A. M. Buchan · U. Ito · F. Colbourne · T. Kuroiwa · I. Klatzo (Eds.)

Maturation Phenomenon in Cerebral Ischemia V

Springer-Verlag Berlin Heidelberg GmbH

A. M. Buchan · U. Ito · F. Colbourne
T. Kuroiwa · I. Klatzo (Eds.)

Maturation Phenomenon in Cerebral Ischemia V

Fifth International Workshop
April 28–May 1, 2002
Banff, Alberta, Canada

With 58 Figures, 15 in Color and 9 Tables

Springer

ALASTAIR M. BUCHAN
Office of Stroke Research
Foothills Medical Centre
Rm 1162, 1403-29 Street N.W.
Calgary, Alberta, T2N 2T9, Canada

UMEO ITO
Department of Neurosurgery
Musashino Red Cross Hospital
1-26-1 Kyonan-cho, Musashino-shu
Tokyo 180-0023, Japan

FRED COLBOURNE
University of Alberta
Department of Psychology
Edmonton, Alberta, T6G 2E9, Canada

TOSHIHIKO KUROIWA
Tokyo Medical and Dental University
Department of Neuropathology
Medical Research Institute
1-5-45 Yushima, Bunkyo-ku
113 Tokyo, Japan

IGOR KLATZO
National Institutes of Health
Laboratory of Neuropathology
and Neuroanatomical Sciences, NINDS
Bethesda, Maryland 20892-4128, USA

Fifth International Workshop, April 28–May 1, 2002
Rimrock Resort Hotel, Banff, Alberta, Canada
Chairmen: A Buchan and U. Ito   **Co-Chairmen:** F. Colbourne and T. Kuroiwa
Secretaries: U. Ito (general), F. Colbourne (local)
International Advisory Board: A. Baethmann, N. Bazan, A. Buchan, D. Choi, G. del Zoppo, C. Fieschi, J. Hallenbeck, K.-A. Hossmann, U. Ito, Y. Katayama, T. Kirino, I. Klatzo, K. Kogure, J. Krieglstein, T. Kuroiwa, J. MacManus, K. Ohno, F. Orzi, F. Plum, F. Sharp, B. Siesjo, A. Tamura, M. Tomita, T. Wieloch
Local Organizers: A. Wilson, A. Kaiser, F. Colbourne

ISBN 978-3-540-40874-1   ISBN 978-3-642-18713-1 (eBook)
DOI 10.1007/978-3-642-18713-1

Cataloging-in-Publication Data applied for
A catalog record for this book is available from the Library of Congress.
Bibliographic information published by Die Deutsche Bibliothek
Die Deutsche Bibliothek lists this publication in the Deutsche Nationalbibliografie; detailed bibliographic data is available in the Internet at <http://dnb.ddb.de>.

springeronline.com

Production: PRO EDIT GmbH, Heidelberg, Germany
Cover-Design: design & production GmbH, Heidelberg, Germany
Typesetting: K + V Fotosatz GmbH, Beerfelden, Germany

Printed on acid-free paper   SPIN 10910597   21/3150/ML   5 4 3 2 1 0

Preface

The Maturation Phenomenon, first described by Ito et al. in 1975, refers to post-ischemic changes that develop hours or days after an ischemic insult. The delayed neuronal death of CA1 pyramidal cells of the hippocampus is a classic example.

The report of the phenomenon boosted research in the field, as it became evident that ischemic damage is not a sudden event but a process potentially susceptible to therapeutic intervention.

In September 1990, Ito and collaborators organized the First International Symposium on "Maturation Phenomenon in Cerebral Ischemia" which was held in Tokyo, Japan. The Second Symposium was organized in Tokyo, Japan in March/April 1996. The Third Symposium was held in Pozzilli, Italy in April 1998 and the Fourth Symposium was held in New Orleans, Louisiana, USA in October/November 1999. This book contains the presentations of the Fifth International Workshop on Maturation Phenomenon in Cerebral Ischemia held at the Rimrock Resort Hotel in Banff, Alberta, Canada on April 28–May 1, 2002. It outlines the present status of investigations and provides further stimulation for research in this field.

The Maturation Phenomenon represents a continuing struggle for survival between the acceleration of tissue or neuronal death and the activation of defense mechanisms leading to neuronal recovery. The elucidation of these mechanisms is important for developing the ability to manipulate them during a long-lasting "therapeutic window".

The book outlines the present status of investigations and provides further stimulation for research in this field. In this current publication, the focus is centered on the elucidation of (1) the Role of Genetic Expression and Neuronal Apoptosis and/or Necrosis, (2) Ischemic Infarction: Inflammation, (3) Clinical Trials, (4) Factors and Mechanisms Enhancing Susceptibility or Tolerance (Glia), (5) Factors Modulating Neuronal Plasticity and the Course of Maturation Phenomenon in Cerebral Ischemia (Metabolic and Inflammatory Factors), and (6) Neurogenesis Stem Cell Activation – Repair and Plasticity.

References

1. Hanyu S, Ito U, Hakamata Y, Yoshida M (1995) Transition from ischemic neuronal necrosis to infarction in repeated ischemia. Brain Res 686:44–48
2. Ito U, Spatz M, Walker J Jr, Klatzo I (1975) Experimental cerebral ischemia in mongolian gerbils. I. Light microscopic obserbations. Acta Neuropathol (Berl) 32:209–223
3. Ito U, Yamaguchi T, Tomita H, Tone O, Shishido T, Hayashi H, Yoshida M (1992) Maturation phenomenon of ischemic injuries observed in Mongolian gerbils: introductory remarks. In: Ito U, Kirino T, Kuroiwa T, Klatzo I (eds) Maturation Phenomenon in Cerebral Ischemia I. Springer, Berlin Heidelberg New York, pp 1–13
4. Ito U, Hanyu S, Hakamata Y, Kuroiwa T, Yoshida M (1997) Features and threshold of infarct development in ischemic maturation phenomenon. In: Ito U, Kirino T, Kuroiwa T, Klatzo I (eds) Maturation Phenomenon in Cerebral Ischemia II. Springer, Berlin Heidelberg New York, pp 115–121
5. Ito U, Hanyu S, Hakamata Y, Arima K, Oyanagi K, Kuroiwa T, Nakano I (1999) Temporal profile of cortical injury following ischemic insult just below and at the threshold level for induction of infarction – light and electron microscopic study. In: Ito U, Orzi F, Kuroiwa T, Fieschi C, Klatzo I (eds) Maturation Phenomenon in Cerebral Ischemia III. Springer, Berlin Heidelberg New York, pp 228–235
6. Kuroiwa T, Yamada I, Hakamata Y, Ohno K, Endo S, Nakano I, Ito U (2001) Time course of postischemic stroke symptoms and delayed infarction after transient cerebral ischemia in gerbils: effect of chemical preconditioning using 3-nitropropionic acid. In: Bazan NG et al (eds) Maturation Phenomenon in Cerebral Ischemia IV. Springer, Berlin Heidelberg New York, pp 141–146
7. Kirino T (1982) Delayed neuronal death in the gerbil hippocampus following ischemia. Brain Res 239:57–69
8. Pulsinelli WA, Brierley JB, Plum F (1982) Temoral profile of neuronal damage in a model of transient forebrain ischemia. Ann Neurol 11:491–498
9. Colbourne F, Li H, Buchan AM (1999) Continuing postischemic neuronal death in CA_1: influence if ischemic duration and cytoprotective doses of NBQX and SNX-111 in rats. Stroke 30:662–668

March 2003   ALASTAIR M. BUCHAN & UMEO ITO

Contents

II Ischemic Infarction: Inflammation

III Clinical Trials/Neuroprotection

IV Factors and Mechanisms Enhancing Susceptibility or Tolerance (Glia)

V Factors Modulating Neuronal Plasticity and the Course of Maturation Phenomenon in Cerebral Ischemia (Metabolic and Inflammatory Factors)

VI Neurogenesis Stem Cell Activation – Repair and Plasticity

VII Abstract and Poster Presentations

VIII Round Table Discussion

List of First-Named Authors

BECKER, K.J.
Box 359775 Harborview Medical Center, 325 Ninth Avenue, Seattle,
WA 98104-2499, USA

BENNETT, M.
Albert Einstein College, Department of Neuroscience, 1300 Morris Park Avenue,
Kennedy 720, Bronx, New York 10461, USA

BRUNEAU, R.
Canadian Center for Behavioral Neuroscience,
Department of Psychology & Neuroscience, University of Lethbridge,
4401 University Drive West, Lethbridge, Alberta, T1K 3M4, Canada

BUCHAN, A.
Professor of Neurology, Department of Clinical Neurosciences,
University of Calgary, Rm 1162, 1403-29 Street N.W., Calgary, Alberta, T2N 2T9,
Canada

CHAN, P.H.
Neurosurgical Laboratories, Stanford University, 1201 Welch Road, #P314,
Stanford, CA 94305, USA

CHOPP, M.
Professor & Vice Chairman, Department of Neurology (E&R3056),
Henry Ford Health Sciences Center, 2799 West Grand Boulevard, Detroit,
MI 48202, USA

COLBOURNE, F.
Department of Psychology, Faculty of Science, University of Alberta, Edmonton,
Alberta, T6G 2E9, Canada

COOPER, N.R.
University of Lethbridge, Department of Psychology & Neuroscience,
4401 University Drive West, Lethbridge, Alberta, T1K 3M4, Canada

DEL ZOPPO, G.J.
Department of Molecular and Experimental Medicine,
The Scripps Research Institute, 10550 North Torrey Pines Road, MEM 132,
La Jolla, CA 92037, USA

DIRNAGL, U.
Department of Neurology, Humboldt-University Berlin, Schumannstr. 20–21,
10098 Berlin, Germany

Dyck, R.
Assistant Professor, University of Calgary, Psychology, 2500 University Drive,
NW, Calgary, Alberta, T2N 1N4, Canada

Ehrenreich, H.
Departments of Neurology and Psychiatry, Georg-August-University, Göttingen,
Germany

Gonzalez, C. L.
Canadian Center for Behavioural Neuroscience, University of Lethbridge,
Department of Psychology & Neuroscience, 4401 University Drive, Lethbridge,
Alberta, T1K 3M4, Canada

Hallenbeck, J. M.
Stroke Branch, National Institute of Neurological Disorders and Stroke,
National Institutes of Health, Building 36, Room 4A03, MSC 4128,
Convent Drive, Bethesda, Maryland 20892-4128, USA

Hossmann, K.-A.
Max-Planck-Institute for Neurological Research,
Department of Experimental Neurology, Gleueler Str. 50, 50931 Cologne,
Germany

Hu, B.
Assistant Professor, Cerebral Vascular Disease Research Center,
Department of Neurology D4–5, University of Miami School of Medicine,
1501 NW 9th Ave, Miami, FL 33136, USA

Ito, U.
4-22-24, Zenpukuji, Suginami-ku, Tokyo 167-0041, Japan

Johansson, B. B.
Professor of Neurology, Wallenberg Neuroscience Center,
Experimental Brain Research, BMC A13, 221 84 Lund, Sweden

Kamiya, T.
Second Department of Internal Medicine, Nippon Medical School, Tokyo,
113-8603, Japan

Kato, H.
Department of Neurology, Tohoku University School of Medicine,
1-1 Seiryo-machi, Aoba-ku, Sendai 980-8574, Japan

Kempski, O.
Institute for Neurosurgical Pathophysiology,
Johannes-Gutenberg-University, Langenbeckstr. 1, 55101 Mainz, Germany

Klatzo, I.
19022 Canadian Court, Gaithersburg, MD, 20886, USA

Kleim, J. A.
AHFMR Medical Scholar, Canadian Centre for Behavioural Neuroscience,
Dept. of Psychology & Neuroscience, University of Lethbridge, Lethbridge, AB,
T1K 3M4, Canada

KRIEGLSTEIN, J.
Institute for Pharmacology and Toxicology, Philipps-Universität,
Ketzerbach 63, 35032 Marburg, Germany

KUBOTA, M.
Department of Neurosurgery, Teikyo University School of Medicine, Tokyo 173,
Japan

KUROIWA, T.
Department of Neuropathology, Medical Research Institute,
Tokyo Medical and Dental University, Yushima 1-5-45, Bunkyo-ku,
Tokyo 113-8510, Japan

MACMANUS, J. P.
Institute for Biological Sciences M54, National Research Council,
Montreal Road Laboratories, Ottawa, Ontario, K1A 0R6, Canada

MIES, G.
Max-Planck-Institut for Neurological Research,
Abteilung für Experimentelle Neurologie, Gleueler Str. 50, 50931 Köln, Germany

NAKANE, M.
Department of Neurosurgery, University Hospital, Mizonokuchi,
Teikyo University School of Medicine, 3-8-3 Mizonokuchi, Takatsu-ku,
Kawasaki, Kanagawa 213-8507, Japan

ORZI, F.
Dipartimento di Scienze Neurologiche, Universita di Roma „La Sapienza",
Il Facolta Di Medicina, Policlinico Sant Andrea, Via di Grottarossa, Roma, Italy

PLESNILA, N.
Institute for Surgical Research,
Klinikum of the University of München-Großhadern, Marchioninistr. 27,
81377 München, Germany

QIAO, M.
Research Technical Officer, Institute of Biodiagnostics, National Research
Council of Canada, 3330 Hospital Drive NW, Calgary, Alberta, T2N 4N1, Canada

SHARP, F. R.
Department of Neurology & Neuroscience Program, University of Cincinnati,
Vontz Center for Molecular Studies, Room 2327, 3125 Eden Avenue,
Cincinnati, OH 45267-0536, USA

SIMON, R. P.
Director and Chair, R.S. Dow Neurobiology Laboratories, Legacy Research,
1225 NE 2nd Avenue, Portland, OR 97232, USA

SNYDER, E. Y.
Harvard Medical School, Harvard Institutes of Medicine, Neurology,
Beth Israel-Deaconess Medical Center, 77 Avenue Louis Pasteur, Room 855,
Boston, MA 02115, USA

SPATZ, M.
  National Institutes of Health, NINDS, Stroke Branch, 36 Convent Drive,
  MSC 4128, Bethesda, Maryland 20892-4128, USA

STANIMIROVIC, D.
  Director, Neurobiology Program, Institute for Biological Sciences,
  National Research Council of Canada, 1200 Montreal Road, Bldg M54, Ottawa,
  Ontario, K1A0R6, Canada

SUBRAMANIAM, S. MD
  Rm 1162, 1403 – 29 Street N.W., Calgary, Alberta, T2N 2T9, Canada

SUTCLIFFE, I.
  National Research Council of Canada, Institute for Biological Sciences,
  1200 Montreal Road, Bldg M54, Ottawa, Ontario, K1A0R6, Canada

TAMURA, A.
  Teikyo University School of Medicine, Department of Neurosurgery,
  2-11-1 Kaga, Itabashi-ku, Tokyo, 173-8605, Japan

TANAKA, K.
  Department of Neurology, School of Medicine, Keio University,
  35 Shinanomachi, Shinjuku-ku, Tokyo 160-8582, Japan

TOMITA, M.
  Department of Neurology, School of Medicine, Keio University, 35 Shinanomachi,
  Shinjuku-ku, Tokyo 160-8582, Japan

TUOR, U.
  Senior Research Professor (Adjunct),
  University of Calgary, Institute for Biodiagnostics (West), 3330 Hospital Drive NW,
  Calgary, Alberta, T2N4N1, Canada

VANDENBERG, P.M.
  Canadian Center for Behavioral Neuroscience, Department of Psychology &
  Neuroscience, University of Lethbridge, 4401 University Drive West, Lethbridge,
  Alberta, T1K3M4, Canada

XUE, J.-H.
  Laboratory for Cerebrovascular Disorders, National Cardiovascular Center and
  Research Institute of NCVC, Suita, 565-8565 Japan

YANAMOTO, H.
  Laboratory for Cerebrovascular Disorders, National Cardiovascular Center and
  Research Institute of NCVC, Suita, 565-8565, Japan

YONG, V.W.
  Department of Clinical Neurosciences, University of Calgary, 1403-29 Street
  N.W., Calgary, Alberta, T2N 2T9, Canada

ZUKIN, R.S.
  Albert Einstein College, Department of Neuroscience, 1300 Morris Park Avenue,
  Bronx, New York 10461, USA

Fifth International Workshop
Maturation Phenomenon in Cerebral Ischemia

Sunday, April 28, to
Wednesday, May 1, 2002

The Rimrock Resort Hotel
Banff, Alberta, Canada

Chairmen:
- Alastair M. Buchan
  Calgary, Alberta, Canada
- Umeo Ito
  Tokyo, Japan

Co-Chairmen:
- Fred Colbourne
  Edmonton, Alberta, Canada
- Toshihiko Kuroiwa
  Tokyo, Japan

Secretaries:
- Fred Colbourne
  (Local Affairs)
  Edmonton, Alberta, Canada
- Umeo Ito
  (General Affairs)
  Tokyo, Japan

Local Organizers:
- Annley Wilson
- Angelika Kaiser
- Fred Colbourne

Alastair M. Buchan
Professor Stroke Research
Department of Clinical Neurosciences
Rm 1162, 1403 - 29 Street N.W.
Calgary, Alberta, Canada   T2N 2T9
Phone: (403) 944-1581 Fax: (403) 944-1602
e-mail: abuchan@ucalgary.ca
Website:
http://www.ucalgary.ca/UofC/conferences/ischemia

International Advisory Board

Alexander Baethmann	Munich, Germany
Nicolas Bazan	New Orleans, Louisana, USA
Alastair Buchan	Calgary, Alberta, Canada
Dennis Choi	Whitehouse Station, NJ, USA
Gregory del Zoppo	LaJolla, California, USA
Cesare Fieschi	Rome, Italy
John Hallenbeck	Bethesda, Maryland, USA
Konstantin Hossmann	Köln, Germany
Umeo Ito	Tokyo, Japan
Yasua Katayama	Tokyo, Japan
Takaaki Kirino	Tokyo, Japan
Igor Klatzo	Gaithersburg, Maryland, USA
Kyuya Kogure	Saitama, Japan
Josef Krieglstein	Marburg, Germany
Toshihiko Kuroiwa	Tokyo, Japan
John MacManus	Ottawa, Ontario, Canada
Kikuo Ohno	Tokyo, Japan
Francesco Orzi	Pozzilli, Italy
Fred Plum	New York, New York, USA
Frank Sharp	Cincinnati, Ohio, USA
Bō Siesjo	Lund, Sweden
Akira Tamura	Tokyo, Japan
Minoru Tomita	Tokyo, Japan
Tadeusz Wieloch	Lund, Sweden

I Role of Genetic Expression and Neuronal Apoptosis and/or Necrosis

Modulation of Neuronal Death
by the Transcription Factor E2F1 in Experimental Stroke

J. P. MacManus, M. Jian, E. Preston, J. Webster, and B. Zurakowski

Summary. Since cultured neurons which are deficient in the transcription factor E2F1 display a resistance to a wide variety of insults, studies were undertaken to see if this resistance extended to a cerebral ischemic insult. Following 2 h of left middle cerebral artery occlusion and 1 d of reperfusion a 33% smaller infarct ($p < 0.05$) was observed by 2,3,5 triphenyltetrazolium staining in the brains of E2F1-null animals compared to E2F1 +/+ and +/− littermate mice. No differences in physiological parameters or cerebrovasculature were observed in the three groups of animals. A milder ischemic insult produced by 20 min of MCA occlusion and 7 d of reperfusion produced a greater difference in the E2F1-null animals with a 71% smaller infarct ($p < 0.001$) compared to littermate controls. A decrease in ischemic injury in E2F1-null mice was also observed by immunohistochemical monitoring of the neuronal-specific MAP2 cytoskeletal protein and TUNEL staining. Finally a battery of six tests of motor function showed both E2F1 +/+ and −/− littermates were similarly impaired at 1 d of reperfusion, but by 7 d the E2F1 −/− mice improved significantly better compared to the wild type mice ($p < 0.01$) and ended up only slightly impaired by that time. It is concluded that the transcription factor E2F1 does modulate neuronal viability in brain after cerebral ischemia.

Key words. E2F1 transcription factor – focal ischemia – infarct– apoptosis – microtubule-associated protein 2

Introduction

Cultured neurons from the cerebellum, cortex or hippocampus isolated from mice missing the transcription factor E2F1 show resistance to a wide variety of cell-death inducers, for example β-amyloid [6], dopamine [11, 12], low potassium [33], oxygen-glucose deprivation [5], or staurosporine [10]. Increases in E2F1 expression are seen in cultured wildtype neurons induced to die by these treatments [5, 6, 10–12, 33, 43]. The inferred involvement of E2F1 in modulation of the cell-death pathway is sup-

Apoptosis Research Group, Institute for Biological Sciences, National Research Council of Canada, Ottawa ON, Canada K1A 0R6

Correspondence to: Dr. John P. MacManus, Institute for Biological Sciences M54, National Research Council, Montreal Road Laboratories, Ottawa, Ontario, Canada, K1A 0R6. Phone: 613/993-9305, Fax: 613/941-4475, E-Mail: john.macmanus@nrc.ca

Maturation Phenomenon in Cerebral Ischemia V
A. M. Buchan et al. (Eds.)
© Springer-Verlag Berlin Heidelberg 2004

ported by the finding of increased apoptosis in cells forced to overexpress this transcription factor, for example in cycling or quiescent cells [14, 16, 21], but also in postmitotic myocardial cells [1, 20] and neurons [10, 11, 33]. In a similar manner to other investigators [13, 35], we have found that the E2F1 protein can induce neuronal cell death in a manner apparently independent of transcriptional activation which may involve inhibition of antiapoptotic signalling pathways [11].

As a transcription factor, E2F1 binds to DNA in a complex with DP-proteins and pocket-proteins, the most renowned of which is the retinoblastoma protein pRb. The actions of E2F1/pRb in control of progression through the cell-cycle by repression or transactivation of target-genes have been well studied, though roles in differentiation, DNA-repair and apoptosis are also described [32]. The mechanism whereby free E2F1 could induce apoptosis is not clear. Three possibilities have been put forward: 1) induction of apoptotic genes, 2) stabilization of the proapoptotic transcription factor p53, and 3) inhibition of anti-apoptotic signalling [31, 37].

In animal models the situation is not so clear. However, in E2F1-null mice decreased cell death in thymus has been observed [4, 46], and in transgenic mice overexpressing E2F1 increased cell death in testes or skin has been reported [9, 38]. Animals deficient in pRb also have excessive E2F1 and it is of interest that such animals have extensive apoptosis during neurogenesis [15, 22, 24]. In light of the above findings with cultured neurons and E2F1-deficient mice, we undertook to examine whether E2F1-null animals had a reduced injury following an episode of focal cerebral ischemia.

Materials and Methods

All procedures using mice were approved by a local committee for the Canadian Council on Animal Care. The E2F1 –/– mice were obtained originally from Jackson Laboratories (Bar Harbor, MA, USA: Stock #2785)[4] and bred locally. We always used F2 generation littermates produced from interbreeding the F1 generation of E2F1 –/– stock crossed with pure-bred C57B/6 mice. F2 generation animals were genotyped by PCR as described [4]. The mice (20 to 23 g) were subjected to occlusion of the left middle cerebral artery (MCA) under isoflurane anaesthesia by an intraluminal filament [26] for 20 min or 2 h of ischemia, the animals briefly reanaesthetized with isoflurane and the filament withdrawn. Regional cerebral blood flow, blood pressure and blood gases were measured as previously described [26].

To measure ischemic damage at 1 or 7 d post MCA occlusion, the brains were removed from euthanized animals, and the cerebrum cut into five 2 mm thick coronal slices which were stained with 2% 2,3,5-triphenyltetrazolium chloride (TTC), photographed with a mm-scale and the images subsequently digitized. Infarct areas were obtained by digital planimetry of the slices using ImagePro software (Media Cybernetics, Silver Spring MD) and normalized for edema as described [26]. Alternatively, ischemic damage was monitored by similar measurement of areas in brain sections that had been stained for the neuronal specific microtubule associated protein 2, MAP2 [3, 19, 28]. Briefly, sections of formalin-fixed brains were incubated overnight with anti-MAP2 (1:100; Sigma), washed and incubated with Cy3-conjugated AffiniPure F(ab)2 fragment goat anti-mouse IgG (1:200, Jackson ImmunoResearch Laboratory) for 4 h. After washing the section was mounted in Vectashield mounting medium (Vector Laboratories).

To assess neurological deficits in motor function in the ischemic E2F1 mutant mice, an expanded six point scale was employed as described [47]. Assessments were made at 1 and 7 d of reperfusion by an individual blinded to the genotype and treatment.

Results

Following 2 h of MCA occlusion and 1 d of reperfusion a similar sized infarct of 60 to 70 mm^3 was observed in the left hemisphere of the brains of E2F1+/+ and +/- littermate mice, but a 33% smaller infarct ($p < 0.05$) was observed in brains of E2F1-null animals (Fig. 1 A). When individual coronal slices of brain from the ischemic littermates were examined, the decrease in damaged tissue in the E2F1-null mice was located primarily in the anterior and mid regions of the brain ($p < 0.001$) (Fig. 1 B). In the slice 4 mm from the frontal pole where the maximum difference in infarcted tissue occurred, this decrease was discernable principally in striatum but also in cortex (Fig. 1 C). It should be stated at this stage that no difference was found in the physiological parameters of blood gases (pO_2, pCO_2, pH) or blood pressure before or after ischemia in all three groups of E2F1-mutant mice (data not shown). The MCA-territory, as visualized by perfusion with india-ink [26], was also similar in all groups of littermate mice (data not shown).

Since the use of littermates eliminated genetic variability as being the cause of the observed decrease in infarct, we were left puzzled at the small effect *in vivo* of a gene whose absence demonstrates greater protection in cultured cortical or cerebellar granule neurons induced to die by a variety of insults [10–12]. We hypothesized that the severity of injury produced by 2 h of MCA occlusion was too great to be easily resisted in the stricken brain tissue, and therefore undertook to repeat our study with a milder insult of 20 min of MCA occlusion followed by 7 d of reperfusion.

The final infarct volume produced at 7 d with the milder chronic ischemic insult was approximately a third of that produced with the 2 h occlusion (Fig. 2 A) and occurred principally in the striatum, although damage was also seen scatted throughout the cortex in about half of the animals. Again the E2F1 +/+ and +/- animals had similar volumes of infarcted tissue of 25 to 30 mm^3, but the E2F1-null littermates had a much smaller amount of damage, approximately 30% of that seen in the wildtype or heterozygote ischemic mice ($p < 0.001$) (Fig. 2 A). In addition to this delineation of infarcted tissue by the monitoring of decreased mitochondrial activity by TTC staining of brain slices, we also undertook delination of the area of neuronal injury by immunohistochemical monitoring of the neuronal-specific MAP2 cytoskeletal protein in brain sections from +/+ and -/- E2F1-mice. In brain sections at the level of the striatum where maximum infarct was observed, areas of MAP2-negative staining were measured. A significantly smaller area of neuronal injury was observed in the E2F1-null mice compared to +/+ animals ($p < 0.001$) (Fig. 2 B).

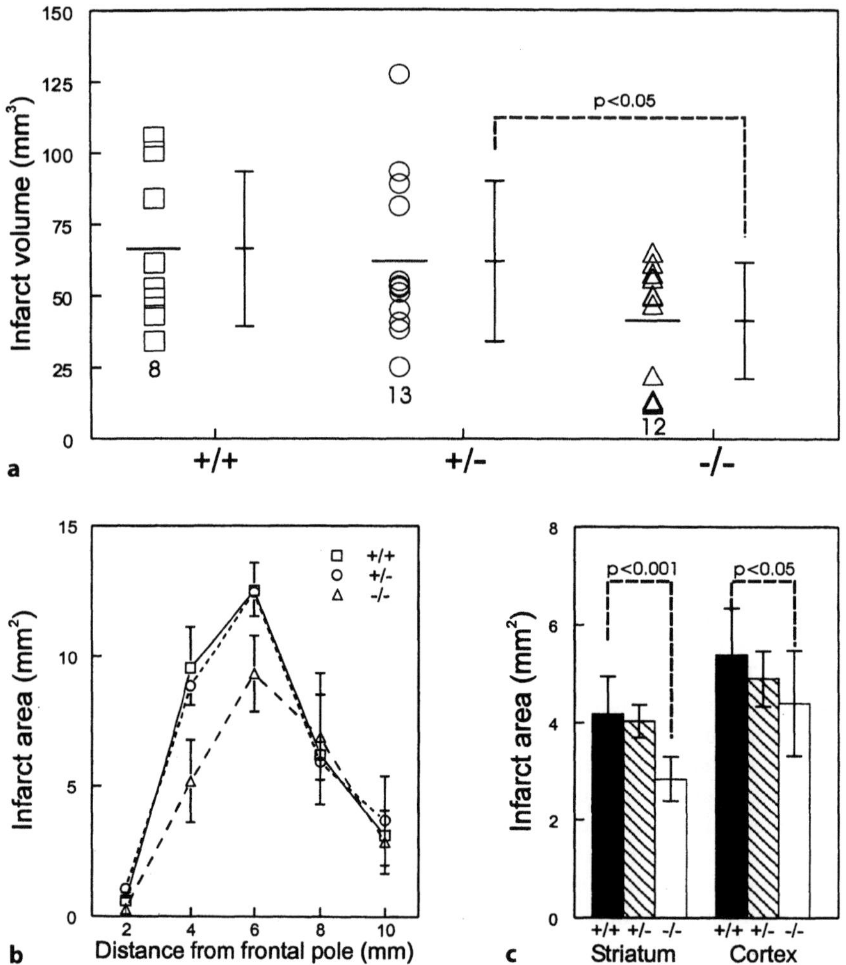

Fig. 1. a Decrease of 33% in infarct volume at 1 d of reperfusion following 2 h of MCA occlusion in E2F1-null mice compared to wild-type and heterozygous littermates (mean ± SD) $p < 0.05$ on ANOVA (n = 8–13). Measurements of infarct volume in individual animals based on TTC staining are shown. **b** Decreased infarct areas in coronal slices of E2F1 −/− mice; $p < 0.001$ on non-paired t test (n = 12) in coronal slice 4 and 6 mm from frontal pole. **c** Decreased infarct areas in both striatum ($p < 0.001$) and cortex ($p < 0.05$) in coronal slice 4 mm from frontal pole in B (n = 12)

We also undertook an assessment of neuronal function in these ischemic E2F1-mutant mice. Using a battery of six tests of motor function [47] both +/+ and −/− littermates were similarly impaired at 1 d of reperfusion after 20 min of MCA occlusion (Fig. 3). The neurological score of both groups of animals showed the expected improvement by 7 d [47], but the E2F1 −/− mice improved significantly more ($p < 0.01$) compared to the wild type mice and were only slightly impaired by that time (Fig. 3).

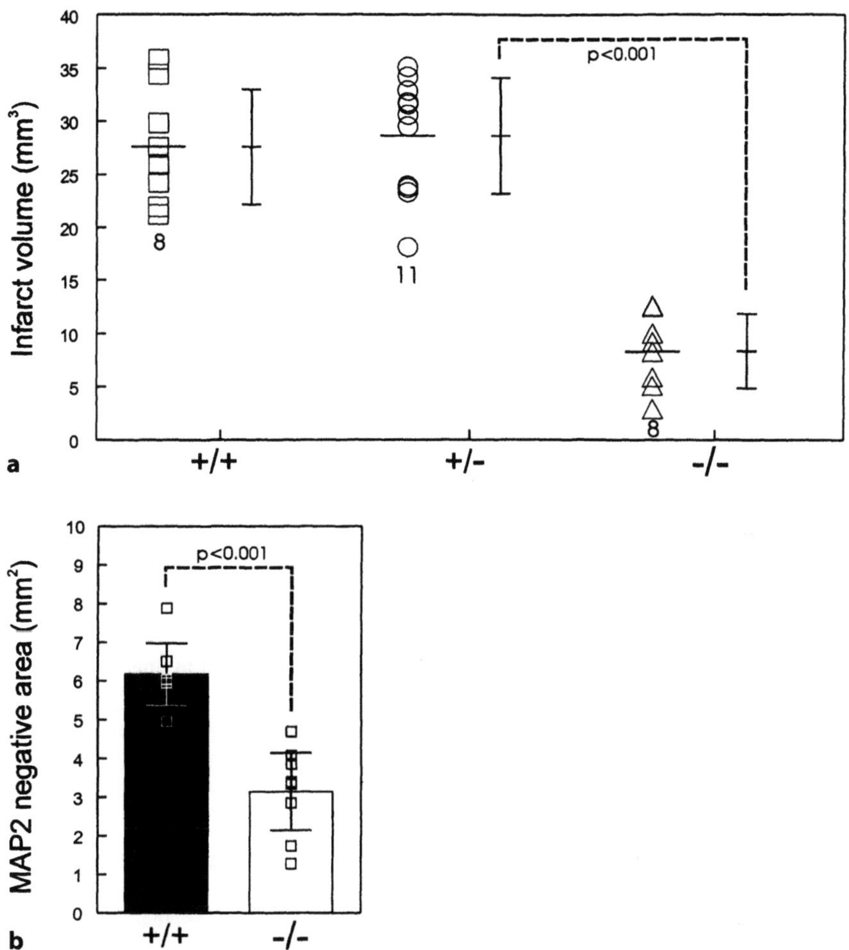

Fig. 2. a Decrease of 71% in infarct volume at 7 d of reperfusion following 20 min of MCA occlusion in E2F1-null mice compared to wild-type and heterozygous littermates (mean ± SD) p < 0.001 on ANOVA (n = 8–11). Measurements of infarct volume in individual animals based on TTC staining are shown. **b** Decreased area free of MAP2 staining in brain sections at the level of striatum in E2F1-null mice compared to E2F1 +/+ littermates (mean ± SD) p < 0.001

Discussion

The E2F1 transcription factor in partnership with the retinoblastoma protein pRb is central in the control of cell-cycle, and modulation of differentiation and apoptosis [32, 37]. However, the E2F1 transcription factor is not essential perhaps due to redundancy in the protein family which has five other members, as evidenced by the fact that E2F1-deficient mice develop and reproduce normally with no histological changes in many tissues including brain for up to six months [4, 46]. Besides the large amount of evidence from cultured neurons cited in the Introduc-

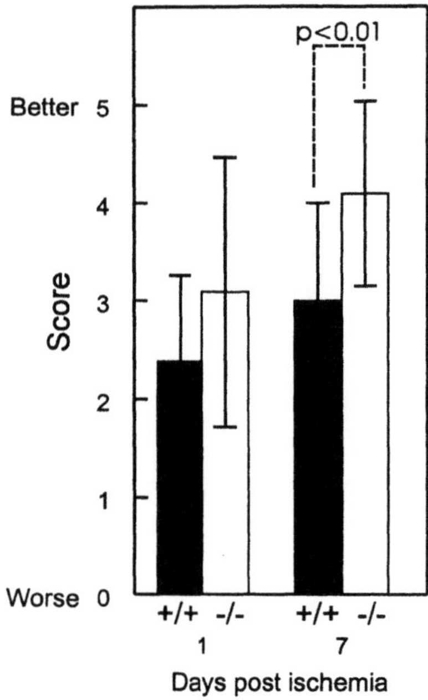

Fig. 3. Less severe behavioural deficit in ischemic E2F1-null mice compared to littermate controls at 7 d of reperfusion (mean ± SD, n = 11–13, p < 0.01). A six point neurological score (see refer. 47) was used from better to worse: 5 = extend both forelimbs when held by tail; 4 = flexion of contralateral forelimb; 3 = reduced resistance to lateral push; 2 = circling towards parietic side when pulled by tail; 1 = spontaneously circling; 0 = without spontaneous motion

tion which indicates a role for E2F1 in cell death of these postmitotic cells, there is also evidence from animal studies that this transcription factor is active in brain. For example, deregulated E2F1 caused by mutation in pRb does lead to massive neuronal apoptosis during development [15, 22, 24]. The mRNA encoding the transcription factor E2F1 increases in brain following either focal (Hou & MacManus, unpublished observations) or global [18] ischemia. In addition, increased E2F1 protein levels have been noted following focal ischemia [34] and also in brains of Down's Syndrome patients [30]. Another indication of deregulation of E2F1 following cerebral ischemia is increased activity of cyclin-dependent kinases (CDK4/6) as seen by increased phosphorylation of pRb following ischemia [8, 34] (see scheme in Fig. 4). This increased kinase activity could occur via the action of increased cyclin D1 which has been reported in both focal [19, 23, 34] and global ischemia [42], although involvement of cyclin D1 is debatable at least following global ischemia [41]. Another possibility for increased kinase activity is via post-ischemic destruction of kinase inhibitors such as p16^{INK4a} [19] which would initiate a cascade of relief from inhibition leading to increased free E2F1 (Fig. 4).

The attenuated ischemic damage in E2F1-null mice compared to their littermate controls (Fig. 1 and 2) is in agreement with our preliminary observations in outbred mice [26], and includes the inexplicable lack of resistance to damage in

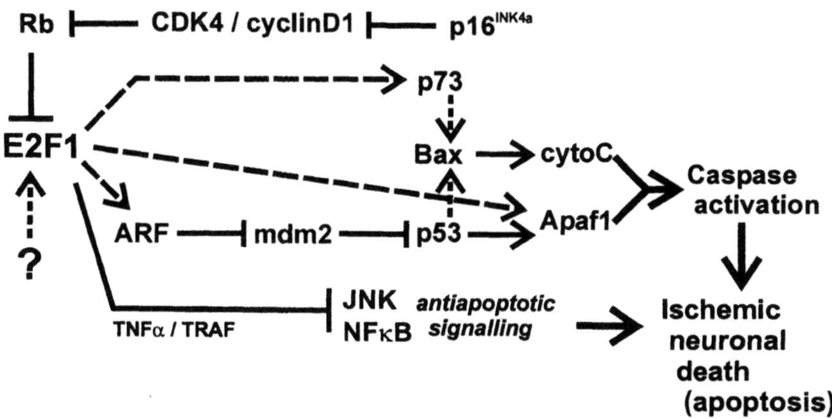

Fig. 4. Putative mechanisms whereby the transcription factor E2F1 might promote neuronal cell death following ischemia. Free E2F1 levels could be increased by release from pRb-binding following phosphorylation by cyclin-dependent kinases caused by increased kinase levels or relief of inhibition by p16^{INK4a}. Alternatively, E2F1 levels could increase due to increased E2F1 gene expression in response to unknown signals. Broken lines indicate transcriptional activation of E2F1 target genes such as p73 and Apaf1 which are involved in ultimate activation of active proteolysis of many cellular constituents by caspase. Another target gene of E2F1 is ARF whose activation via mdm2 leads to stabilization of p53. Finally E2F1 might act by inhibiting anti-apoptotic signaling pathways, for example involving NFκB. Any one, or combination of such events could explain the potent proapoptotic actions of E2F1

the hindmost areas of the brain (Fig. 1b). This decreased damage in littermates is not due to differences in cerebral vasculature nor physiological parameters such as blood pressure or blood flow [26]. The resistance to ischemic injury is greater following a brief ischemic episode of 20 min (Fig. 2a) compared to the standard more severe 2 h of MCA occlusion (Fig. 1a).

A loss of the cytoskeletal neuronal protein MAP2 has been touted by many investigators as an early indicator of neuronal injury following either global [41, 45] or focal ischemia [23, 28], and this loss has been observed in mice following mild ischemia [3, 19]. The decreased ischemic damage in E2F1-null mice was also seen in the less severe loss of neuronal MAP2 staining (Fig. 2b). It has been reported that areas depleted of MAP2 contain cells with DNA fragments as seen by TUNEL staining following global [45] or focal [19] ischemia. Following the 20 min of focal ischemia in our present study, we observed increased caspase activity and TUNEL-positive cells at 7 d of reperfusion in the area of negative MAP2 staining (not shown). This increase in neurons with DNA fragmentation was also attenuated in the E2F1-null ischemic mice. The decrease of 50 to 70% in the extent of infarction in the absence of the E2F1 transcription factor also manifested itself in the fact that the E2F1-null ischemic mice had less behavioral impairment than littermate controls (Fig. 3). Some investigators claim that the mild ischemia produced by 20 to 30 min of MCA occlusion results in delayed neuronal death at 7 d of reperfusion that has more apoptotic characteristics than the more severe 2 h of occlusion which leads to a massive necrotic lesion [2, 3, 19]. Thus the ability of E2F1 loss to more successfully attenuate damage after the mild ischemic episode compared to the prolonged 2 h period may be due to a different mode of cell death predominating. However it must be said that we have not observed a qualitative difference between the cell death following mild versus severe ischemia in mice which would indicate a more apoptotic form after 20 min of ischemia using any of the apoptotic criteria such as DNA fragmentation [27].

There are three potential mechanisms described whereby E2F1 may influence the cell death pathway (Fig. 4): 1) Induction of proapoptotic genes, 2) stabilization of the proapoptotic transcription factor p53, and 3) inhibition of anti-apoptotic signaling [32, 37]. To elaborate, firstly, a genome-wide survey using micro array techniques unearthed 494 of 35000 genes which were responsive to the transcription factor E2F1 so there are plenty of possibilities for the induction of proapoptotic genes [31]. The most appealing genes at the moment are the proapoptotic transcription factor p73, and several genes involved in the execution machinery such as Apaf1 and both caspase 3 and 7 [29, 31]. There are as yet no indications in gene profiling studies following cerebral ischemia of increases in p73 or Apaf1 (reviewed in ref. 39), but of course activation of caspase 3 and subsequent neuronal apoptosis has been widely implicated in both global and focal cerebral ischemia (reviewed in refs. 7, 25, 27, 39).

Secondly, several studies have identified Arf as an E2F1 target gene and so stabilization of the proapoptotic transcription factor p53 by decreased interaction with mdm2 is a definite possibility for the death-inducing actions of E2F1 [31, 37]. There are numerous studies implicating p53 in neuronal apoptosis in culture [17, 40, 44], and in ischemic brain [7, 25]. Since increased Arf alone is not sufficient to induce cell death, more than one E2F1-evoked signal appears necessary and thus these putative mechanisms may be mutually dependent [31, 37].

Thirdly, in addition to these transcriptional mechanisms whereby E2F1 could modulate apoptosis, the inhibition of anti-apoptotic signaling is also suggested as an alternate mechanism [36]. It has been reported that E2F1 can induce apoptosis without transcriptional transactivation in several cell types [13, 35] including neurons [11]. Interference with the anti-apoptotic signaling involving NFκB has been described as the means whereby E2F1 acts in such a direct way [36] and we have obtained some evidence to support this idea in dopamine-evoked neuronal apoptosis [11].

All of the above potential mechanisms whereby E2F1 might influence the cell death pathway are not necessarily mutually exclusive, and may indeed be interdependent (Fig. 4). Whether such mechanisms are at work in ischemic brain, or in injured cultured neurons, must await further work. It is hoped that publication of the ongoing gene profiling studies following cerebral ischemia will show the way towards pinpointing the death-promoting actions of E2F1 in brain.

Acknowledgements. Our thanks are due to Jack Daoust and his team for animal management. This study was supported in part by the Heart and Stroke Foundation of Canada (grants # A3457 & T4427 to JPM).

References

1. Agah R, Kirshenbaum LA, Abdellatif M, Truong LD, Chakraborty S, Michael LH, Schneider MD (1997) Adenoviral delivery of E2F1 directs cell cycle reentry and p53 independent apoptosis in postmitotic adult myocardium *in vivo*. J Clin Invest 100:2722–2728
2. Du C, Hu R, Csernansky CA, Hsu CY, Choi DW (1996) Very delayed infarction after mild focal cerebral ischemia: a role for apoptosis? J Cerebral Blood Flow Metab 16:195–201
3. Endres M, Namur S, Shimizu-Sasamata M, Waeber C, Zhang L, Gomez-Isla T, Hyman BT, Moskowitz MA (1998) Attenuation of delayed neuronal death after mild focal ischemia in mice by inhibition of the caspase family. J Cerebral Blood Flow Metab 18:238–247
4. Field SJ, Tsai FY, Kuo F, Zubiaga AM, Kaelin WG, Livingston DM, Orkin SH, Greenberg ME (1996) E2F1 functions in mice to promote apoptosis and suppress proliferation. Cell 85:549–561
5. Gendron TF, Mealing GAR, Paris J, Lou A, Edwards A, Hou ST, MacManus JP, Hakim AM, Morley P (2001) Attenuation of neurotoxicity in cortical cultures and hippocampal slices from E2F1 knockout mice. J Neurochem 78:316–324
6. Giovanni A, Keramaris E, Morris EJ, Hou ST, O'Hare M, Dyson N, Robertson GS, Slack RS, Park DS (2000) E2F1 mediates death of B-amyloid-treated cortical neurons in a manner independent of p53 and dependent on Bax and caspase 3. J Biol Chem 275:11553–11560
7. Graham SH, Chen J (2001) Programmed cell death in cerebral ischemia. J Cerebral Blood Flow Metab 21:99–109
8. Hayashi T, Sakai K, Sasaki C, Zhang WR, Abe K (2000) Phosphorylation of retinoblastoma protein in rat brain after transient middle cerebral artery occlusion. Neuropathol & Applied Neurobiol 26:390–397
9. Holmberg C, Helin K, Sehested M, Karlstrom O (1998) E2F-1-induced p53-independent apoptosis in transgenic mice. Oncogene 17:143–155
10. Hou ST, Callaghan D, Fournier MC, Hill I, Kang L, Massie B, Morley P, Murray C, Rasquinha I, Slack R, MacManus JP (2000) The transcription factor E2F1 modulates apoptosis of neurons. J Neurochem 75:91–100
11. Hou ST, Cowan, E, Walker T, Ohan N, Dove M, Rasquinha I, MacManus JP (2001) The transcription factor E2F1 promotes dopamine-evoked neuronal apoptosis by a mechanism independent of transcriptional activation. J Neurochem 78:1–12
12. Hou ST, Cowan E, Dostanic S, Rasquinha I, Comas T, Morley P, MacManus JP (2001) Increased expression of E2F1 during dopamine-evoked caspase-3-mediated apoptosis in rat cortical neurons. Neurosci Lett 306:153–156
13. Hsieh JK, Fredersdorf S, Kouzarides T, Martin K, Lu X (1997) E2F1-induced apoptosis requires DNA binding but not transactivation and is inhibited by the retinoblastoma protein through direct interaction. Genes & Dev 11:1840–1852
14. Hunt KK, Deng J, Liu TJ, Wilson-Heiner J, Swisher SG, Clayman G, Hung MC (1997) Adenovirus-mediated over expression of the transcription factor E2F1 induces apoptosis in human breast and ovarian carcinoma cell lines and does not require p53. Cancer Res 57:4722–4726
15. Jacks T, Fazeli A, Schmitt EM, Bronson RT, Goodell MA, Weinberg RA (1992) Effects of an Rb mutation in the mouse. Nature 359:295–300
16. Johnson DG, Schwarz JK, Cress WD, Nevins JR (1993) Expression of transcription factor E2F1 induces quiescent cells to enter S phase. Nature 36:349–352
17. Johnson MD, Kinoshita Y, Xiang H, Ghatan S, Morrison RS (1999) Contribution of p53-dependent caspase activation to neuronal cell death declines with neuronal maturation. J Neurosci 19:2996–3006
18. Jin K, Mao XO, Eshoo MW, Nagayama T, Minami M, Simon RP, Greenberg DA (2001) Microarray analysis of hippocampal gene expression in global cerebral ischemia. Ann Neurol 50:93–103
19. Katchanov J, Harms C, Gertz K, Hauch L, Waeber C, Hirt L, Priller J, von Harsdorf R, Bruck W, Hortnagl H, Dirnagl U, Bhide PG, Endres M (2001) Mild cerebral ischemia induces loss of cyclin-dependent kinase inhibitors and activation of cell cycle machinery before delayed neuronal cell death. J Neurosci 21:5045–5053
20. Kirshenbaum LA, Abdellatif S, Chakraborty S, Schneider MD (1996) Human E2F-1 reactivates cell cycle progression in ventricular myocytes and represses cardiac gene transcription. Dev Biol 179:402–411
21. Kowalik TF, DeGregori J, Schwarz JK, Nevins JR (1995) E2F1 overexpression in quiescent fibroblasts leads to induction of cellular DNA synthesis and apoptosis. J Virol 69:2491–2500
22. Lee EYHP, Chang CY, Hu N, Wang YCJ, Lai CC, Herrup K, Lee WH, Bradley A (1992) Mice deficient for Rb are nonviable and show defects in neurogenesis and haematopoiesis. Nature 359:388–394

23. Li Y, Chopp M, Powers C, Jiang N (1997) Immunoreactivity of cyclin D1/cdk4 in neurons and oligodendrocytes after focal cerebral ischemia in rat. J Cerebral Blood Flow Metab 17:846–856
24. MacLeod KF, Hu Y, Jacks T (1996) Loss of Rb activates both p53-dependent and independent cell death pathways in the developing mouse nervous system. EMBO J 15:6178–6188
25. MacManus JP, Buchan AM (2000) Apoptosis after experimental stroke: fact or fashion? J Neurotrauma 17:899–914
26. MacManus JP, Koch CJ, Jian M, Walker T, Zurakowski B (1999) Decreased brain infarct following focal ischemia in mice lacking the transcription factor E2F1. NeuroReport 10:2711–2714
27. MacManus JP, Rasquinha I (2000) Ischemic Neuronal death is not by classic apoptosis: are all DNA-ladders the same? In: *Apoptosis in Health and Disease*, New Horizons in Therapeutics V3, Harwood Academic Publishers, Amsterdam, pp 33–46
28. Matsushita K, Matsuyama T, Kitagawa K, Matsumoto M, Yanagihara T, Sugita M (1998) Alterations of Bcl-2 family proteins precede cytoskeletal proteolysis in the penumbra, but not in infarct centers following focal cerebral ischemia in mice. Neurosci 83:439–448
29. Moroni MC, Hickman ES, Denchi EL, Caprara G, Colli E, Cecconi F, Muller H, Helin K (2001) Apaf-1 is a transcriptional target for E2F and p53. Nature Cell Biol 3:552–558
30. Motonaga K, Itoh M, Hirayama A, Hirano S, Becker LE, Goto Y, Takashima S (2001) Up-regulation of E2F-1 in Down's Syndrome brain exhibiting neuropathological features of Alzheimer-type dementia. Brain Res 905:250–253
31. Muller H, Bracken AP, Vernell R, Moroni MC, Christians F, Grassilli E, Prosperini E, Vigo E, Oliner JD, Helin K (2001) E2Fs regulate the expression of genes involved in differentiation, development, proliferation, and apoptosis. Genes & Dev 15:267–285
32. Muller H, Helin K (2001) The E2F transcription factors: key regulators of cell proliferation. Biochim Biophys Acta 1470:M1–M12
33. O'Hare MJ, Hou ST, Morris EJ, Cregan SP, Xu Q, Slack RS, Park DS (2000) Induction and modulation of cerebellar granule neuron death by E2F-1. J Biol Chem 275:25358–25364
34. Osuga H, Osuga S, Wang F, Fetni R, Hogan MJ, Slack RS, Hakim AM, Ikeda JE, Park DS (2000) Cyclin-dependent kinases as a therapeutic target for stroke. Proc Natl Acad Sci USA 97:10254–10259
35. Phillips AC, Bates S, Ryan KM, Helin K, Vousden KH (1997) Induction of DNA synthesis and apoptosis are separable functions of E2F-1. Genes & Dev 11:1853–1863
36. Phillips AC, Ernst MK, Bates S, Rice NR, Vousden KH (1999) E2F-1 potentiates cell death by blocking antiapoptotic signaling pathways. Mol Cell 4:771–781
37. Phillips AC, Vousden KH (2001) E2F-1 induced apoptosis. Apoptosis 6:173–182
38. Pierce AM, Gimenez-Conti IB, Schneider-Broussard R, Martinez LA, Conti CJ, Johnson DG (1998) Increased E2F1 activity induces skin tumors in mice heterozygous and nullizygous for p53. Proc Natl Acad Sci USA 95:8858–8863
39. Read SJ, Parsons AA, Harrison DC, Philpott K, Kabnick K, O'Brien S, Clark S, Brawner M, Bates S, Gloger I, Legos JJ, Barone FC (2001) Stroke genomics: approaches to identify, validate and understand ischemic stroke gene expression. J Cerebral Blood Flow Met 21:755–778
40. Slack RS, Belliveau DJ, Rosenberg M, Atwal J, Lochmuller H, Aloyz R, Haghighi A, Lach B, Seth P, Cooper E, Miller FD (1996) Adenovirus-mediated gene transfer of the tumor suppressor, p53, induces apoptosis in postmitotic neurons. J Cell Biol 135:1085–1095
41. Small DL, Monette R, Fournier MC, Zurakowski B, Fiander H, Morley P (2001) Characterization of cyclin D1 expression in a rat global model of cerebral ischemia. Brain Res 900:26–37
42. Timsit S, Rivera S, Ouaghi P, Guischard F, Tremblay E, Ben-Ari Y, Khrestchatisky M (1999) Increased cyclin D1 in vulnerable neurons in the hippocampus after ischaemia and epilepsy: a modulator of in vivo programmed cell death? Eur J Neurosci 11:253–263
43. Trinh E, Boutillier AL, Loeffler JP (2001) Regulation of the retinoblastoma-dependent Mdm2 and E2F-1 signaling pathways during neuronal apoptosis. Mol Cell Neurosci 17:342–353
44. Xiang H, Hochman DW, Saya H, Fujiwara T, Schwartzkroin PA, Morrison RS (1996) Evidence for p53-mediated modulation of neuronal viability. J Neurosci 16:6753–6765
45. Yagita Y, Matsumoto M, Kitagawa K, Mabuchi T, Ohtsuki T, Hori M, Yanagihara T (1999) DNA cleavage and proteolysis of microtubule-associated protein 2 after cerebral ischemia of different severity. Neurosci 92:1417–1424
46. Yamasaki L, Jacks T, Bronson R, Goillot E, Harlow E, Dyson NJ (1996) Tumor induction and tissue atrophy in mice lacking E2F1. Cell 85:537–548
47. Zausinger S, Hungerhuber E, Baethmann A, Reulen HJ, Schmid-Elsaesser R (2000) Neurological impairment in rats after transient middle cerebral artery occlusion: a comparative study under various treatment paradigms. Brain Res 863:94–105

Brain Genomic Responses to Ischemic Stroke, Hemorrhage, Seizures, Hypoglycemia and Hypoxia

Y. Tang, A. Lu, B. J. Aronow, K. R. Wagner, and F. R. Sharp

Summary. RNA expression in rat brain was examined one day following ischemic stroke, intracerebral hemorrhage, kainate-induced seizures, insulin-induced hypoglycemia, and hypoxia. Rat oligonucleotide microarrays were used to compare expression of over 8000 transcripts for these brain injuries compared to sham operated and untouched controls. Of the genes induced one day following ischemia, hemorrhage, and hypoglycemia, approximately half were unique for each condition suggesting unique components of the responses to each of the injuries. A significant number of the genes were immune-related, likely representing responses to dying cells in ischemia, to blood in hemorrhage, and to neuronal death following hypoglycemia. All of the genes induced by kainate were also induced either by ischemia, hemorrhage or hypoglycemia. This strongly supports the concept that excitotoxicity plays a role in ischemia, intracerebral hemorrhage and hypoglycemia. In contrast, there was only a single gene that was down-regulated by all of the injury conditions suggesting there is not a common gene down-regulation response to injury.

Key words. Genomics – brain – immune reactions – stroke – hemorrhage – hypoglycemia – gene regulation – hypoxia – excitotoxicity

Introduction

Decreased delivery of glucose and oxygen during cerebral ischemia eventually results in tissue infarction called a stroke. The relative role of decreased glucose and oxygen delivery on the damage is unsettled, though sustained severe hypoglycemia produces diffuse neuronal death without infarction [9]. With DNA microarray technology it is possible to monitor gene expression patterns on a global scale [17,

Yang Tang, Aigang Lu, Bruce J. Aronow[1], Kenneth R. Wagner, Frank R. Sharp
Department of Neurology and Neuroscience Program, University of Cincinnati and Divisions of Molecular Developmental Biology and Informatics, [1]Children's Hospital Research Foundation, University of Cincinnati, Cincinnati, Ohio 45267, USA

Corresponding author: Dr. Frank R. Sharp, Department of Neurology & Neuroscience Program, University of Cincinnati, Vontz Center for Molecular Studies, Room 2327, 3125 Eden Avenue, Cincinnati, OH 45267-0536, USA, Tel.: 513/558-7082, Fax: 513/558-7009, E-Mail: frank.sharp@uc.edu

Maturation Phenomenon in Cerebral Ischemia V
A. M. Buchan et al. (Eds.)
© Springer-Verlag Berlin Heidelberg 2004

66, 69] and study a variety of diseases [80, 116, 138]. In this study we examined the relative roles of hypoxia and low glucose on stroke by examining gene expression using microarrays in the ischemic brain and compared this to the patterns of gene expression following hypoxia alone or insulin-induced hypoglycemia alone.

Moreover, though glutamate-induced excitotoxic injury is thought to play a central role in ischemia-induced brain injury [26], the role of excitotoxic injury in human stroke is less clear because of the failure of glutamate receptor antagonists to improve stroke outcome in the clinical setting [25, 32, 33, 62]. Therefore, we also assessed the role of excitotoxicity in ischemic brain by comparing the patterns of gene expression in the ischemic brain to that following kainic acid-induced excitotoxic damage [77]. All of the genes induced by kainate-induced seizures were induced following ischemic stroke and/or following hypoglycemia and hemorrhage. The finding that there were no genes specifically induced by kainate suggests that excitotoxicity-mediated injury is common to all of these acute injuries [15, 25, 33, 57].

We also applied the approach of comparative gene expression to an injury state where the mechanisms of tissue injury are unclear intracerebral hemorrhage. Though there is a decreased blood flow [146], marked edema [130] and apoptotic cell death around intracerebral hemorrhages [74], the mechanisms of injury are unknown. Therefore, the brain genomic response around an intracerebral hemorrhage was compared to that following ischemic stroke, hypoglycemia, kainate-induced seizures and hypoxia. Since there were a substantial number of genes induced around hemorrhages that were also induced following hypoglycemia and kainate, this suggests similar mechanisms of injury around hemorrhage related to altered glucose delivery and excitotoxicity.

A surprising finding was that there was only a single gene that was down-regulated following all of the injury conditions. This suggests that decreased gene expression is unlikely to be a regulated process common to many injury conditions. More importantly it suggests that searching for specifically down-regulated genes is unlikely to provide therapeutic gene targets that would be generally useful.

Methods

Male Sprague-Dawley rats weighing 250 to 300 g were studied. Six rats were used for each group. Brain samples from each animal in each group were taken from the same region of parietal neocortex or striatum. Therefore, there were three separate RNA samples placed on three separate microarrays. We used triplicate microarrays because of recent reports of greater reliability with triplicate samples [63]. To produce brain ischemia, rats ($n = 6$) were anesthetized with isoflurane (3% in 21% oxygen and 76% nitrogen), the neck skin and muscle incised, and the left common carotid artery isolated. Body temperature was maintained at $37.0 \pm 0.2\,^{\circ}\mathrm{C}$ with a rectal thermistor connected to a feedback controller. The external carotid and pterygopalatine arteries were ligated. A 3-0 monofilament nylon suture was threaded through the external carotid artery stump into the internal carotid artery and up to the stem of the middle cerebral artery (MCA). The suture was then anchored in place with 4-0 silk to produce a permanent MCA occlusion. The muscle and skin were sutured, and once animals recovered, they were returned to their home cages with food and water available ad libitum. This "suture" or "thread" model of MCA occlusion produces reliable infarction in the distribution of the MCA artery [150], and has been used by many groups for a variety of gene regulation studies in rat and mouse brains [97]. For sham ischemia, these animals (n = 6) were treated similarly to those animals that had the brain ischemia performed. They were anesthetized and neck and muscle incisions performed. The common carotid was isolated, but no suture was inserted into the carotid and no vessels were occluded. The wounds were sutured. Once animals recovered from anesthesia, they were returned to their home cages where food and water were available ad libitum. For intracerebral hemorrhage adult rats (n = 6) were anesthetized with isoflurane using methods identical to those used for the stroke group. The scalp was incised and a burr hole drilled 0.5 mm anterior and 4 mm lateral to bregma. A 25 gauge needle was used to deliver 50 µl of lysed autologous blood 4 mm deep into the right striatum. The wound was cleaned and sutured. Animals were allowed to recover in their home cages with food and water ad libitum for a period of 24 h. This hemorrhage model results in moderate, isolated cell death around the margins of the hemorrhage [74]. Sham hemorrhage animals (n = 6) were treated similarly to those that had the ICH performed. They were anesthetized, the scalp incised and a burr hole drilled. Instead of blood, they were injected with 50 µl of saline. The wounds were sutured. Once animals recovered from anesthesia, they were returned to their home cages where food and water were available ad libitum. To produce seizures, rats (n = 6) were injected subcutaneously with 10 mg/ kg of kainic acid dissolved in 0.9% sterile saline (Sigma). Only animals that demonstrated evidence of severe, and prolonged generalized seizures were studied [152]. The seizures generally occur repeatedly for several hours and then stop spontaneously. Animals were injected in their home cages and remained there for 24 h – the entire experiment – with food and water available ad libitum. The prolonged seizures result in reproducible injury to large numbers of neurons in hippocampus, cortex, entorhinal cortex and many other brain regions [151, 152]. To produce hypoglycemia, adult rats (n = 6) were injected with 10 U/kg regular insulin subcutaneously. One of the three subjects became obtunded, and the other two became obtunded and had repeated seizures. Approximately 4–6 h after the insulin all animals were given 20 cc of 20% sterile glucose intraperitoneally that was repeated every hour for two hours. All animals were allowed to survive for 24 h following the insulin injections. Though neuronal injury was not assessed, animals that have an isoelectric EEG for over one hour following insulin administration generally have significant neuronal injury [10]. To produce hypoxia, adult rats (n = 6), while still in their home cages, were placed in a large plexiglass chamber (Reming Bioinstruments, Redfield, New York) through which 8% oxygen was circulated for a period of 6 h. The oxygen concentration was monitored continuously in the chamber during that time, as was the carbon dioxide concentration. After 6 h of hypoxia, the animals while still in their home cages were returned to the animal room and normoxia (20.8% oxygen in room air) for a period of 18 h. This degree and duration of hypoxia is not known to produce any injury in brain, but is known to induce the hypoxia-inducible factor (HIF-1) in brain [13]. Rats that had not been handled in any way were used as untouched controls (n = 6). These animals were allowed access to food and water ad libitum and were exposed to a 12-h light and 12-h dark cycle just as was the case for the all of the other animals in this study. In addition, all experimental and control animals were housed in the same room prior to and at the conclusion of the study.

At one day (24 h) after brain ischemia, intracerebral hemorrhage, sham ischemia surgery, sham hemorrhage surgery, kainic acid injection, hypoxia, insulin/glucose injection, or being assigned as an untouched control, all subjects were anesthetized with ketamine 100 mg/kg and xylazine 20 mg/kg. Immediately following this the animal was decapitated and the brain removed as rapidly as possible. The brain was sectioned at the level of bregma. For ischemia, sham-ischemia, hemorrhage, kainate, insulin-glucose, hypoxia and untouched animals, parietal cortex was taken for microarray analysis. For sham-hemorrhage animals, striatum was taken and for hemorrhage animals, striatum and the over-lying cortex were dissected separately and both taken for microarray analysis.

Total RNA was isolated with TRIZOL Reagent (Life Technology, Rockville, Maryland). Briefly, the brain tissue pellets were lysed in Trizol Reagent using a homogenizer or repetitive pipetting. After extraction with chloroform, RNA was precipitated by isopropyl alcohol and subjected to further purification using a RNeasy mini kit (Qiagen, Valencia, California).

Expression analysis was performed according to the Affymetrix expression analysis technical manual. Double-stranded cDNA is synthesized from total RNA. To synthesize the first cDNA strand, a HPLC-purified oligo-dT primer was annealed to the RNA and extension by reverse transcriptase was performed in the presence of deoxy-oligonucleotides. The second strand was synthesized using DNA polymerase I. Double-stranded cDNA was purified using a modified phenol/chloroform extraction procedure (Eppendorf-5 Prime, Boulder, Colorado), followed by ethanol precipitation. An *in vitro* transcription was performed to produce biotin-labeled cRNA from the cDNA. cRNA was synthesized using T7 RNA polymerase and biotin-labeled ribonucleotides, and purified with affinity columns (Qiagen, Valencia, California) followed by ethanol precipitation. No amplification procedure was performed to produce the final cRNA. The amount of product was quantified by spectrophotometric analysis, and the quality of the cRNA was assessed by gel electrophoresis.

Once prepared, cRNA was hybridized to Affymetrix U34A rat arrays (Affymetrix, Santa Clara, California). The hybridization cocktail, containing the cRNA sample, BSA, and herring sperm DNA, was incubated at 99 °C, transferred to 45 °C, and then injected into the microarray chamber (Affymetrix, Santa Clara, California). Hybridization was performed automatically in a chamber at 45 °C for 16 h while being rotated at 60 rpm. The hybridized probe array was then automatically washed, dried, and scanned two times at an excitation wavelength of 488 nm.

Affymetrix rat U34A arrays analyze approximately 7000 full-length sequences and approximately 1000 EST clusters (referred to as genes in the following text). Each gene on the array is assessed using 16 probe pairs. Each probe pair consists of an oligomer (25 base long) that is designed to be perfectly complementary to a particular message (called the perfect match or PM) and a companion oligomer that is identical to PM probe except for a single base difference in a central position (called the mismatch or MM probe). The mismatch probe serves as a control for hybridization specificity and helps subtract non-specific hybridization. After hybridization intensity data is captured, the Affymetrix Genechip software MAS 4.0 automatically calculates intensity values for each probe cell and uses these probe cell intensities to calculate an average intensity for each gene (called average difference), which directly correlates with mRNA abundance. The software also gives each gene a qualitative assessment of "present" or "absent" based on a "voting scheme", with the number of instances in which the PM signal is significantly larger than the MM signal across the whole probe set. Prior to comparing any two measurements, a scaling procedure is performed so that all signal intensities on an array are multiplied by a factor that makes the mean PM-MM value for each array equal to a preset value 1500. The scaling corrects for any inter-array differences or small differences in sample concentration, labeling efficiency or fluorescence detection and makes inter-array comparisons possible. In the case of a pair-wise comparison of two array results, the patterns of change of the whole probe set (with consistent voting) is used to make a qualitative call of "Increase", "Decrease", "Marginally increase", "Marginally decrease" or "No change". The fold change is derived by the ratio of Average Differences from one experimental compared to a control array.

We compared the RNA abundance of parietal cortex from ischemia and intracerebral hemorrhage to that of parietal cortex of sham-ischemia. The RNA abundance of parietal cortex from kainate, insulin-glucose and hypoxia groups was compared to parietal cortex of the untouched control group. Lastly, the RNA abundance of striatum from the intracerebral hemorrhage group was compared to that of striatum of the sham-hemorrhage group. For each of these comparisons, there were three arrays used for each of the three animals in each experimental group and three arrays used for each of the three control animals in the control groups (triplicate arrays for each group).

To obtain differentially expressed genes for each condition, Affymetrix Genechip software MAS 4.0 was first used to compare three experimental arrays to three control arrays. For example, ischemia 1 was compared to sham-ischemia 1, 2, 3, respectively, then ischemia 2 was compared to sham-ischemia 1, 2, 3, then ischemia 3 is compared to sham-ischemia 1, 2, 3. As a result, there were 9 comparisons for each condition. By combining the fold change and the difference calls derived from the 9 comparisons, we obtained a stringent list and less stringent list for each condition. The criteria for the stringent lists were: (1) Average of the three experimental samples was at least two fold greater or less than the average of the three control samples. (2) The fold change for each of the 9 comparisons was at least 1.3 fold. (3) There were at least 7 Increase or 7 Decrease calls as determined by the Affymetrix Genechip software (Marginally Increase and Marginally Decrease are treated as Increase or Decrease). The criteria for the less stringent list were average of the three experimental samples was at least two fold greater or less than the average of three control samples, and at least 5 Increase or 5 Decrease calls in the experimental compared to control group. Classifying regulated genes into stringent and less stringent lists can provide an index of the reliability of data. Genes in the stringent lists are more reliable than genes in the less stringent list as based on the Affymetrix algorithm. The numbers of genes regulated by each condition based solely on these criteria are listed in Table 1.

Table 1. Numbers of genes that were regulated 24 h following brain ischemia, brain hemorrhage, kainate-induced seizures, insulin-glucose, and hypoxia. For the brain ischemia group, the parietal cortex was compared to that of sham-ischemia animals. For kainate-induced seizures, insulin-glucose and hypoxia, the parietal cortex was compared to the parietal cortex of untouched controls. For the brain hemorrhage group, the blood-injected striatum was compared to saline-injected striatum controls, and the parietal cortex over the striatal hemorrhage was compared to cortex of the sham-ischemia animals. There were three animals per group. The "Stringent" and "Less stringent" criteria for this table are discussed in the methods, but include an average of two-fold change or greater in the experimental compared to controls, and at least 5 out of 9 "Difference calls" for the comparisons of the three experimental samples to the three control samples using the Affymetrix Genechip software MAS 4.0

Condition		Stringent	Less stringent	Total
Brain ischemia vs sham (cortex)	up-regulated	331	84	415
	down-regulated	84	74	158
Brain hemorrhage vs sham (cortex)	up-regulated	116	89	205
	down-regulated	19	65	73
Brain hemorrhage vs sham (striatum)	up-regulated	82	40	122
	down-regulated	69	68	137
Kainate-induced seizure vs untouch (cortex)	up-regulated	104	83	187
	down-regulated	40	49	89
Insulin-glucose vs untouch (cortex)	up-regulated	91	211	302
	down-regulated	83	72	155
Hypoxia vs untouch (cortex)	up-regulated	1	14	15
	down-regulated	2	9	11

Results

Only part of the genome is expressed in a specific tissue, with the makeup of a given tissue being determined by the spectrum of expressed genes. Of the 8740 genes surveyed by the Affymetrix rat U34A chips, transcripts of 3869 genes 44.3% were detected in parietal cortex and transcripts of 4121 genes were detected in striatum as defined by criteria using the Affymetrix software (with 3 present calls in parietal cortex of untouched controls and 3 present calls in striatum of sham-hemorrhage, respectively). The most abundant genes, including mitochondrial cytochrome oxidase subunit (J01435), 14-3-3 protein eta-subtype (D17445), thy-1 (EST AA874848), cyclophilin A (EST AI228674), hsc-70 (EST AI234604), alpha-tubulin (EST AI169370), aldolase A (M12919), ribosomal protein L9 (X51706), and hsp-90 (S45392), were expressed at similar levels in cortex and striatum. Of interest, there were a significant number of genes that were differentially expressed between cortex and striatum, including such genes as oxytocin/neurophysin (K01701), melanin-concentrating hormone (M62641), and many others that were expressed at much higher levels in striatum compared to cortex.

Each condition addressed in this study produced a distinct genomic response in brain. Table 1 shows the numbers of genes that were up-regulated or down-regulated for each condition. For example, if stringent criteria were used 331 transcripts were up-regulated by brain ischemia and if less stringent criteria were used 84 transcripts were up-regulated by brain ischemia. Similarly, 82 genes were up-regulated by brain hemorrhage if stringent criteria were used compared to 40 genes if less stringent criteria were used (Table 1).

Brain ischemia regulated more genes than any other condition (415 up-regulated genes and 158 down-regulated genes), possibly because ischemia/stroke damages all cellular elements including neurons, glia, axons/white matter and the vessels. Of note, exposure of animals to 8% hypoxia for 6 h produced relatively few changes of gene expression assessed 18 h later (Table 1: 15 up-regulated genes and 11 down-regulated genes). This suggests that any effects of "systemic stress" as would be produced by global hypoxia resulted in relatively few changes of gene expression.

It is notable that for most conditions, the number of up-regulated genes in the stringent list was greater than the number in the less stringent list. This probably reflects the consistency of the fold change and difference calls using the Affymetrix algorithm/software. The only exception was insulin-glucose treatment, in which the number of less stringent genes (211 up-regulated) was greater than that of stringent genes (91 up-regulated). This was likely due to the fact that one animal in this group did not develop seizures whereas the two others had persistent seizures. Therefore, it is likely that there was considerable variation in the brain injury and in the brain genomic response in the animals given insulin; and this is reflected in a greater number of genes in the less stringent category.

Many genes that were up-regulated by each experimental condition were modulated in two or more of the groups. For example, of 122 genes up-regulated by intracerebral hemorrhage, 82 were also induced by brain ischemia and 53 were induced by kainate-induced seizures. Similarly, of 302 genes up-regulated by insulin-glucose, 152 were also up-regulated by ischemia and 111 were up-regulated by kainate.

The genes up-regulated in common by the various conditions might provide an index of common mechanisms of injury. For example, there were 52 genes induced in common by ischemia, kainate and hemorrhage. There were 98 genes induced in common by ischemia, kainate and hypoglycemia. Of the genes up-regulated by ischemia (415), kainate (187), hemorrhage (122), and by insulin-glucose (302), there were 46 genes that were up-regulated by all of these injury conditions (Table 2). Since these genes were not up-regulated by hypoxia, they could serve as markers of neuronal injury, indices of common mechanisms of injury and/or common responses to injury. For example, the up-regulation of several of these genes including GFAP, vimentin, Hsp27, heme oxygenase, and S-100 would be expected to occur mainly in astrocytes and microglia following ischemia, hemorrhage, kainate and hypoglycemia [32, 110, 120]. This would suggest that these are glial responses that are common to all of these injury conditions.

Using a cluster approach, additional genes were identified that are likely to be regulated by brain injury. These included ras-related protein (U12187), cytochrome P450 (U09540), platelet-activating factor acetylhydrolase alpha 1 subunit (AF016047), thyrotropin-releasing hormone (M23643), lysyl hydrolase (S66184), transcriptional repressor CREM (S66024), small proline-rich protein (L46593),

Table 2. Genes induced (up-regulated) to the greatest extent by brain ischemia, brain hemorrhage, kainate-induced seizures and hypoglycemia. The fold change was obtained by comparing ischemia to cortex-sham, hemorrhage to striatum-sham, and kainate and hypoglycemia to untouched controls. When the baseline level was less than 25, it was rounded to 25 to facilitate the computation of fold change and denoted with "~"

Classification	Accession #	Name	Fold change			
			Ischemia	Hemorrhage	Kainate	Hypo-glycemia
Heat shock protein	M86389	heat shock protein (HSP27)	~946.6	11.4	~734.3	~362.7
	J02722	heme oxygenase	~267.7	25	~39.8	~111.7
	Z75029	heat shock protein 70 (HSP70)	50.5	3.2	5.2	2.2
DNA damage-repairing protein	L32591	GADD45	7.3	2.5	3.1	1.4
Metal ion chelator	L33869	ceruloplasmin	7.2	2.5	4	2.4
	M11794	metallothionein-1 and metallothionein-2	6.3	3.1	4.3	4.5
Protease and protease inhibitor	D00753	contrapsin-like protease inhibitor-related protein (CPi-26)	~164.8	18.1	~99.7	~101.8
	AI169327	tissue inhibitor of metalloproteinase-1 (TIMP1)	88.7	6.4	255.9	170.5
	D90404	cathepsin C	4.3	3	3.1	1.8
	AA892775	lysozyme	2.8	3.2	2.4	3
Matrix-modeling molecules	M14656	osteopontin	56.5	5.7	41.3	37.4
	AI012030	Matrix Gla protein (Mgp)	3.1	2.8	4.5	2.6
	S61865	syndecan	5.3	1.5	2.8	1.8
	S61868	ryudecan	2.1	1.1	2	1.6
Cytoskeleton	AF028784	glial fibrillary acidic protein alpha (GFAP)	~71.5	5.8	~71.5	~83.8
	AA892333	tubulin alpha 6 (Tuba6)	15	6	10.5	9.4
	X62952	vimentin	3.3	3.3	4	4.1
	AF004811	moesin	20.7	2.3	~25.8	~9.0
Receptor	M61875	CD44	53.7	10.8	~93	~45.9
	AF087943	CD14	16.6	6.6	4.7	3.5
	J05122	peripheral-type benzodiazepine receptor (PKBS)	9	3.8	4.3	5.6

Table 2 (continued)

Classification	Accession #	Name	Fold change			
			Ischemia	Hemorrhage	Kainate	Hypo-glycemia
Molecules with immune function	AF075383	suppressor of cytokine signaling-3 (SOCS-3)	~55.5	9.2	~13.4	~13.7
	X17053	immediate-early serum-responsive JE gene	24.4	14.4	15.5	8.3
	J02962	IgE-binding protein	21.4	6.1	48.8	49.3
	X13044	MHC-associated invariant chain gamma	11.7	13.5	5	33
	X73371	Fc gamma receptor	8.1	10.3	5.5	6.2
	U31599	MHC class II-like beta chain (RT1.DMb)	3.4	2	6.8	5.4
Others	AA946503	alpha-2u globulin-related protein	~52.1	20.5	~18.6	~147.4
	S74141	hck = tyrosine kinase	~8.3	5.2	~13.6	~26.9
	U18729	cytochrome b558 alpha-subunit	29.3	3.6	~29.1	~64.6
	L05489	heparin-binding EGF-like growth factor	11.4	2.3	7	4.5
	M65149	CELF	10.9	3.4	8.1	8.4
	U92081	epithelial cell trans-membrane protein antigen precursor (RTI40)	10.9	3.2	10.2	6.5
	X59864	ASM15	9	4.3	7.3	2.2
	D10729	proteasome subunit RC1	5	2.5	6.1	3.7
	L13039	annexin II	4.9	3	2.5	2.3
	J03627	S-100-related protein	4.7	3.6	4.7	2.7
	X06916	protein p9Ka homologous to calcium-binding protein	4.1	5.8	3.6	4.7
	AF083269	p41-Arc	3.8	3.4	1.9	1.6
	J04792	ornithine decarboxylase	3.4	2	2.1	2.2
	S69874	C-FABP = cutaneous fatty acid-binding protein	2.7	1.8	2.5	2.1
	X59375	ribosomal protein S27	2.6	2.2	2.5	2.4
	M24604	proliferating cell nuclear antigen (PCNA/cyclin)	2.5	1.9	1.9	1.7
	AA944422	acidic calponin	2.5	1.6	2.9	2.6
	AF036537	homocysteine-respondent protein HCYP2	2	2	4.1	4

brain glucose transporter protein 1 (M13979), plasminogen activator inhibitor-1 (M24067), and chemokine ST38 precursor (AF053312) (not shown). Genes that are listed in Table 2, and were detected in this cluster analysis included Hsp70 heat shock protein and GADD45. These two genes would appear to be among the most reliable predictors of severe neuronal injury as suggested from previous studies because they met the stringent criteria for Table 2 and they were also detected in the cluster analysis [24, 105, 111].

Genes that were identified according to the criteria in Table 2, and that were also identified by cluster analysis to be induced by all injury conditions included but were not limited to: CELF, SOCS-3, GFAP, ASM15, CD44, immediate-early serum-responsive JE, Hsp27, heme oxygenase (Hsp32), alpha-2u globulin-related protein, and TIMP-1. These would appear to be among the most reliable, and yet sensitive injury-related genes since they were detected using two completely different methods of analysis.

The genes induced by all of the injury conditions (Table 2) are of particular interest since they suggest common pathways mediating injury or repair. CELF is a family of RNA binding proteins that are implicated in cell-specific alternative splicing which may be important for brain-specific alternative splicing [60]. Heparin-binding EGF-like growth factor is an anti-apoptotic factor that promotes neuronal and glial survival [143], and is induced in ischemic brain [122]. Annexin II, one of a family of calcium-binding proteins found in a variety of tissues, is induced in astrocytes and endothelial cells following ischemia [34]. P41-Arc is part of a complex involved in cross-linking of actin filaments [153] with an unknown function in brain. Ornithine decarboxylase (ODC) is the rate-limiting enzyme involved in the synthesis of polyamines. ODC is found in neurons and is induced in glia following ischemia, seizures, and other stressful conditions [14]. Though the polyamines appear to modulate NMDA and other receptors [54, 86], their precise role in brain injury has not been elucidated. Cutaneous fatty acid-binding protein upregulates VEGF in tumor cells [53, 65], but has not been reported in brain. The ribosomal protein S27 is found throughout normal brain, particularly in hypothalamus, is induced during the visual critical period, and has zinc finger motifs [21, 94, 123, 140]. PCNA is induced in all dividing cells following all types of injury [32, 64, 67, 68]. Acidic calponin, is an actin-, tropomyosin- and Ca^{2+} calmodulin-binding protein that inhibits MgATPase, is a calpain substrate, and is expressed in neurons, synapses and some glia in the brain [3, 39, 90, 149]. Nothing is reported on the homocysteine-respondent protein HCYP2 other than its sequence and homocysteine inducibility. Homocysteine metabolism is of interest because homocysteinemia is associated with increased stroke and myocardial infarction risk [44].

Proteases and their inhibitors are also induced by all of these acute injury conditions (Table 2) including tissue inhibitor of metalloproteinase-1 (TIMP-1), cathepsin C, lysozyme, and contrapsin-like protease inhibitor related protein (Cpi-26). TIMP-1 [132] and cathepsin C [144] are induced following brain ischemia. Cpi-26, an inhibitor of proteases including trypsin [93], and lysozyme, a white blood cell lysozomal enzyme that increases following myocardial infarction [135], have not been studied in brain.

P8 is a notable gene induced by most of the injury conditions. P8 appears to play a role in signaling TGF activation via p8 to Smad proteins [41]. The Smad proteins are transcription factors that modulate a number of target genes, particularly those involved with immune regulation that may play an important role in a variety of tissue injuries [81]. TGF and Smads are increased during scar formation after myocardial infarction [45]. Moreover, TGF is induced following stroke and may serve to protect or enhance recovery following brain ischemia [4, 85]. It is possible that different TGF and p8-mediated Smads are induced with different types of injury, and with induction of different down-stream genes in the different types of injury [49, 81, 104]. For example, TGF signaling through Smad3 may be important for induction of TIMP-1, c-jun and Fos (all induced by all acute injury conditions examined here), whereas signaling through other Smads might induce other target genes [101, 128].

The RET ligand was also induced by most of the injury conditions. RET receptor tyrosine kinase is a functional receptor for GDNF (glial cell-derived neurotrophic factor) that requires calcium [5]. GDNF binds to GDNF family receptor alpha (GFR alpha) which stimulates autophosphorylation of RET and downstream signaling [103], where the docking protein FRS2 links RET with the nitrogen-activated protein kinase-signaling cascade [78]. GDNF is a potent survival factor for specific neurons including dopamine, noradrenaline and sympathetic neurons and acts in concert with TGF to support survival of other neurons [103]. The current study suggests that GDNF-Ret signaling could be important for survival of cells following many types of acute neuronal injury.

Though the large majority of the genes induced by kainate were also induced by focal ischemia, there were a few genes induced by kainate that were also induced by hemorrhage or hypoglycemia and not by ischemia. These included C4 complement protein (U42719), macrophage metalloelastase (X98517) and proteasome activator PA28 subunit beta (D45250). These genes are of interest since they may point to excitotoxic mechanisms specific for hemorrhage and hypoglycemia injury that may not necessarily be shared with ischemia.

For example, the macrophage metalloelastase was induced by kainate and hypoglycemia but not ischemia. The macrophage metalloelastase is a matrix metalloproteinase, MMP12, one of a family of metalloproteinases that can degrade all components of the extracellular matrix [134, 136]. MMP12 appears to exacerbate lung injury [134], is not expressed in normal astrocytes [2], but is expressed to high levels in astrocytic brain tumors [55]. TGF beta inhibition of MMP12 is mediated by Smad3 [136]. The role of macrophage metalloelastase in kainate and hypoglycemic injury is not clear, but could be related to macrophage removal of dying neurons in specific regions of brain in both conditions.

The proteasome activator PA28 was also induced by kainate and hypoglycemia, and not by ischemia. Proteasome activator PA28 is a gamma interferon-inducible complex [38]. PA28 (also called 11S REG) is one of two protein complexes that have been found to bind the ends of the proteasome and activate it. There are three PA28 subunits, alpha, beta and gamma; alpha and beta being found in the cytosol, and gamma in the nucleus [147]. PA28 is required for the presentation of certain major histocompatibility (MHC) class I antigens that allows cytotoxic T lymphocytes to lyse the cells expressing cell-surface MHC Class I molecules complexed with foreign peptides [99, 118]. Again, PA28 upregulation by kainate and hypoglycemia suggests specific mechanisms of cell mediated immune injury in these conditions. However, it has also been found that Hsc70, Hsp40 and PA28 are necessary and sufficient to fully reconstitute Hsp90-mediated refolding of partially denatured firefly luciferase [79], suggesting a possible protective role for PA28.

The data also show that though there were many genes induced in common by ischemia, hemorrhage, hypoglycemia, and kainate, there were a number of genes induced that were specific for each of these conditions except for kainate. Of the 415 genes induced by ischemia, 236 were only induced by ischemia and not by kainate or hemorrhage. Of the 122 genes induced by hemorrhage, 39 were induced only by hemorrhage and not induced by ischemia or kainate. Of the 302 genes induced by insulin, 132 were only induced by insulin and not ischemia or kainate. Kainate, however, was the only condition in which there were no specifically up-regulated genes. Of the 187 genes induced by kainate, 150 were induced by ischemia and/or hemorrhage. Of the same 187 genes induced by kainate, 162 were induced by ischemia and/or insulin-glucose. All of the 187 genes induced by kainate were induced either by ischemia, hemorrhage and/or insulin. As discussed below, this data suggests that excitotoxicity is an important mechanism of injury in ischemia, hemorrhage and hypoglycemia and that there is not likely to be a mechanism of injury specific for kainate.

Genes that are markedly regulated at 24 h after ischemia included the following: inflammatory, immune and signaling molecules like JAK2 (U13396), MAP-kinase phosphatase (cpg-21) (AF013144), dual specificity phosphatase (U42627), lipopoly-saccaride-binding protein (L32132), interleukin-6 (M26745), decay accelerating factor GPI-form precursor (DAF) (AF039583), major acute phase alpha-1 protein (K02814), eNOS (AJ011116), CC chemokine ST38 precursor (AF053312), intercellular adhesion molecule-1 (ICAM-1) (D00913), interleukin 1 receptor antagonist (M63101), and integrin-alpha 1 (X52140). Hypoxia-inducible factor 1 (Y09507) and some of its target genes like hexokinase II (S56464), VEGF (L20913), angiopoietin (AF030378), adrenomedullin (D14069), and brain glucose-transporter protein (M13979) were all induced. Of note is the finding that the DNA damaging-inducible protein GADD153 (U30186) is induced at 24 h after ischemia but not in the other conditions examined here.

There were a number of genes that were induced mainly following hemorrhage. Many of these genes were related to inflammatory or immune responses, such as pre-pro-complement C3 (X52477), macrophage inflammatory protein-2 precursor (U45965), CD38 (D29646), intercellular calcium binding protein (macrophage-related protein 8) (EST AA957003), alpha-1 acid glycoprotein (V01216) and Gly-cam-1 (glycosylation-dependent cell adhesion molecule 1 L08100). Others included MHC class II MHC RT1-B region Ia antigen (M15562), leukocyte antigen MRC-OX44 (M57276), MRC OX-45 surface antigen (X13016), and interleukin-1 receptor type 2 (Z22812). Additional genes up-regulated by hemorrhage included CC chemokine receptor (E13732), DORA protein (AJ223184), apolipoprotein B mRNA-editing protein (L07114), and interleukin-1 beta converting enzyme (U14647).

There were also some genes induced specifically by insulin/glucose. These included genes such as MHC class I RT1.C/E mRNA (EST AI235890), presenilin-2 (X99267), plakoglobin (U58858) and EST AA799766. It is notable that the rat U34A chips only include a few of the known, cloned glucose-regulated proteins (GRPs) that are known to be glucose responsive [72].

A cautionary note should be made about the patterns of genes regulated for a specific condition. The problem of the time course of gene expression has not been dealt with at all here since all samples were compared at 24 h. For example, c-fos (X06769), BDNF (S71196, S76758, D10938, X67108), and cyclooxygenase-2 (COX2) (L25925) all are induced in ischemic brain at 24 h but are not up-regulated in the other injury conditions at 24 h. Similarly, some immediate-early genes like NGFI-A (N18416), NGFI-B (U17254), NGFI-C (N92433), Krox-24 (Y753397), and Arc (U19866) are up-regulated in ischemia, but down-regulated by kainate and probably hemorrhage, hypoxia, insulin-glucose and even sham when measured at 24 h after treatments. However, all of these molecules, including c-fos, BDNF, COX2, NGFI-A, NGFI-B, NGFI-C are induced in ischemic brain and following seizures within a few hours and return to normal or decrease below normal levels by 24 h after kainate [58, 110]. Hence, ischemia differs from these other injury conditions in that there is persistent induction of these early genes in ischemic brain compared to the other conditions examined here.

Like up-regulated genes, some genes are down-regulated following one or more conditions. The number of down-regulated genes for each condition tended to be less than the number of up-regulated genes. Among 646 up-regulated genes in parietal cortex by any condition, 130 genes were not expressed in untouched brain and were induced by one or more injury conditions. In comparison, among 381 cortex genes down-regulated by any condition, only 20 genes were completely suppressed by any of the injury conditions.

Like up-regulated genes, some down-regulated genes are modulated by multiple conditions. There were 42 genes down-regulated by both ischemia and kainate, 12 genes down-regulated by both kainate and hemorrhage and 10 genes down-regulated by both insulin-glucose and kainate. Compared to the up-regulated genes, the numbers of shared genes were fewer. One of the most unexpected results was that only a single gene – calmodulin-dependent protein kinase (CaM kinase, M63333) – was down-regulated in common by all of the injury conditions including ischemia, kainate, hemorrhage and insulin. The failure to detect a group of common genes that were down-regulated in all of the injury conditions suggests that transcriptional down-regulation is probably not a major part of the genomic response to brain injury. Thus, protein phosphorylation and de-phosphorylation may play the major role in down-regulating signaling cascades. The current results suggest that targeting down-regulated genes for therapeutic interventions is not likely to yield a generally useful therapeutic strategy.

Discussion

Many of the genes induced by ischemia were also induced following kainate-induced seizures, intracerebral hemorrhage and hypoglycemia including heat shock proteins, DNA damage repair proteins, proteases and protease inhibitors, metal ion and free-radical scavengers, and calcium-binding proteins. The induction of heme-oxygenase (Hsp32, microglia), Hsp27 (astrocytes), ceruloplasmin (astrocytes), GFAP (astrocytes), vimentin (astrocytes), matrix glia protein, and S-100 (astrocytes) following ischemia, hemorrhage, kainate and hypoglycemia points to a generalized glial response to all of these injury conditions [7, 36, 46, 52, 88, 120, 139]. The induction of the heat shock proteins/chaperones, including Hsp27, Hsp32 (heme oxygenase), and Hsp70, following these injuries points to a general role of these stress genes to respond to denatured proteins within cells, metabolism of heme proteins, and cytoskeletal stress [1, 110, 111]. All of these heat shock proteins are induced following ischemia and hemorrhage and kainate, but their role in hypoglycemia was previously unknown [1, 59, 83, 91, 110, 113, 125].

The induction of CD44, CD14, MHC, and other immune-related molecules reinforces the concept that brain injuries of all types activate an immune/inflammatory response, with certain aspects of this immune response being common to all injuries. CD44, a transmembrane glycoprotein involved in endothelial cell recognition and trafficking of lymphocytes and regulation of cytokine gene expression, is induced in microglia, macrophages, and vessels following stroke [131]. This study shows similar up-regulation following neuronal injury caused by hemorrhage, hypoglycemia, and seizures. MHC molecules are up-regulated on microglia/macrophages following ischemia and kainate [16, 40, 56, 73] and all of the injuries studied here and would appear to play a central role in the immune-mediated removal of necrotic and apoptotic cells.

The induction of ceruloplasmin and metallothionein-1 and -2 emphasizes the importance of iron and other ions following acute injury. Though induced by ischemia [19, 145], there was little previous evidence for the role of metallothioneins in hemorrhage, hypoglycemia and seizures. Metallothioneins protect against stroke [127] and may protect against other causes of acute injury. Ceruloplasmin is important in the metabolism and transport of both iron and copper [137, 148], so that a prominent role for this metal ion-related protein can be suggested in many acute neuronal injuries.

Matrix and cytoskeletal molecules, including osteopontin, are also induced by all of the injury conditions. Osteopontin and its integrin receptor alpha vs beta 3 are induced in macrophages and microglia in the region around a stroke, with subsequent up-regulation of its integrin receptor on astrocytes [36, 133]. These data suggest that osteopontin released from microglia/macrophages at the edge of an infarct modulate the astrocytic scar/response following stroke [36]. The current findings suggest that this also occurs in injury states where there is selective neuronal cell death, including following hemorrhage, hypoglycemia and seizures. The above genes are believed to have particular significance because they are all induced following ischemia, hemorrhage, seizures, and hypoglycemia, and represent common pathways for injury or repair in the acutely injured brain.

Excitotoxicity plays a key role in the pathogenesis of many neurological disorders, and especially in ischemic stroke and seizures [15, 26, 76, 114]. Though there have been several suggestions that excitotoxicity plays a role in neuronal damage following ischemia, seizures, and hypoglycemia [77, 114, 115], there has not been data suggesting that damage around hemorrhages might be related to excitotoxicity. The finding that all of the genes induced following kainate are also induced following hemorrhage, hypoglycemia and ischemia, suggests a prominent role for excitotoxicity in each of these conditions.

Though there has been little previous evidence that glutamate might mediate injury around an intracerebral hemorrhage, in vitro studies suggest interactions between blood/hemoglobin and excitotoxicity [42, 100]. Importantly, glutamate could increase around intracerebral hemorrhages (ICH) from multiple sources including blood plasma, lysed red blood cells and injured brain cells. Plasma, which has a high concentration of glutamate [124], diffuses into brain around a hemorrhage [129, 130]. The intracellular concentrations of glutamate in red blood cells are very high, and are in the range that are neurotoxic in vitro and are in the range of the extracellular glutamate following cerebral ischemia [77, 84, 124]. The lysis of red blood cells would markedly increase glutamate around hemorrhages. Moreover, hemoglobin released by red blood cells can exacerbate excitotoxic injury in vitro and thrombin released in regions of ICH can potentiate NMDA function [42, 100].

An intense inflammatory response is a well-documented process after various brain injuries [43]. Despite the overwhelming evidence of the existence of inflammation following brain injury, many of the signaling mechanisms remain to be elucidated [30]. The induction of immune-related molecules, such as MHC antigens, IgE-binding protein, and Fc gamma receptor by all injury conditions support the involvement of an immune response that could contribute to secondary brain injury and/or contribute to repair.

The current data provides some new insights into common mechanisms of immune responses to these different causes of injury. CD14 is induced by all of the injury conditions here, as is the lipopolysaccharide (LPS) receptor present both in plasma and at the surface of myeloid cells. It is considered to be the key player in the induction of septic shock provoked by gram-negative bacteria [95]. LPS is first aggregated by LPS-binding protein (LBP, which is also induced following ischemia) and then binds to CD14. The LPS-CD14 interaction can lead to the activation of NF-κB through a transduction pathway in which Toll-like 4 receptor is involved [92, 106, 141, 142]. Through this transduction pathway, LPS causes dramatic transcriptional regulation of a wide range of pro-inflammatory genes including TNF, IL-1, IL-6, IL-8, ICAM-1, E-selectin and others [141]. CD14 may be involved in the regulation of a brain inflammatory response via an autocrine/paracrine loop [82]. A low dose of LPS dramatically sensitizes the immature brain to injury following short periods of hypoxia-ischemia that by themselves cause little injury and this effect has been associated with an enhanced expression of CD14 mRNA in the brain [35]. Although no exogenous LPS is present in brain following injury, the heat shock protein Hsp70 can serve as an endogenous ligand to activate CD14 and stimulate the production of IL-1 beta, IL-6 and TNF-alpha [8]. The discovery of the induction of CD14 may provide further clues about the induction and regulation of inflammatory reactions in injured brain, and could encourage further examination of the potentially harmful role of secondary infections in acute brain injury. Moreover, since CD14 polymorphisms are associated with increased risk of myocardial infarction [48, 112, 126], they could have a role in stroke.

The immediate-early JE gene is an analog of murine monocyte chemoattractant protein-1, which is responsible for recruiting hematogenous macrophages and is induced after peripheral nerve injury [20]. It was induced by all of the injury conditions examined here, and probably plays a central role in recruiting macrophages into acutely injured brain. The peripheral-type benzodiazepine receptor

(PTBR) is found on the outer mitochondrial membranes of peripheral tissues and cells. In mammalian CNS, the peripheral-type benzodiazepine receptor is localized within the astrocytes and microglia. PTBR transports cholesterol to the site of neurosteroid biosynthesis, and is induced following brain trauma and global ischemia [96, 98]. Suppressor of cytokine signaling (SOCS) is a family of molecules induced by the activation of JAK-STAT pathway following cytokine stimulation and that negatively regulate cytokine signaling. SOCS-3 has been shown to negatively regulate fetal liver erythropoiesis by inhibiting JAK2 activity, although its function in brain injury is not defined [70].

Some immune-related molecules such as cytokine-induced neutrophil chemoattractant (CINC/gro D11445) and macrophage inflammatory protein-1 beta (U06434) are induced following ICH and ischemia but not in kainate or hypoglycemia. This indicates neutrophil and probably macrophage infiltration may play a greater role in the development of injury following ischemia and hemorrhage than in kainate. This is also consistent with reports that there is no neutrophil recruitment and a 2-day delay in the increase of macrophage-microglial cell numbers following kainate-induced seizures [6].

Ischemia and brain hemorrhage appeared to induce the greatest number of immune-mediated molecules. Genes induced specifically by brain hemorrhage and not by the other conditions included complement component C3, macrophage inflammatory protein-2 precursor, CD38, intercellular calcium-binding protein (macrophage-related protein 8), Glycam1 (glycosylation-dependent cell adhesion molecule 1) and MHC RT1-B region Ia antigen. Only a few of these genes have been studied following brain hemorrhage [75]. The large number of immune/cytokine genes induced by hemorrhage likely relate to the unique molecules presented to the brain following this type of injury: blood plasma proteins; red blood cells and hemoglobin; clotting factors; and white blood cells themselves within the hemorrhage. Most of these molecules and cells would not be present with the other types of injury, and hence likely stimulate a unique and possibly aggressive immune reaction [130].

Similarly, there are a significant number of immune-related genes that are induced specifically in ischemic brain. These included lipopolysaccaride-binding protein, interleukin-6, decay-accelerating factor, GPI-form precursor (DAF), major acute phase alpha-1 protein, CC chemokine ST38 precursor, intercellular adhesion molecule-1 (ICAM-1), integrin-1 and ICE-like cysteine protease (Lice). Though IL-6, ICAM-1, IL-1, and intergrin-1 are known to be induced following ischemia, many of the others were not [11, 16, 29, 61, 102, 119]. The unique feature of ischemic infarction compared to kainate, hypoglycemia and hemorrhage is that infarction involves death of not only neurons but also death of glia and vascular elements. Hence, the immune response to these dying cells/cellular elements would be different in ischemic brain compared to hemorrhage and compared to the selective neuronal cell death that occurs with kainate seizures and hypoglycemic neuronal injury.

Liver arginase, a urea cycle and nitric oxide synthase-regulating enzyme that is also expressed in CNS [50, 117], was induced following intracerebral hemorrhage (ICH). Arginase inhibits nitric oxide generation and excitotoxic necrosis in cortical neurons exposed to glutamate by depleting arginine and preventing it from being oxidized by nitric oxide synthase to nitric oxide [28]. Furthermore, intracellular arginine depletion results in an accumulation of uncharged tRNAs, leading to eIF-2 phosphorylation and repression of global protein synthesis and eventually suppression of apoptosis in response to some stimuli [37]. The induction of liver arginase following ICH could be protective.

There were more genes specifically regulated following focal ischemia than in any other condition. This probably relates in part to injury to neurons, glia and vascular elements that is not a feature of the other injury models studied here. For example, the DNA damage-repairing protein GADD45 was induced in all of the injury conditions (ischemia, hemorrhage, kainate and hypoglycemia – Table 2), whereas the growth and DNA-damage-inducible protein GADD153 is induced only following ischemia. This would suggest that GADD45 could be involved in DNA repair common to all of the injury conditions, and this DNA repair would likely be occurring mainly in neurons since this is the only cell type injured in all of these conditions. In fact, current data demonstrate induction of GADD45 in neurons following focal and global ischemia [22, 24, 51, 105]. In addition, the above findings suggest that GADD153 might be involved in DNA repair in glia and/or vascular elements in the ischemic brain and that it is not induced in the other conditions because of the relative absence of glial and vascular injury. Though GADD45 is induced in neurons following ischemia, it is not known if it is induced in neurons following the other conditions examined here. In addition, though GADD153 is induced in ischemic brain [7, 52, 88], it is not known if it is induced in glial or vascular elements following ischemia. This seems likely since GADD153 is induced in injured heart and liver, and it is present at high levels in gliomas [23, 27]. Further studies will be needed to determine if there are different DNA repair genes in different cell types in brain, and whether this might contribute to selective vulnerability or resistance of certain cell populations.

A family of hypoxia-responsive genes is also induced specifically in the ischemic brain and not following kainate, hemorrhage or hypoglycemia. These include hypoxia inducible factor-1 and its target genes like VEGF, brain-glucose transporter, glycolytic enzymes, angiopoietin and adrenomedullin [108]. HIF-1 and its target genes are induced following ischemia in brain and other tissues [13, 108, 109]. It is likely that the HIF-1 induction is protective since pharmacological induction of HIF-1 with cobalt chloride protects neonatal brain against ischemia [12]. The finding that six hours of hypoxia produced only minimal changes of gene expression 24 h later was somewhat of a surprise. However, once hypoxia is reversed there is rapid ubiquitin-mediated degradation of HIF-1 and the stimulus for the induction of hypoxia-inducible genes would decrease rapidly [107].

These genomic studies point to some expected and unexpected genomic responses following ischemic stroke, hemorrhage, kainate-induced seizure, and hypoglycemia. However, gene expression after neuronal injury cannot be studied in isolation. Genes expressed at different time points and in different cell types can have different implications for injury and repair. It will be important to characterize the spatial, temporal and cellular expression before speculations regarding therapeutic potential can be addressed. Lastly, proteins provide the framework for the working cell, with some RNAs not being transcribed, and with massive protein regulation unrelated to transcription but related to phosphorylation and alterations of cellular compartments and rates of degradation. Another major complicating factor following many types of injury is that though many genes may be transcribed into RNA, many proteins are not translated in injured tissues including brain [47]. The role of this translation block in protecting injured tissue or promoting injury is still unknown [18, 31, 71, 87, 89, 110, 121].

Acknowledgements. This study was supported by ES 08822/ES/NIEHS (BJA), and NS28167, AG19561 and a Bugher Award (FRS) from the American Heart Association (National).

References

1. Abe H, Nowak TS Jr (1996) The stress response and its role in cellular defense mechanisms after ischemia. Adv Neurol 71:451–466
2. Agapova OA, Ricard CS, Salvador-Silva M, Hernandez MR (2001) Expression of matrix metalloproteinases and tissue inhibitors of metalloproteinases in human optic nerve head astrocytes. Glia 33:205–216
3. Agassandian C, Plantier M, Fattoum A, Represa A, der Terrossian E (2000) Subcellular distribution of calponin and caldesmon in rat hippocampus. Brain Res 887:444–449
4. Ali C, Docagne F, Nicole O, Lesne S, Toutain J, Young A, Chazalviel L, Divoux D, Caly M, Cabal P, Derlon JM, MacKenzie ET, Buisson A, Vivien D (2001) Increased expression of transforming growth factor-beta after cerebral ischemia in the baboon: an endogenous marker of neuronal stress? J Cereb Blood Flow Metab 21:820–827
5. Anders J, Kjar S, Ibanez CF (2001) Molecular modeling of the extracellular domain of the RET receptor tyrosine kinase reveals multiple cadherin-like domains and a calcium-binding site. J Biol Chem 276:35808–35817
6. Andersson PB, Perry VH, Gordon S (1991) The CNS acute inflammatory response to excitotoxic neuronal cell death. Immunol Lett 30:177–181
7. Anguelova E, Boularand S, Nowicki JP, Benavides J, Smirnova T (2000) Up-regulation of genes involved in cellular stress and apoptosis in a rat model of hippocampal degeneration. J Neurosci Res 59:209–217
8. Asea A, Kraeft SK, Kurt-Jones EA, Stevenson MA, Chen LB, Finberg RW, Koo GC, Calderwood SK (2000) HSP70 stimulates cytokine production through a CD14-dependant pathway, demonstrating its dual role as a chaperone and cytokine. Nat Med 6:435–442
9. Auer RN, Kalimo H, Olsson Y, Siesjo BK (1985) The temporal evolution of hypoglycemic brain damage. I. Light- and electron-microscopic findings in the rat cerebral cortex. Acta Neuropathol (Berl) 67:13–24
10. Auer RN, Olsson Y, Siesjo BK (1984) Hypoglycemic brain injury in the rat. Correlation of density of brain damage with the EEG isoelectric time: a quantitative study. Diabetes 33:1090–1098
11. Barone FC, Feuerstein GZ (1999) Inflammatory mediators and stroke: new opportunities for novel therapeutics. J Cereb Blood Flow Metab 19:819–834
12. Bergeron M, Gidday JM, Yu AY, Semenza GL, Ferriero DM, Sharp FR (2000) Role of hypoxia-inducible factor-1 in hypoxia-induced ischemic tolerance in neonatal rat brain. Ann Neurol 48:285–296
13. Bergeron M, Yu AY, Solway KE, Semenza GL, Sharp FR (1999) Induction of hypoxia-inducible factor-1 (HIF-1) and its target genes following focal ischaemia in rat brain. Eur J Neurosci 11:4159–4170
14. Bernstein HG, Muller M (1999) The cellular localization of the L-ornithine decarboxylase/polyamine system in normal and diseased central nervous systems. Prog Neurobiol 57:485–505
15. Bittigau P, Ikonomidou C (1997) Glutamate in neurologic diseases. J Child Neurol 12:471–485
16. Bona E, Andersson AL, Blomgren K, Gilland E, Puka-Sundvall M, Gustafson K, Hagberg H (1999) Chemokine and inflammatory cell response to hypoxia-ischemia in immature rats. Pediatr Res 45:500–509
17. Brown PO, Botstein D (1999) Exploring the new world of the genome with DNA microarrays. Nat Genet 21:33–37

18. Burda J, Martin ME, Gottlieb M, Chavko M, Marsala J, Alcazar A, Pavon M, Fando JL, Salinas M (1998) The intraischemic and early reperfusion changes of protein synthesis in the rat brain. eIF-2 alpha kinase activity and role of initiation factors eIF-2 alpha and eIF-4E. J Cereb Blood Flow Metab 18:59–66
19. Campagne MV, Thibodeaux H, van Bruggen N, Cairns B, Lowe DG (2000) Increased binding activity at an antioxidant-responsive element in the metallothionein-1 promoter and rapid induction of metallothionein-1 and -2 in response to cerebral ischemia and reperfusion. J Neurosci 20:5200–5207
20. Carroll SL, Frohnert PW (1998) Expression of JE (monocyte chemoattractant protein-1) is induced by sciatic axotomy in wild type rodents but not in C57BL/Wld(s) mice. J Neuropathol Exp Neurol 57:915–930
21. Chan YL, Suzuki K, Olvera J, Wool IG (1993) Zinc finger-like motifs in rat ribosomal proteins S27 and S29. Nucleic Acids Res 21:649–655
22. Charriaut-Marlangue C, Richard E, Ben-Ari Y (1999) DNA damage and DNA damage-inducible protein Gadd45 following ischemia in the P7 neonatal rat. Brain Res Dev Brain Res 116:133–140
23. Chen BP, Wolfgang CD, Hai T (1996) Analysis of ATF3, a transcription factor induced by physiological stresses and modulated by gadd153/Chop10. Mol Cell Biol 16:1157–1168
24. Chen J, Uchimura K, Stetler RA, Zhu RL, Nakayama M, Jin K, Graham SH, Simon RP (1998) Transient global ischemia triggers expression of the DNA damage-inducible gene GADD45 in the rat brain. J Cereb Blood Flow Metab 18:646–657
25. Choi DW (1998) Antagonizing excitotoxicity: a therapeutic strategy for stroke? Mt Sinai J Med 65:133–138
26. Choi DW (1988) Glutamate neurotoxicity and diseases of the nervous system. Neuron 1:623–634
27. Collins VP (1995) Gene amplification in human gliomas. Glia 15:289–296
28. Dawson VL, Dawson TM, London ED, Bredt DS, Snyder SH (1991) Nitric oxide mediates glutamate neurotoxicity in primary cortical cultures. Proc Natl Acad Sci USA 88:6368–6371
29. del Zoppo GJ, Ginis I, Hallenbeck JM, Iadecola C, Wang X, Feuerstein GZ (2000) Inflammation and stroke: putative role for cytokines, adhesion molecules and iNOS in brain response to ischemia. Brain Pathol 10:95–112
30. del Zoppo GJ, Becker KJ, Hallenbeck JM (2001) Inflammation after stroke: is it harmful? Arch Neurol 58:669–672
31. Dever TE (1999) Translation initiation: adept at adapting. Trends Biochem Sci 24:398–403
32. Dirnagl U, Iadecola C, Moskowitz MA (1999) Pathobiology of ischaemic stroke: an integrated view. Trends Neurosci 22:391–397
33. Doble A (1999) The role of excitotoxicity in neurodegenerative disease: implications for therapy. Pharmacol Ther 81:163–221
34. Eberhard DA, Brown MD, Van den Berg SR (1994) Alterations of annexin expression in pathological neuronal and glial reactions. Immunohistochemical localization of annexins I, II (p36 and p11 subunits), IV, and VI in the human hippocampus. Am J Pathol 145:640–649
35. Eklind S, Mallard C, Leverin AL, Gilland E, Blomgren K, Mattsby-Baltzer I, Hagberg H (2001) Bacterial endotoxin sensitizes the immature brain to hypoxic-ischaemic injury. Eur J Neurosci 13:1101–1106
36. Ellison JA, Barone FC, Feuerstein GZ (1999) Matrix remodeling after stroke. *De novo* expression of matrix proteins and integrin receptors. Ann NY Acad Sci 890:204–222
37. Esch F, Lin KI, Hills A, Zaman K, Baraban JM, Chatterjee S, Rubin L, Ash DE, Ratan RR (1998) Purification of a multipotent antideath activity from bovine liver and its identification as arginase: nitric oxide-independent inhibition of neuronal apoptosis. J Neurosci 18:4083–4095
38. Fabunmi RP, Wigley WC, Thomas PJ, DeMartino GN (2001) Interferon gamma regulates accumulation of the proteasome activator PA28 and immunoproteasomes at nuclear PML bodies. J Cell Sci 114:29–36
39. Ferhat L, Charton G, Represa A, Ben-Ari Y, der Terrossian E, Khrestchatisky M (1996) Acidic calponin cloned from neural cells is differentially expressed during rat brain development. Eur J Neurosci 8:1501–1509
40. Finsen BR, Jorgensen MB, Diemer NH, Zimmer J (1993) Microglial MHC antigen expression after ischemic and kainic acid lesions of the adult rat hippocampus. Glia 7:41–49
41. Garcia-Montero AC, Vasseur S, Giono LE, Canepa E, Moreno S, Dagorn JC, Iovanna JL (2001) Transforming growth factor beta-1 enhances Smad transcriptional activity through activation of p8 gene expression. Biochem J 357:249–253

42. Gingrich MB, Junge CE, Lyuboslavsky P, Traynelis SF (2000) Potentiation of NMDA receptor function by the serine protease thrombin. J Neurosci 20:4582–4595
43. Hallenbeck JM (1996) Significance of the inflammatory response in brain ischemia. Acta Neurochir Suppl (Wien) 66:27–31
44. Hankey GJ, Eikelboom JW (2001) Homocysteine and stroke. Curr Opin Neurol 14:95–102
45. Hao J, Ju H, Zhao S, Junaid A, Scammell-La Fleur T, Dixon IM (1999) Elevation of expression of Smads 2, 3, and 4, decorin and TGF-beta in the chronic phase of myocardial infarct scar healing. J Mol Cell Cardiol 31:667–678
46. Hertz L, Yu AC, Kala G, Schousboe A (2000) Neuronal-astrocytic and cytosolic-mitochondrial metabolite trafficking during brain activation, hyperammonemia and energy deprivation. Neurochem Int 37:83–102
47. Hossmann KA (1994) Viability thresholds and the penumbra of focal ischemia. Ann Neurol 36:557–565
48. Hubacek JA, Rothe G, Pit'ha J, Skodova Z, Stanek V, Poledne R, Schmitz G (1999) C(-260)→ T polymorphism in the promoter of the CD14 monocyte receptor gene as a risk factor for myocardial infarction. Circulation 99:218–220
49. Itoh S, Itoh F, Goumans MJ, Ten Dijke P (2000) Signaling of transforming growth factor-beta family members through Smad proteins. Eur J Biochem 267:6954–6967
50. Jenkinson CP, Grody WW, Cederbaum SD (1996) Comparative properties of arginases. Comp Biochem Physiol B Biochem Mol Biol 114:107–132
51. Jin K, Chen J, Kawaguchi K, Zhu RL, Stetler RA, Simon RP, Graham SH (1996) Focal ischemia induces expression of the DNA damage-inducible gene GADD45 in the rat brain. Neuroreport 7:1797–1802
52. Jin K, Mao XO, Eshoo MW, Nagayama T, Minam M, Simon RP, Greenberg DA (2001) Microarray analysis of hippocampal gene expression in global cerebral ischemia. Ann Neurol 50:93–103
53. Jing C, Beesley C, Foster CS, Chen H, Rudland PS, West DC, Fujii H, Smith, PH, Ke Y (2001) Human cutaneous fatty acid-binding protein induces metastasis by up-regulating the expression of vascular endothelial growth factor gene in rat Rama 37 model cells. Cancer Res 61:4357–4364
54. Johnson TD (1998) Polyamines and cerebral ischemia. Prog Drug Res 50:193–258
55. Kachra Z, Beaulieu E, Delbecchi L, Mousseau N, Berthelet F, Moumdjian R, Del Maestro R, Beliveau R (1999) Expression of matrix metalloproteinases and their inhibitors in human brain tumors. Clin Exp Metastasis 17:555–566
56. Kato H, Kogure K, Liu XH, Araki T, Itoyama Y (1996) Progressive expression of immunomolecules on activated microglia and invading leukocytes following focal cerebral ischemia in the rat. Brain Res 734:203–212
57. Kaul M, Garden GA, Lipton SA (2001) Pathways to neuronal injury and apoptosis in HIV-associated dementia. Nature 410:988–994
58. Koistinaho J, Hokfelt T (1997) Altered gene expression in brain ischemia. Neuroreport 8:1–8
59. Krueger AM, Armstrong JN, Plumier J, Robertson HA, Currie RW (1999) Cell specific expression of Hsp70 in neurons and glia of the rat hippocampus after hyperthermia and kainic acid-induced seizure activity. Brain Res Mol Brain Res 71:265–278
60. Ladd AN, Charlet N, Cooper TA (2001) The CELF family of RNA binding proteins is implicated in cell-specific and developmentally regulated alternative splicing. Mol Cell Biol 21:1285–1296
61. Lebel E, Vallieres L, Rivest S (2000) Selective involvement of interleukin-6 in the transcriptional activation of the suppressor of cytokine signaling-3 in the brain during systemic immune challenges. Endocrinology 141:3749–3763
62. Lee JM, Zipfel GJ, Choi DW (1999) The changing landscape of ischaemic brain injury mechanisms. Nature 399:7–14
63. Lee ML, Kuo FC, Whitmore GA, Sklar J (2000) Importance of replication in microarray gene expression studies: statistical methods and evidence from repetitive cDNA hybridizations. Proc Natl Acad Sci USA 97:9834–9839
64. Lehrmann E, Christensen T, Zimmer J, Diemer NH, Finsen B (1997) Microglial and macrophage reactions mark progressive changes and define the penumbra in the rat neocortex and striatum after transient middle cerebral artery occlusion. J Comp Neurol 386:461–476
65. Leo CP, Pisarska MD, Hsueh AJ (2001) DNA array analysis of changes in preovulatory gene expression in the rat ovary. Biol Reprod 65:269–276
66. Lipshutz RJ, Fodor SP, Gingeras TR, Lockhart DJ (1999) High density synthetic oligonucleotide arrays. Nat Genet 21:20–24

67. Liu J, Bartels M, Lu A, Sharp FR (2001) Microglia/macrophages proliferate in striatum and neocortex but not in hippocampus after brief global ischemia that produces ischemic tolerance in gerbil brain. J Cereb Blood Flow Metab 21:361–373
68. Liu J, Solway K, Messing RO, Sharp FR (1998) Increased neurogenesis in the dentate gyrus after transient global ischemia in gerbils. J Neurosci 18:7768–7778
69. Lockhart DJ, Barlow C (2001) Expressing what's on your mind: DNA arrays and the brain. Nat Rev Neurosci 2:63–68
70. Marine JC, McKay C, Wang D, Topham DJ, Parganas E, Nakajima H, Pendeville H, Yasukawa H, Sasaki A, Yoshimura A, Ihle JN (1999) SOCS3 is essential in the regulation of fetal liver ery-thropoiesis. Cell 98:617–627
71. Martin De La Vega C, Burda J, Nemethova M, Quevedo C, Alcazar A, Martin ME, Danielisova V, Fando JL, Salinas M (2001) Possible mechanisms involved in the down-regulation of transla-tion during transient global ischaemia in the rat brain. Biochem J 357:819–826
72. Massa SM, Swanson RA, Sharp FR (1996) The stress gene response in brain. Cerebrovasc Brain Metab Rev 8:95–158
73. Matsuoka Y, Kitamura Y, Okazaki M, Sakata M, Tsukahara T, Taniguchi T (1998) Induction of heme oxygenase-1 and major histocompatibility complex antigens in transient forebrain isch-emia. J Cereb Blood Flow Metab 18:824–832
74. Matsushita K, Meng W, Wang X, Asahi M, Asahi K, Moskowitz MA, Lo EH (2000) Evidence for apoptosis after intercerebral hemorrhage in rat striatum. J Cereb Blood Flow Metab 20:396–404
75. Mayne M, Ni W, Yan HJ, Xue M, Johnston JB, Del Bigio MR, Peeling J, Power C (2001) Anti-sense oligodeoxynucleotide inhibition of tumor necrosis factor-alpha expression is neuropro-tective after intracerebral hemorrhage. Stroke 32:240–248
76. Meldrum BS (1993) Excitotoxicity and selective neuronal loss in epilepsy. Brain Pathol 3:405–412
77. Meldrum BS (1994) The role of glutamate in epilepsy and other CNS disorders. Neurology 44:14–23
78. Melillo RM, Santoro M, Ong SH, Billaud M, Fusco A, Hadari YR, Schlessinger J, Lax I (2001) Docking protein FRS2 links the protein tyrosine kinase RET and its oncogenic forms with the mitogen-activated protein kinase signaling cascade. Mol Cell Biol 21:4177–4187
79. Minami Y, Kawasaki H, Minami M, Tanahashi N, Tanaka K, Yahara I (2000) A critical role for the proteasome activator PA28 in the Hsp90-dependent protein refolding. J Biol Chem 275:9055–9061
80. Mirnics K, Middleton FA, Marquez A, Lewis DA, Levitt P (2000) Molecular characterization of schizophrenia viewed by microarray analysis of gene expression in prefrontal cortex. Neuron 28:53–67
81. Miyazono K, ten Dijke P, Heldin CH (2000) TGF-beta signaling by Smad proteins. Adv Immu-nol 75:115–157
82. Nadeau S, Rivest S (2000) Role of microglial-derived tumor necrosis factor in mediating CD14 transcription and nuclear factor kappa B activity in the brain during endotoxemia. J Neurosci 20:3456–3468
83. Nimura T, Weinstein PR, Massa SM, Panter S, Sharp FR (1996) Heme oxygenase-1 (HO-1) pro-tein induction in rat brain following focal ischemia. Brain Res Mol Brain Res 37:201–208
84. Obrenovitch TP, Richards DA (1995) Extracellular neurotransmitter changes in cerebral isch-aemia. Cerebrovasc Brain Metab Rev 7:1–54
85. Pang L, Ye W, Che XM, Roessler BJ, Betz AL, Yang GY (2001) Reduction of inflammatory re-sponse in the mouse brain with adenoviral-mediated transforming growth factor-ss1 expres-sion. Stroke 32:544–552
86. Paschen W (1992) Polyamine metabolism in different pathological states of the brain. Mol Chem Neuropathol 16:241–271
87. Paschen W (1996) Disturbances of calcium homeostasis within the endoplasmic reticulum may contribute to the development of ischemic-cell damage. Med Hypotheses 47:283–288
88. Paschen W, Gissel C, Linden T, Althausen S, Doutheil J (1998) Activation of gadd153 expres-sion through transient cerebral ischemia: evidence that ischemia causes endoplasmic reticulum dysfunction. Brain Res Mol Brain Res 60:115–122
89. Pestova TV, Kolupaeva VG, Lomakin IB, Pilipenko EV, Shatsky IN, Agol VI, Hellen CU (2001) Molecular mechanisms of translation initiation in eukaryotes. Proc Natl Acad Sci USA 98:7029–7036
90. Plantier M, Fattoum A, Menn B, Ben-Ari Y, der Terrossian E, Represa A (1999) Acidic calponin immunoreactivity in postnatal rat brain and cultures: subcellular localization in growth cones, under the plasma membrane and along actin and glial filaments. Eur J Neurosci 11:2801–2812

91. Plumier JC, David JC, Robertson HA, Currie RW (1997) Cortical application of potassium chloride induces the low-molecular weight heat shock protein (HSP27) in astrocytes. J Cereb Blood Flow Metab 17:781–790

92. Poltorak A, He X, Smirnova I, Liu MY, Huffel CV, Du X, Birdwell D, Alejos E, Silva M, Galanos C, Freudenberg M, Ricciardi-Castagnoli P, Layton B, Beutler B (1998) Defective LPS signaling in C3H/HeJ and C57BL/10ScCr mice: mutations in Tlr4 gene. Science 282:2085–2088

93. Potempa J, Enghild JJ, Travis J (1995) The primary elastase inhibitor (elastasin) and trypsin inhibitor (contrapsin) in the goat are serpins related to human alpha 1-anti-chymotrypsin. Biochem J 306:191–197

94. Prasad SS, Cynader MS (1994) Identification of cDNA clones expressed selectively during the critical period for visual cortex development by subtractive hybridization. Brain Res 639:73–84

95. Pugin J, Heumann ID, Tomasz A, Kravchenko VV, Akamatsu Y, Nishijima M, Glauser MP, Tobias PS, Ulevitch RJ (1994) CD14 is a pattern recognition receptor. Immunity 1:509–516

96. Raghavendra Rao VL, Dogan A, Bowen KK, Dempsey RJ (2000) Traumatic brain injury leads to increased expression of peripheral-type benzodiazepine receptors, neuronal death, and activation of astrocytes and microglia in rat thalamus. Exp Neurol 161:102–114

97. Rajdev S, Hara K, Kokubo Y, Mestril R, Dillmann W, Weinstein PR, Sharp FR (2000) Mice overexpressing rat heat shock protein 70 are protected against cerebral infarction. Ann Neurol 47:782–791

98. Rao VL, Bowen KK, Rao AM, Dempsey RJ (2001) Up-regulation of the peripheral-type benzodiazepine receptor expression. J Neurosci Res 64:493–500

99. Rechsteiner M, Realini C, Ustrell V (2000) The proteasome activator 11 S REG (PA28) and class I antigen presentation. Biochem J 345 Pt 1:1–15

100. Regan RF, Panter SS (1996) Hemoglobin potentiates excitotoxic injury in cortical cell culture. J Neurotrauma 13:223–231

101. Roberts AB, Piek E, Bottinger EP, Ashcroft G, Mitchell JB, Flanders KC (2001) Is smad3 a major player in signal transduction pathways leading to fibrogenesis? Chest 120:S43–S47

102. Rothwell NJ, Luheshi GN (2000) Interleukin 1 in the brain: biology, pathology and therapeutic target [In Process Citation]. Trends Neurosci 23:618–625

103. Saarma M (2000) GDNF – a stranger in the TGF-beta superfamily? Eur J Biochem 267:6968–6971

104. Schiffer M, von Gersdorff G, Bitzer M, Susztak K, Bottinger EP (2000) Smad proteins and transforming growth factor-beta signaling. Kidney Int 58 Suppl 77:S45–52

105. Schmidt-Kastner R, Zhao W, Truettner J, Belayev L, Busto R, Ginsberg MD (1998) Pixel-based image analysis of HSP70, GADD45 and MAP2 mRNA expression after focal cerebral ischemia: hemodynamic and histological correlates. Brain Res Mol Brain Res 63:79–97

106. Schumann RR, Leong SR, Flaggs GW, Gray PW, Wright SD, Mathison JC, Tobias PS, Ulevitch RJ (1990) Structure and function of lipopolysaccharide binding protein. Science 249:1429–1431

107. Semenza GL (1999) Perspectives on oxygen sensing. Cell 98:281–284

108. Semenza GL (1999) Regulation of mammalian O_2 homeostasis by hypoxia-inducible factor 1. Annu Rev Cell Dev Biol 15:551–578

109. Semenza GL (2000) Expression of hypoxia-inducible factor 1: mechanisms and consequences. Biochem Pharmacol 59:47–53

110. Sharp FR, Lu A, Tang Y, Millhorn DE (2000) Multiple molecular penumbras after focal cerebral ischemia. J Cereb Blood Flow Metab 20:1011–1032

111. Sharp FR, Massa SM, Swanson RA (1999) Heat-shock protein protection. Trends Neurosci 22:97–99

112. Shimada K, Watanabe Y, Mokuno H, Iwama Y, Daida H, Yamaguchi H (2000) Common polymorphism in the promoter of the CD14 monocyte receptor gene is associated with acute myocardial infarction in Japanese men. Am J Cardiol 86:682–684, A8

113. Simon RP, Cho H, Gwinn R, Lowenstein DH (1991) The temporal profile of 72-kDa heat-shock protein expression following global ischemia. J Neurosci 11:881–889

114. Simon RP, Schmidley JW, Meldrum BS, Swan JH, Chapman AG (1986) Excitotoxic mechanisms in hypoglycaemic hippocampal injury. Neuropathol Appl Neurobiol 12:567–576

115. Simon RP, Swan JH, Griffiths T, Meldrum BS (1984) Blockade of N-methyl-D-aspartate receptors may protect against ischemic damage in the brain. Science 226:850-852

116. Soriano MA, Tessier M, Certa U, Gill R (2000) Parallel gene expression monitoring using oligonucleotide probe arrays of multiple transcripts with an animal model of focal ischemia. J Cereb Blood Flow Metab 20:1045–1055

117. Spector EB, Kern RM, Haggerty DF, Cederbaum SD (1985) Differential expression of multiple forms of arginase in cultured cells. Mol Cell Biochem 66:45–53

118. Stohwasser R, Salzmann U, Giesebrecht J, Kloetzel PM, Holzhutter HG (2000) Kinetic evidences for facilitation of peptide channelling by the proteasome activator PA28. Eur J Biochem 267:6221–6230

119. Stoll G, Jander S, Schroeter M (1998) Inflammation and glial responses in ischemic brain lesions. Prog Neurobiol 56:149–171

120. Streit WJ (2000) Microglial response to brain injury: a brief synopsis. Toxicol Pathol 28:28–30

121. Sullivan JM, Alousi SS, Hikade KR, Bahu NJ, Rafols JA, Krause GS, White BC (1999) Insulin induces dephosphorylation of eukaryotic initiation factor 2alpha and restores protein synthesis in vulnerable hippocampal neurons after transient brain ischemia. J Cereb Blood Flow Metab 19:1010–1019

122. Tanaka N, Sasahara M, Ohno M, Higashiyama S, Hayase Y, Shimada M (1999) Heparin-binding epidermal growth factor-like growth factor mRNA expression in neonatal rat brain with hypoxic/ischemic injury. Brain Res 827:130–138

123. Thomas EA, Alvarez CE, Sutcliffe JG (2000) Evolutionarily distinct classes of S27 ribosomal proteins with differential mRNA expression in rat hypothalamus. J Neurochem 74:2259–2267

124. Tsai PJ, Huang PC (2000) Circadian variations in plasma and erythrocyte glutamate concentrations in adult men consuming a diet with and without added monosodium glutamate. J Nutr 130:1002S–1004S

125. Turner CP, Bergeron M, Matz P, Zegna A, Noble LJ, Panter SS, Sharp FR (1998) Heme oxygenase-1 is induced in glia throughout brain by subarachnoid hemoglobin. J Cereb Blood Flow Metab 18:257–273

126. Unkelbach K, Gardemann A, Kostrzewa M, Philipp M, Tillmanns H, Haberbosch W (1999) A new promoter polymorphism in the gene of lipopolysaccharide receptor CD14 is associated with expired myocardial infarction in patients with low atherosclerotic risk profile. Arterioscler Thromb Vasc Biol 19:932–938

127. Van Lookeren Campagne M, Thibodeaux H, Van Bruggen N, Cairns B, Gerlai R, Palmer JT, Williams SP, Lowe DG (1999) Evidence for a protective role of metallothionein-1 in focal cerebral ischemia. Proc Natl Acad Sci USA 96:12870–12875

128. Verrecchia F, Chu ML, Mauviel A (2001) Identification of novel TGF-beta/Smad gene targets in dermal fibroblasts using a combined cDNA microarray/promoter transactivation approach. J Biol Chem 276:17058–17062

129. Wagner KR, Xi G, Hua Y, Kleinholz M, de Courten-Myers GM, Myers RE (1998) Early metabolic alterations in edematous perihematomal brain regions following experimental intracerebral hemorrhage. J Neurosurg 88:1058–1065

130. Wagner KR, Xi G, Hua Y, Kleinholz M, de Courten-Myers GM, Myers RE, Broderick JP, Brott TG (1996) Lobar intracerebral hemorrhage model in pigs: rapid edema development in perihematomal white matter. Stroke 27:490–497

131. Wang H, Zhan Y, Xu L, Feuerstein GZ, Wang X (2001) Use of suppression subtractive hybridization for differential gene expression in stroke: discovery of CD44 gene expression and localization in permanent focal stroke in rats. Stroke 32:1020–1027

132. Wang X, Barone FC, White RF, Feuerstein GZ (1998) Subtractive cloning identifies tissue inhibitor of matrix metalloproteinase-1 (TIMP-1) increased gene expression following focal stroke. Stroke 29:516–520

133. Wang X, Louden C, Yue TL, Ellison JA, Barone FC, Solleveld HA, Feuerstein GZ (1998) Delayed expression of osteopontin after focal stroke in the rat. J Neurosci 18:2075–2083

134. Warner RL, Lewis CS, Beltran L, Younkin EM, Varani J, Johnson KJ (2001) The role of metalloelastase in immune complex-induced acute lung injury. Am J Pathol 158:2139–2144

135. Welman E, Colbeck JF, Selwyn AP, Fox KM, Orr I (1980) Plasma lysosomal enzyme activity in acute myocardial infarction and the effects of drugs. Adv Myocardiol 2:359–369

136. Werner F, Jain MK, Feinberg MW, Sibinga NE, Pellacani A, Wiesel P, Chin MT, Topper JN, Perrella MA, Lee ME (2000) Transforming growth factor-beta 1 inhibition of macrophage activation is mediated via Smad3. J Biol Chem 275:36653–36658

137. Wessling-Resnick M (1999) Biochemistry of iron uptake. Crit Rev Biochem Mol Biol 34:285–314

138. Whitney LW, Becker KG, Tresser NJ, Caballero-Ramos CI, Munson PJ, Prabhu VV, Trent JM, McFarland HF, Biddison WE (1999) Analysis of gene expression in multiple sclerosis lesions using cDNA microarrays. Ann Neurol 46:425–428

139. Wilson JX (1997) Antioxidant defense of the brain: a role for astrocytes. Can J Physiol Pharmacol 75:1149–1163

140. Wong JM, Mafune K, Yow H, Rivers EN, Ravikumar TS, Steele GD Jr, Chen LB (1993) Ubiquitin-ribosomal protein S27a gene overexpressed in human colorectal carcinoma is an early growth response gene. Cancer Res 53:1916–1920
141. Wright SD (1999) Toll, a new piece in the puzzle of innate immunity. J Exp Med 189:605–609
142. Wright SD, Ramos RA, Tobias PS, Ulevitch RJ, Mathison JC (1990) CD14, a receptor for complexes of lipopolysaccharide (LPS) and LPS binding protein. Science 249:1431–1433
143. Xian CJ, Zhou XF (2000) Roles of transforming growth factor-alpha and related molecules in the nervous system. Mol Neurobiol 20:157–183
144. Yamashima T (2000) Implication of cysteine proteases calpain, cathepsin and caspase in ischemic neuronal death of primates. Prog Neurobiol 62:273–295
145. Yanagitani S, Miyazaki H, Nakahashi Y, Kuno K, Ueno Y, Matsushita M, Naitoh Y, Taketani S, Inoue K (1999) Ischemia induces metallothionein III expression in neurons of rat brain. Life Sci 64:707–715
146. Yang GY, Betz AL, Chenevert TL, Brunberg JA, Hoff JT (1994) Experimental intracerebral hemorrhage: relationship between brain edema, blood flow, and blood-brain barrier permeability in rats. J Neurosurg 81:93–102
147. Yawata M, Murata S, Tanaka K, Ishigatsubo Y, Kasahara M (2001) Nucleotide sequence analysis of the approximately 35-kb segment containing interferon-gamma-inducible mouse proteasome activator genes. Immunogenetics 53:119–129
148. Yonekawa M, Okabe T, Asamoto Y, Ohta M (1999) A case of hereditary ceruloplasmin deficiency with iron deposition in the brain associated with chorea, dementia, diabetes mellitus and retinal pigmentation: administration of fresh-frozen human plasma. Eur Neurol 42:157–162
149. Yoshimoto R, Hori M, Ozaki H, Karaki H (2000) Proteolysis of acidic calponin by mu-calpain. J Biochem (Tokyo) 128:1045–1049
150. Zarow GJ, Karibe H, States BA, Graham SH, Weinstein PR (1997) Endovascular suture occlusion of the middle cerebral artery in rats: effect of suture insertion distance on cerebral blood flow, infarct distribution and infarct volume. Neurol Res 19:409–416
151. Zhang X, Boulton AA, Yu PH (1996) Expression of heat shock protein-70 and limbic seizure-induced neuronal death in the rat brain. Eur J Neurosci 8:1432–1440
152. Zhang X, Gelowitz DL, Lai CT, Boulton AA, Yu PH (1997) Gradation of kainic acid-induced rat limbic seizures and expression of hippocampal heat shock protein-70. Eur J Neurosci 9:760–769
153. Zhao X, Yang Z, Qian M, Zhu X (2001) Interactions among subunits of human Arp2/3 complex: p20-Arc as the hub. Biochem Biophys Res Commun 280:513–517

Temporal Profile of Gene Induction After Venous Ischemia and Effects of Spreading Depression

T. Kaido, Y. Kamada, T. Nishioka, C. Heers, D. Bartsch, A. Heimann, and O. Kempski

Summary. Occlusion of two adjacent cortical veins is followed by a widespread reduction of rCBF and the occurrence of small infarcts, which become larger if spreading depression (SD) occurs. The infarct matures over time with TUNEL-positive cells seen in the penumbra up to 4 days after vein occlusion. Caspase inhibition with zVAD.fmk reduces infarct size. Here, the time course of gene expression in the penumbra is compared to that induced by SD alone.

In rats an occlusion of two cortical veins was induced by I.V. rose bengal and fiberoptic illumination. Ten SDs were induced in 7 min intervals. Changes of gene expression after 2 h, 8 h, 24 h and 72 h were analysed for 13 genes by RT-PCR. The housekeeping gene GAPDH was used for normalization. Expression patterns were studied by cluster analysis, c-fos, cox-2 and bcl-2 responded early (2 h) and far more after two-vein occlusion plus SD than after SD alone. Secondly, HO-1 and c-myc were increased 24 h after SD in animals without ischemia. These increases were specific for SD. bcl-xL, bad and bax had minor and late increases. Cyclin D1 showed a moderately increased expression 72 h after ischemia. A cluster analysis confirmed related expression patterns of (1) HO-1 and c-myc 24 h after SD, and (2) c-fos and cox-2 early after two-vein occlusion.

Key words. Heme oxygenase-1 – cyclooxygenase-2 – penumbra – gene expression – spreading depression – delayed neuronal cell death

Introduction

Treatment of stroke is to a large part an attempt to save the ischemic penumbra [1] which, when untreated, sooner or later will die. In the rat MCA-occlusion model most of the penumbra turns into infarct within 6 h [8]. A possible reason for this second-

[1] Takanobu Kaido MD, [1] Yoshitaka Kamada, [1] Toshikazu Nishioka, [2] Cara Heers, PhD;
[2] Daniela Bartsch PhD, [1] Axel Heimann DVM, [1] Oliver Kempski MD, PhD
[1] Institute for Neurosurgical Pathophysiology, Johannes Gutenberg-University Mainz, Germany,
[2] Janssen Research Foundation, Neuss, Germany

Corresponding author: Univ.-Prof. Dr. med. Oliver Kempski, Institute for Neurosurgical Pathophysiology, Johannes-Gutenberg-University, Langenbeckstr. 1, 55101 Mainz, Germany
Tel.: +49-61 31/17 36 36, Fax: +49-61 31/17 66 40, E-Mail: kempski@nc-patho.klinik.uni-mainz.de

Maturation Phenomenon in Cerebral Ischemia V
A. M. Buchan et al. (Eds.)
© Springer-Verlag Berlin Heidelberg 2004

ary decay is an increase of the metabolic needs of the tissue as a consequence of metabolic activation by periinfarct depolarizations – spreading depression (SD) – which go along with a complete breakdown of the ion gradients and, hence, increased metabolic demands for ion pumping [2, 13]. Another cause is the persisting low flow conditions in the penumbra which is tolerated for several hours but is insufficient to support tissue survival over longer intervals. The individual significance of these mechanisms – persisting low flow and metabolic activation – remains to be determined.

The occlusion of two adjacent cortical veins is followed by a more widespread reduction of regional CBF than after MCA occlusion and the occurrence of comparatively small infarcts, which also become larger if SD occurs spontaneously or is induced [18]. Moreover, in this model maturation of the infarct continues longer than in rat MCA occlusion, with infarct growth and TUNEL-positive cells found for at least four days after vein occlusion [Kamada *et al.*, in preparation]. In addition the intraventricular injection of the caspase inhibitor zVAD.fmk after two-vein occlusion reduces infarct size [Kamada *et al.*, in preparation, Nishioka et al., in preparation]. Therefore the model appears useful for studying processes in penumbra zones, mechanisms of secondary tissue damage in the penumbra in particular. One way to detect such mechanisms is to study changes of the expression of potentially harmful or protective genes. The aim of the current study, therefore, was to determine the time course of gene expression changes in the ischemic/periischemic low blood flow zone as compared to respective changes induced by spreading depression alone.

Material and Methods

Male Wistar rats were anesthetized with chloral hydrate. Rectal temperature was kept close to 37.0 °C throughout the experiment by a feedback-controlled heating pad. Polyethylene catheters were inserted into the right femoral artery and vein for continuous registration of mean arterial pressure, for arterial blood gas sampling and for administration of fluid and drugs. Rats were mounted in a stereotaxic frame. After a 2.0 cm midline skin incision, a right fronto-parietal cranial window was made for access to the brain surface. An occlusion of two adjacent cortical veins was induced by I.V. rose bengal and fiberoptic illumination [14–16, 18]. The diameter of the occluded veins was approximately 100 mm. Rose bengal (50 mg/kg body wt) was injected slowly without effect on the systemic arterial pressure, and care was taken to avoid illumination of tissue and other vessels near the target vein. The vein was illuminated for 10 min via the micromanipulator-assisted light guide. To occlude the second selected vein, half of the initial rose bengal dose was injected intravenously before the illumination was repeated with the new target. After two-vein occlusion, ten SDs were induced with cortical microinjections of KCl (150 mM) in 7 min intervals. Spreading depressions were monitored by cortical impedance measurements using a precision LCR monitor (4282A, Hewlett-Packard, USA). After 2, 8, 24 or 72 h cortical tissue was dissected immediately (ipsi- vs. contralateral tissue between the two occluded veins, i.e. in the territory with low rCBF) and frozen in liquid nitrogen. Three to five animals were investigated for each condition. mRNA levels were determined by semi-quantitative RT-PCR for the following genes: heme oxygenase-1 (HO-1), the immediate early genes c-fos, c-jun and c-myc, the bcl-2 family members bcl-2, bcl-xL, bax and bad, superoxide dismutases 1 and 2, and the cell cycle-related gene cyclin D1. The housekeeping gene glyceraldehyde-3-phosphate dehydrogenase (GAPDH) was used for normalization. The ratio of the relative amounts of mRNA (normalized to GAPDH) from the operated to the contralateral side was calculated. Related expression patterns were evaluated by cluster analysis [4].

Results

There were no statistical differences in physiological data which all were within normal range. Venous occlusion evolved with the well-described typical time course [14–16]. We were able to relate selected genes to four major groups because of their different expression patterns in the tissue. First, we found a group of genes which respond early (2h) and strong, but far more after two-vein occlusion plus SD than after SD alone (c-fos, cox-2, bcl-2). Cox-2 was stimulated 13.6-fold by vein occlusion + SD but only 1.55-fold by SD alone. The mean stimulation of c-fos expression was 69.4-fold whereas with SD alone the stimulation was more moderate (10.8-fold). These effects were transient. After 24 h, there was a strong decrease to a mRNA level which was significantly lower than in SD-animals. Similar, but less strong, changes were observed in bcl-2 expression which was also elevated (2.7-fold) 2 h after two-vein occlusion and then normalized again. Secondly, we could detect changes in gene expression at 24 h after induction of SD in animals without vein occlusion (HO-1, c-myc). These increases (19.0-fold and 5.4-fold, respectively) were specific for SD, whereas a combination of two-vein occlusion plus SD after an initial moderate stimulation rather led to a suppression at that time point. Later, i.e. at 72 h after vein occlusion, these genes developed moderately increased expression patterns. The third group comprises the remaining tested members of the bcl-2 family (bcl-xL, bad and bax). In general, these expression changes were small: bax, bad and bcl-xL increased late, at 72 h after two-vein occlusion, with bax showing the most conspicuous changes (2-fold increase). Finally, cyclin D1 showed a moderately (5.6-fold) increased expression late (72 h) after ischemia.

Cluster analysis confirmed these expression patterns: expression of (a) HO-1 and c-myc were related 24 h after SD, and (b) c-fos and cox-2, and also bcl-2 were coexpressed early after two-vein occlusion in combination with SD.

Discussion

The two-vein occlusion model permits to discriminate changes in gene expression in the vicinity of focal ischemia from those elicited by spreading depression. SD is known to occur spontaneously in the ischemic penumbra and tends to worsen the metabolic situation as well as outcome [2, 13]. On the other hand, SD is known to precondition the brain, i.e. to induce tolerance for a subsequent episode of ischemia [11]. Therefore the current study and others were performed to discriminate tolerance inducing from damaging alterations. Among the changes observed many have been described before in other models (e.g. immediate early gene induction after ischemia). It was also known that SD and also hypoxia (which can also induce tolerance) induce HO-1 [6, 12]. Other authors have described an increase of HO-1 protein by focal and global ischemia [7, 17]. However a direct comparison between focal ischemia and SD so far has not been made. Our data clearly indicate that only SD but not ischemia (here even in combination with SD) caused an enhanced expression of HO-1. HO-1 is interesting as a potentially beneficial gene/protein since animals overexpressing the protein have a better outcome from focal ischemia [19] and neurons from those animals resist oxidative stress-mediated cell death [3]. HO-1 inhibits inflammatory reactions (selectin expression [20]) and catalyzes the formation of the vasodilator CO and, thereby, also suppresses the hypoxic induction of the gene-encoding plasminogen activator inhibitor-1 (PAI-1) [5]. Further research is necessary to evaluate the role of HO-1 in tolerance induction. The lack of HO-1 upregulation in subacute stages of the penumbra may contribute to the decay of penumbra tissue.

Another significant result of this study was the early and more than moderate increase of COX-2 expression after venous ischemia but not after SD. COX-2 not only catalyzes the production of vasoactive prostaglandins from arachidonic acid but also the formation of free radicals during that process. COX-2 is also blamed for a potentiation of excitotoxicity [9]. Hence, the increased expression of COX-2 mRNA may reflect an attempt of penumbra tissue to improve collateral perfusion by vasodilation, i.e. by prostacyclin formation [10] which may be accompanied by the cited deleterious side effects of an increased production of reactive oxygen species and aggravation of excitotoxicity via COX-2 in the penumbra.

So far unpublished results of our own group demonstrate a maximum of TUNEL-positive cells between days 2 and 4 after vein occlusion. Moreover, the intraventricular injection of the caspase inhibitor zVAD.fmk after two-vein occlusion reduces infarct size [Kamada et al., in preparation, Nishioka et al., in preparation]. Therefore the rather modest alterations of the bcl-2 family genes are somewhat puzzling. Hence it cannot be excluded that the time frame of the current study with samples taken at one and 4 days after occlusion prevented the detection of related changes.

In conclusion we here present evidence that mechanisms initiated by brain activation by SD as well as the loss of CBF control in the penumbra can be adequately studied by assessing gene expression changes in the two-vein occlusion model. The potentially protective nature of HO-1 as well as a postulated unfavorable role of COX-2 are supported by our data.

References

1. Astrup J, Siesjö BK, Symon L (1981) Thresholds in cerebral ischemia – the ischemic penumbra. Stroke 12:723–725
2. Back T, Ginsberg M, Dietrich WD, Watson BD (1996) Induction of spreading depression in the ischemic hemisphere following experimental middle cerebral artery occlusion: effect on infarct morphology. J Cereb Blood Flow Metab 16:202–213

3. Chen K, Gunter K, Maines MD (2000) Neurons overexpressing heme oxygenase-1 resist oxidative stress-mediated cell death. J Neurochem 75(1):304–313

4. Eisen MB, Spellman PT, Brown PO, Botstein D (1998) Cluster analysis of genome-wide expression patterns. PNAS 95:14863–14868

5. Fujita T, Toda K, Karimova A, Yan SF, Naka Y, Yet SF, Pinsky DJ (2001) Paradoxical rescue from ischemic lung injury by inhaled carbon monoxide driven by derepression of fibrinolysis. Nat Med 7(5):598–604

6. Garnier P, Demougeot C, Bertrand N, Prigent-Tessier A, Marie C, Beley A (2001) Stress response to hypoxia in gerbil brain: HO-1 and Mn SOD expression and glial activation. Brain Research 893:301–309

7. Geddes JW, Pettigrew LC, Holtz ML, Craddock SD, Maines MD (1996) Permanent focal and transient global cerebral ischemia increase glial and neuronal expression of heme oxygenase-1, but not heme oxygenase-2, protein in rat brain. Neuroscience Letters 210:205–208

8. Ginsberg MD (1997) Injury mechanisms in the ischemic penumbra – approaches to neuroprotection in acute ischemic stroke. Cerebrovasc Dis 7(suppl 2):7–12

9. Kelley KA, Ho L, Winger D, Freire-Moar J, Borelli CB, Aisen PS, Pasinetti GM (1999) Potentiation of excitotoxicity in transgenic mice overexpressing neuronal cyclooxygenase-2. Am J Pathology 155(3):995–1004

10. Kempski O, Shohami E, von Lubitz D, Hallenbeck JM, Feuerstein G (1987) Postischemic production of eicosanoids in gerbil brain. Stroke 18:111–119

11. Kobajashi S, Harris VA, Welsh FA (1995) Spreading depression induces tolerance of cortical neurons to ischemia in rat brain. J Cereb Blood Flow Metab 15:721–727

12. Koistinaho J, Pasonen S, Yrjänheikki J, Chan PH (1999) Spreading depression-induced gene expression is regulated by plasma glucose. Stroke 30:114–119

13. Mies G, Iijima T, Hossmann K-A (1993) Correlation between periinfarct DC shifts and ischemic neuronal damage in rat. Neuroreport 4:709–711

14. Nakase H, Heimann A, Kempski O (1996) Local cerebral blood flow in a rat cortical occlusion model. J Cereb Blood Flow Metabol 16:720–728

15. Nakase H, Kempski O, Heimann A, Takeshima T, Tintera J (1997) Microcirculation following cerebral venous occlusions assessed by laser Doppler scanning. J Neurosurg 87:307–314

16. Nakase H, Nagata K, Otsuka H, Sakaki T, Kempski O (1998) Local cerebral blood flow autoregulation following asymptomatic cerebral venous occlusion in the rat. J Neurosurg 89:118–124

17. Nimura T, Weinstein PR, Massa SM, Panter S, Sharp FR (1996) Heme oxygenase-1 (HO-1) protein induction in rat brain following focal ischemia. Molecular Brain Research 37:201–208

18. Otsuka H, Ueda K, Heimann A, Kempski O (2000) Effects of cortical spreading depression on cortical blood flow, impedance, DC-potential, and infarct size in a rat venous infarct model. Experimental Neurology 162:201–214

19. Panahian N, Yoshiura M, Maines MD (1999) Overexpression of heme oxygenase-1 is neuroprotective in a model of permanent middle cerebral artery occlusion in transgenic mice. J Neurochem 72(3):1187–1203

20. Vachharajani TJ, Work J, Issekutz AC, Granger DN (2000) Heme oxygenase modulates selectin expression in different regional vascular beds. Am J Physiol Heart Circ Physiol 278:H1613–H1617

The Role of Protein Phosphatases Type 2C in Neuronal Apoptosis

J. Krieglstein, D. Selke, Y. Zhu, and S. Klumpp

Summary. In previous work we found out that the protein phosphatase type 2C (PP2C) could be activated by low molecular weight compounds which had to fulfill special requirements. The activators had to be lipophilic, acidic and oxidizable and they had to have a special configuration, for instance only the *cis*- but not the *trans*-configuration of oleic acid was effective. The position of the double bond was crucial as well.

In addition, we could also demonstrate that the activation was true only for PP2C. Other phosphatases tested such as PP1, PP2A, PP2B, acid and alkaline phosphatases and a tyrosine phosphatase were not activated or even inhibited by these compounds. Surprisingly, the compounds which massively activated PP2C also significantly induced apoptotic damage of cultured neurons obtained from embryonic chick telencephalon. There was a striking correlation between activation of PP2C and induction of apoptosis and the question arose whether and how PP2C could act on the apoptotic cascades. Which protein from the signaling transduction of apoptosis could be dephosphorylated by this phosphatase with the consequence to cause apoptosis? In principle, there are several proteins which could be looked for. The pro-apoptotic oncogene Bad is a candidate of the first choice because it is distinctly involved in neuronal damage and protection.

Key words. Apoptosis – Bad – oleic acid – protein phosphatases – transforming growth factor (TGF-β1)

Josef Krieglstein[1], Dagmar Selke[2], Yuan Zhu[1], Susanne Klumpp[2]
[1] Institut für Pharmakologie und Toxikologie, Philipps-Universität, Marburg, Germany;
[2] Institut für Pharmazeutische und Medizinische Chemie, Westfälische Wilhelms-Universität, Münster, Germany

Corresponding author: Professor Dr. Dr. Josef Krieglstein, Institut für Pharmakologie und Toxikologie, Philipps-Universität, Ketzerbach 63, D-35032 Marburg, Germany,
Tel.: +49-6421/2821311, Fax: +49-6421/2828918, E-Mail: krieglst@mailer.uni-marburg.de

Maturation Phenomenon in Cerebral Ischemia V
A. M. Buchan et al. (Eds.)
© Springer-Verlag Berlin Heidelberg 2004

Introduction

Oleic Acid Activates PP2C and Causes Apoptosis in Cultured Rat Hippocampal Cells

The effect of fatty acids on cell viability varies and seems to be cell type-specific and concentration-dependent. In a previous study we demonstrated oleic acid to specifically activate PP2C under in vitro conditions [3]. We evaluated the apoptotic damage of cultured rat hippocampal cells by Hoechst-staining 24 h after adding oleic acid. At a concentration of 1 µM oleic acid the percentage of apoptotic cells was significantly increased and the number of apoptotic cells further increased concentration-dependently up to 300 µM oleic acid in the culture medium. A plateau was achieved at about 60% of apoptotic cells which was 3-fold higher than that in the vehicle-treated control. The damaged cells showed the characteristic apoptotic features including condensed chromatin, shrunken and fragmented nuclei as well as apoptotic bodies. Next, we established the time-course of induction of apoptosis at 300 µM oleic acid. An increase in the apoptotic damage was already seen 2 h after adding oleic acid, and became significant at 4 h. The maximum of cell damage (50%) was almost achieved 24 h after the addition of oleic acid. Because both neurons (about 60%) and astrocytes (about 40%) were present in these primary cultures, it was interesting to identify which type of cells was more vulnerable to oleic acid. We could observe that 24 h after oleic acid treatment mainly neurons were damaged. This finding was confirmed by double staining of the cells with Hoechst and the neuronal marker NeuN, where almost all NeuN-positive cells exhibited apoptotic nuclear morphology. Furthermore, double staining of the cells with Hoechst 33258 and the astrocyte marker GFAP indicated that most of the astrocytes remained normal in morphology up to 24 h and were damaged 48 h after the exposure to oleic acid. Thus, we can conclude that neurons in the primary cultures of rat hippocampal cells are more vulnerable to oleic acid than astrocytes.

The Role of Bad in Neuronal Apoptosis and Neuroprotection

Interactions between the pro- and anti-apoptotic members of the Bcl-2 family play an important role in the regulation of programmed cell death. Bad is a pro-apoptotic member of the Bcl-2 family. Like other death promoters, Bad can heterodimerize with anti-apoptotic Bcl-2 or Bcl-X_L thereby inducing apoptosis [9]. In previous work, we demonstrated that the level of Bad in the mouse brain increased tremendously after ischemia and this increase could be almost completely abolished by TGF-β1 (Fig. 1). Those experiments were performed by occlusion of the middle cerebral artery (tMCAO) for 30 min in mice which were transduced with either lacZ (control) or the TGF-β1 gene. The Bad level was examined in the lesioned striatum 8 h, 24 h and 48 h after tMCAO by Western blotting using an antibody recognizing Bad independent of its phosphorylation state. Strikingly, ischemia led to a 30-fold increase in Bad expression 8 h after reperfusion accompanied by a 5-fold activation of caspase-3. The level of Bad and the activation of caspase-3 in those lacZ-transduced mice decreased 24 h and 48 h after tMCAO,

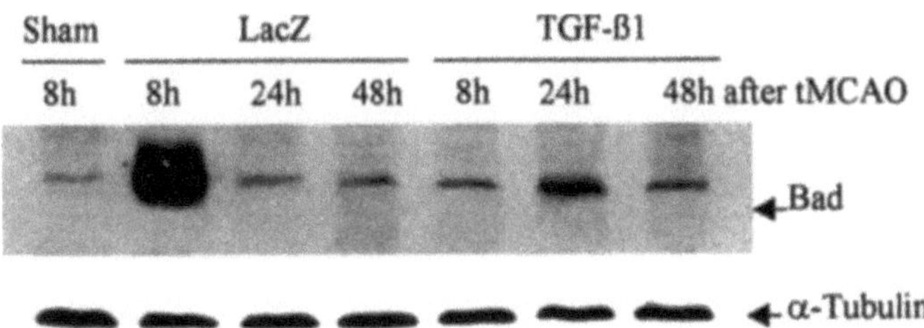

Fig. 1. Ischemia increases and TGF-β1 inhibits Bad expression in mouse brain. Focal ischemia was performed by occlusion of the middle cerebral artery for 30 min. The adenovirus rectors AdRSV-Lac2 and AdRSVahTGF-β were injected into the cerebroventrical 5 d before ischemia. Proteins were extracted from the striatum at the given time points after ischemia. 60 µg of total protein were loaded to each lane for Western blotting of Bad and a-tubulin

but remained higher than that in sham-operated mice. This dramatically increased Bad level was significantly reduced in TGF-β1-transduced mice, where activation of caspase-3 was also diminished. These findings were confirmed by immunostaining of Bad and active caspase-3 in mouse brain sections prepared 8 h after tMCAO. An enhanced Bad immunoreactivity and a significant activation of caspase-3 were detected in the ipsilateral striatum. In contrast, Bad immunoreactivity was only weakly detected in the contralateral side and no active caspase-3 was found. Double-staining indicated co-localization of Bad with the neuronal marker neurofilament 200, as well as co-localization of Bad with active caspase-3. The ischemia-induced increase in Bad immunoreactivity and activation of caspase-3 were inhibited in the striatum of TGF-β1-transduced mice. These findings were in agreement with the results obtained from Western blotting.

We further tested whether inhibition of apoptotic cascades by TGF-β1 contributed to neuroprotection and improvement of the neurological outcome of the ischemic mice. High numbers of TUNEL-positive cells with shrunken nuclei and condensed chromatin were detected in the lesioned striatum of lacZ-transduced mice, whereas the number of TUNEL-positive cells was considerably reduced in the striatum of TGF-β1-transduced mice. In comparison with saline and lacZ-controls, the infarct volume in TGF-β1-transduced mice was reduced by 40% at day 1 after tMCAO. Evaluation of the neurological outcome by the beam test showed that tMCAO markedly increased the number of falls from the beam in mice subjected to either saline- or lacZ-injection, but to a lesser extent in TGF-β1-transduced mice. From these results a neuroprotective potency of TGF-β1 mediated by an inhibition of Bad expression can be assumed.

We further attempted to find out whether the elevation of Bad level also occurred in cultured neurons after apoptotic stimuli. It has been previously shown that staurosporine induces neuronal apoptosis in primary cultures of rat hippocampal neurons [12]. This upregulation of the Bad level was diminished by TGF-β1 in a concentration-dependent manner. Here we determined Bad expression in cultured rat hippocampal cells which were treated with 100 nM staurosporine in the presence and absence of TGF-β1 (1 and 10 ng/ml). The level of Bad increased about 2-fold at 6 h and 24 h after the exposure of the cells to staurosporine, compared with the corresponding vehicle-treated controls. TGF-β1 added to the culture medium to give a final concentration of 1 ng/ml 24 h prior to staurosporine treatment, effectively inhibited the elevation of Bad level, and 10 ng/ml TGF-β1 further reduced Bad expression. These results suggest that the inhibited increase in the Bad level crucially contributes to the neuroprotective effect of TGF-β1.

Bad: Phosphorylation, BH3 Domain, and Interaction with 14-3-3

The phosphorylation status of Bad has crucial functional implications. In its unphosphorylated state, Bad forms heterodimers with anti-apoptotic Bcl-2 homologs such as Bcl-X$_L$ and promotes cell death. The binding of Bad to Bcl-X$_L$ involves BH3 domains. In its phosphorylated state, Bad associates in the cytoplasm with 14-3-3 proteins. This interaction requires certain phosphoserine containing sequence motifs. The freed Bcl-X$_L$ then is capable of preventing cell death.

According to current knowledge Bad undergoes reversible phosphorylation at Ser-112, Ser-136 and Ser-155 (Fig. 2). The amino acid numbers refer to the mouse protein. The three phosphorylation sites comprise two different phosphorylation-dependent mechanisms that both inhibit the death-promoting activity of Bad. Phosphorylation of Ser-112 and Ser-136 touches on the interaction with 14-3-3 proteins [10] whereas phosphorylation of Ser-155 affects the BH3 domain conformation and dimerization with Bcl-X$_L$ [11].

The 14-3-3 proteins were discovered in 1967, and were given their name on the basis of the fraction number on DEAE-cellulose chromatography and the migration position in starch-gel electrophoresis. The 14-3-3 proteins are a family of closely related highly acidic homodimeric intracellular proteins which were first identified as being very abundant in brain tissues and located preferentially in neurons. The 14-3-3 proteins participate in apoptosis and other pathways by altering the subcellular localization of their numerous binding partners (for review see [5]). Ser-112 and Ser-136 of Bad are nested within such consensus motifs for 14-3-3. Mutation of Ser-112 and Ser-136 of Bad to alanine residues that can not be phosphorylated – (1) prevents complexation of Bad with 14-3-3, (2) enhances binding of Bad to Bcl-2 and Bcl-X$_L$, and (3) increases pro-apoptotic activity.

BH domains refer to *Bcl-2 homology*. Proteins belonging to the Bcl-2 family share a number of such homology domains (BH1, BH2, BH3 and BH4). The anti-apoptotic Bcl-2 family members possess all four BH domains. All pro-apoptotic members contain a BH3 domain – except for Bad – necessary for dimerization with other proteins of the Bcl-2 family and crucial for their killing activity. The binding of Bad to Bcl-X$_L$ is mediated by a BH3-like domain of Bad dimerizing with the canonical BH3 domain of Bcl-X$_L$. The very rudimentary BH3-like domain of Bad is located at residues 151–157 (LxxxxDE). One of the phosphorylation sites, Ser-155, lies in there.

Current thinking is that phosphorylation at Ser-155, similar to phosphorylation at Ser-112 and Ser-136, inhibits Bad-induced death-promoting activity but does so in a different manner. Phosphorylation of both Ser-112 and Ser-136 is required for cooperative high affinity 14-3-3 binding. When the 2×30 kD 14-3-3 dimer binds to the tiny 20 kD Bad via Ser-112, Ser-136 or both, one can envisage that the BH3-like domain of Bad may be obscured. In this way, 14-3-3 affects the subcellular

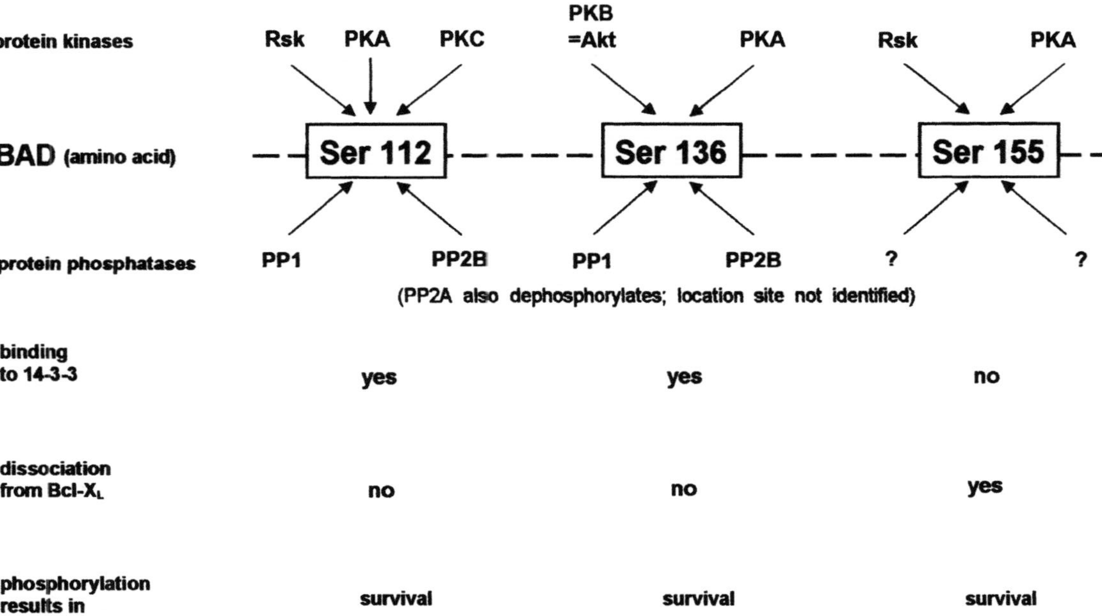

Fig. 2. Scheme summarizing the current knowledge on phosphorylation sites of Bad, the protein kinases involved in the phosphorylation reactions, and protein phosphatases responsible for dephosphorylation. Amino acid positions refer to the mouse protein. Interaction sites for 14-3-3 and Bcl-X$_L$ are indicated

localization by providing molecular interference that prevents the association of the target protein (Bad) with a third protein (Bcl-X_L). Complete cytosolic sequestration of Bad may additionally involve the Ser-155 phosphorylation pathway. Phosphorylation of Ser-155 of Bad is sufficient to induce a conformational change that enhances dissociation of Bad from Bcl-X_L. Whether or not phosphorylation of Ser-155 is crucial for binding of 14-3-3 is unclear at present. Tan et al. [6] reported that phosphorylation at Ser-155 does not induce 14-3-3 binding. In contrast, Lizcano et al. [4] described that phosphorylation of Bad at Ser-155 promotes its interaction with 14-3-3 proteins. One has to be aware that the interaction models so far are based on yeast two-hybrid assays, coimmunoprecipitation experiments, overlay technique and competition studies using peptides. They all remain unproven because Bad has not been co-crystallized with Bcl-X_L or 14-3-3.

Ser-112 of Bad is reported to be phosphorylated by mitogen-activated protein (MAP) kinase-activated protein kinase A (MAPKAP-K1, also called Rsk for ribosomal S6 kinase) and by protein kinase C (PKC) ε and θ that are diacylglycerol-dependent but calcium-independent. The identification of Ser-112 as phosphorylation site for PKC is somewhat intriguing. Running a motif scan for PKC phosphorylation sites in Bad points to Thr-3, Thr-60 and Ser-161 without mentioning Ser-112. Ser-136 of Bad is thought to be phosphorylated by protein kinase B (PKB, also called Akt), which is activated when cells are exposed to agonists that stimulate phosphatidylinositol 3-kinase (PI3K). To some extent and only at high levels of ATP, Ser-112 and Ser-136 can also be phosphorylated by cAMP-dependent protein kinase A (PKA). The major site on Bad phosphorylated by PKA, however, is Ser-155. Phosphorylation at Ser-112 and Ser-136 creates binding sites for the phosphoserine-specific interaction of 14-3-3 proteins, which retain Bad in the cytoplasm and prevent cytotoxic interactions with Bcl-X_L at the mitochondrial membrane.

Ser-155 of Bad can also be phosphorylated. It is embedded in a canonical PKA motif (RRxS) and located within the hydrophobic amphipathic face of the BH3-like α helix, crucial for dimerization with Bcl-X_L. Regulation of BH3 domain accessibility by covalent modification may be a general feature of the death-promoting "BH3 domain only" proteins. Although the precise mechanism is still unknown, deletion analysis of Bad has demonstrated the importance of the BH3-like domain in mediating both its heterodimerization with Bcl-X_L and its death-promoting activity. Ser-155 can be phosphorylated by PKA and Rsk. Such phosphorylation of Bad at Ser-155 is predicted to introduce a charged phosphate directly between the BH3 hydrophobic face and the Bcl-X_L hydrophobic pocket, resulting in dissociation of the two molecules. Bad lacks the hydrophobic C-terminal sequence that, as in the case of Bcl-2, is required for its targeting to mitochondria.

Dephosphorylation of Bad

The general role of Bad phosphorylation in the maintenance of cell survival is well accepted. The mechanism by which Bad is phosphorylated is more complicated than was previously thought, and the literature is controversial in some points. In general, however, most studies have focussed on the kinase site. Much less is known about dephosphorylation of Bad (for review on ser/thr protein phosphatases see [8]).

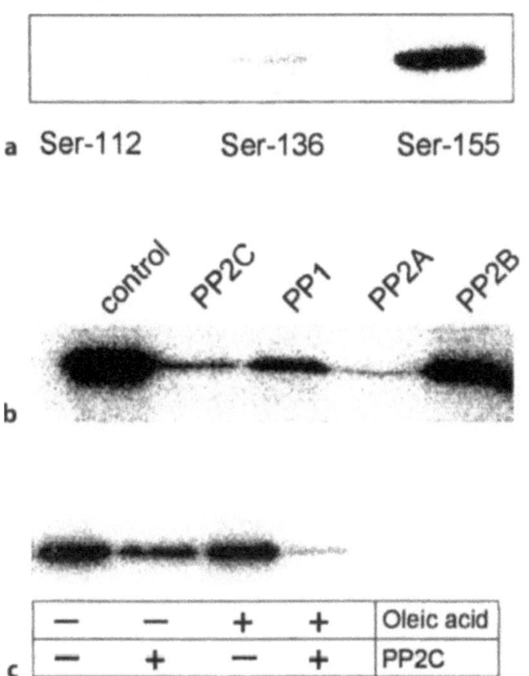

Fig. 3a–c. Phosphorylation and dephosphorylation of Bad. **a** Western-blot. GST-Bad was phosphorylated by PKA. Antibodies directed against peptides containing phosphoserine 112, 136 or 155 were used to identify Ser-155 as the predominant phosphorylation site. **b** Autoradiogram. GST-Bad (9 µg) phosphorylated by PKA at Ser-155 was incubated at 37 °C for 30 min with the protein phosphatases as indicated (300 ng, respectively). Dephosphorylation by PP2A and PP2C was most efficient. **c** Autoradiogram. Dephosphorylation of Bad (phosphorylated at Ser-155) by PP2C in the absence and presence of 0.6 mM oleic acid. Incubations contained 0.5 mM Mg^{2+}. Oleic acid stimulated dephosphorylation of Bad by PP2C

The Ser/Thr-protein phosphatases type-1 (PP1) and type-2B (PP2B, calcineurin) were reported to dephosphorylate Bad at phosphoserines 112 and 136 [1, 7]. Type-2A phosphatase (PP2A) was also described to dephosphorylate Bad [2]. Studies to locate such dephosphorylation sites have not been published.

Phosphatases acting on phosphoserine 155 of Bad have not been identified yet. We addressed the question which of the serine/threonine protein phosphatases would be able to reverse the phosphate introduced to Bad at Ser-155 by PKA and Rsk. Our findings as described above – correlation of PP2C activation with induction of apoptosis – furthermore prompted us to investigate a possible involvement of PP2C in the dephosphorylation of Bad. This has never been examined before. To test the hypothesis of PP2C exerting its apoptotic effect via dephosphorylation of Bad we once more concentrated on Ser-155 because the surrounding amino acids suggest dealing with a sequence motif PP2C might act on: arginine at position -3 and lack of adjacent C-terminal proline residues.

Phosphorylation and dephosphorylation of Bad was performed using bacterially expressed GST-Bad fusion protein kindly provided by Sir Philip Cohen, Dundee, UK. Bad was used as a substrate in an *in vitro* kinase reaction using PKA, Mg^{2+}

and $[\gamma\text{-}^{32}P]$ATP. Subsequent SDS-PAGE and autoradiography revealed heavily labeled Bad. Western blot analysis using polyclonal antibodies raised against phosphopeptides that recognize either phosphoserine 112, 136 or 155 of Bad (Cell Signal Technology) was performed to identify the phosphorylation sites. Incubation of Bad with PKA in the presence of 100 μM ATP specifically resulted in phosphorylation of Ser-155 (Fig. 3 A).

In order to examine which protein phosphatases are able to dephosphorylate phosphoserine 155 of Bad, unincorporated ATP had to be removed after the kinase reaction using spin columns. $[^{32}P]$Bad was then incubated with an equivalent amount of various protein phosphatases. Autoradiograms showed that PP1, PP2A and PP2C were all able to dephosphorylate phosphoserine 155 of Bad (Fig. 3 B). For PP2B higher concentrations were required (data not shown). Tyrosine phosphatase was taken as control for hydrolysis due to time and temperature. As expected, it had no effect (data not shown).

Dephosphorylation of phosphoserine 155 of Bad by PP2C was studied in more detail. This dephosphorylation reaction showed features characteristic for PP2C: requirement of Mg^{2+}-ions and insensitivity to okadaic acid (data not shown). Oleic acid is known to activate dephosphorylation of the artificial substrate for PP2C, $[^{32}P]$casein, and known to induce apoptosis in neuronal cell cultures. Interestingly, oleic acid was found to also stimulate dephosphorylation of phosphoserine 155 of Bad by PP2C (Fig. 3 C).

Conclusion

The present results suggest PP2C to be significantly involved in neuronal apoptosis. An inhibition of PP2C, unknown till now, may possess neuroprotective potency and could play a role in therapy of neurodegenerative diseases and stroke.

References

1. Ayllon V, Martinez AC, Garcia A, Cayla X, Rebollo A (2000) Protein phosphatase 1α is a ras-activated bad phosphatase that regulates interleukin-2 deprivation-induced apoptosis. EMBO J 19:2237–2246
2. Chiang CW, Harris G, Ellig C, Masters SC, Subramanian R, Shenolikar S, Wadzinski BE, Yang E (2001) Protein phosphatase 2A activates the proapoptotic function of Bad in interleukin-3-dependent lymphoid cells by a mechanism requiring 14-3-3 dissociation. Blood 97:1289–1297
3. Klumpp S, Selke D, Krieglstein J (2000) Protein phosphatases and neuronal apoptosis. In: Krieglstein J, Klumpp S (eds) Pharmacology of cerebral ischemia. medpharm Scientific Publishers, Stuttgart, pp 95–104
4. Lizcano JM, Morrice N, Cohen P (2000) Regulation of Bad by cAMP-dependent protein kinase is mediated via phosphorylation of a novel site, Ser[155]. Biochem J 349:547–557
5. Muslin AJ, Xing H (2000) 14-3-3 proteins: regulation of subcellular localization by molecular interference. Cellular Signal 12:703–709
6. Tan Y, Demeter MR, Ruan H, Comb MJ (2000) Bad ser-155 phosphorylation regulates Bad/Bcl-XL interaction and cell survival. J Biol Chem 275:25865–25869
7. Wang HG, Pathan N, Ethell IM, Krajewski S, Yamaguchi Y, Shibasaki F, McKeon F, Bobo T, Franke TF, Reed JC (1999) Ca^{2+}-induced apoptosis through calcineurin dephosphorylation of Bad. Science 284:339–343
8. Wera S, Hemmings BA (1995) Serine/threonine protein phosphatases. Biochem J 311:17–29

9. Yang E, Zha J, Jockel J, Boise LH, Thompson CB, Korsemeyer SJ (1995) Bad, a heterodimeric partner for Bcl-xl and Bcl-2, displaces Bax and promotes cell death. Cell 80:285–291
10. Zha J, Harada H, Yang E, Jocker J, Korsmeyer SJ (1996) Serine phosphorylation of death agonist Bad in response to survival factor results in binding to 14-3-3 not Bcl-X. Cell 87:619–628
11. Zha J, Harada H, Osipov K, Jockel J, Waksman G, Korsemeyer SJ (1997) BH3 domain of Bad is required for heterodimerization with Bcl-X and apoptotic activity. J Biol Chem 272:24101–24104
12. Zhu Y, Ahlemeyer B, Bauerbach E, Krieglstein J (2001) TGF-β1 prevents neuronal apoptosis in rat hippocampal cultures involving inhibition of caspase-3 activation. Neurochem Int 38:227–235

Mitochondrial Cytochrome C- and Smac-Dependent Apoptosis in Cerebral Ischemia: Role of Oxidative Signaling

T. Sugawara, M. Ferrand-Drake, F. Yu, C. Maier, E. E. Hoyte, and P. H. Chan

Summary. Mitochondria are known to be involved in the early stage of apoptosis by releasing cytochrome c, caspase-9 and second mitochondria-derived activator of caspases (Smac). We have reported that overexpression of copper/zinc superoxide dismutase (SOD1) reduced superoxide production and ameliorated neuronal injury in the hippocampal CA1 subregion after global ischemia. However, the role of oxygen-free radicals produced after ischemia/reperfusion in the mitochondrial signaling pathway has not been clarified. Five minutes of global ischemia was induced in male SOD1 transgenic (Tg) and wild-type (Wt) littermate rats. In the Wt animals, early superoxide production, mitochondrial release of cytochrome c, Smac and cleaved caspase-9 were observed after ischemia. Active caspase-3 was subsequently increased and 85% of the hippocampal CA1 neurons showed apoptotic DNA damage 3 days after ischemia. Tg animals showed less superoxide production and cytochrome c and Smac release. These results suggest that overexpression of SOD1 reduced oxidative stress, thereby attenuating the mitochondrial release of cytochrome c and Smac, resulting in less caspase activation and apoptotic cell death. Oxygen-free radicals may play a pivotal role in the mitochondrial signaling pathway of apoptotic cell death in the hippocampal CA1 neurons after global ischemia.

Key words. Superoxide dismutase – oxidative stress – global cerebral ischemia – neuron – apoptosis – cytochrome c – second mitochondrial activator of caspases (Smac) – caspase

Introduction

It has been demonstrated in numerous studies that oxygen radicals are directly involved in oxidative damage of cellular macromolecules such as lipids, proteins, and nucleic acids in ischemic tissues, which lead to cell death. Recent studies have

Taku Sugawara, Michel Ferrand-Drake, Fengshan Yu, Carolina Maier, Elizabeth E. Hoyte, Pak H. Chan
Department of Neurosurgery, Department of Neurology and Neurological Sciences, and Program in Neurosciences, Stanford University School of Medicine, Stanford, California

Corresponding author: Pak H. Chan, Ph.D., Neurosurgical Laboratories, Stanford University, 1201 Welch Road, #P314, Stanford, CA 94305, USA
Tel.: 650-498/44 57, Fax: 650-498/45 50, E-Mail: phchan@leland.stanford.edu

Maturation Phenomenon in Cerebral Ischemia V
A. M. Buchan et al. (Eds.)
© Springer-Verlag Berlin Heidelberg 2004

provided evidence that indirect signaling pathways by oxygen radicals can also cause cellular damage and death in cerebral ischemia and reperfusion [4].

Mitochondria are assumed to be involved in apoptosis by releasing cytochrome c from their intermembrane space to the cytoplasm. If ATP or deoxy-ATP is present, cytochrome c binds to the *C. elegans* gene ced-3 (CED) homolog, Apaf-1, and subsequently, Apaf-1 binds to procaspase-9, resulting in activation of caspase-9, which is an initiator of the cytochrome c-dependent caspase cascade [16, 17, 31]. Activated caspase-9 directly cleaves procaspase-3, and active caspase-3 triggers activation of additional caspases and leads to apoptosis [18, 26]. On the other hand, the inhibitor of apoptosis (IAP) family proteins negatively regulate caspase activation. IAPs suppress apoptosis by inhibiting the enzymatic activity of active caspases [9, 21]. In the early stage of apoptosis, a newly identified apoptosis regulator, second mitochondria-derived activator of caspases (Smac), is released from mitochondria into the cytosol concurrently with cytochrome c. Smac eliminates the inhibitory effects of many IAPs and promotes caspase activation [10, 30].

We have shown evidence that copper/zinc-SOD (SOD1), a cytosolic antioxidant, plays a protective role against focal [5, 14] and global [6] cerebral ischemia. Our studies showed that the early release of cytochrome c from mitochondria to the cytosol and subsequent DNA-fragmented cell death were attenuated in transgenic (Tg) mice that overexpress SOD1 after transient focal cerebral ischemia [12], and that cytochrome c release corresponded to the selective vulnerability of the rats hippocampal CA1 neurons after global ischemia [28]. However, whether SOD1 can affect the mitochondrial signaling pathway after transient global ischemia has not been studied. Using SOD1-Tg rats, we provide evidence that the delayed death of vulnerable hippocampal CA1 neurons is partly mediated by the superoxide radical-mitochondrial signaling pathway.

Materials and Methods

SOD1 Tg Rats

Heterozygous SOD1 Tg rats of the SOD1 with Sprague-Dawley background, carrying human SOD1 genes with a 4- to 6-fold increase in SOD1, were derived from the founder stock described previously [6]. They were further bred with wild-type (Wt) Sprague Dawley rats to generate heterozygous rats. The SOD1 Tg rats were identified by isoelectric focusing gel electrophoresis, as described [6]. There were no observable phenotypic differences, including cerebral vasculature, between the Tg rats and their Wt littermates [6].

Surgery

Five minutes of transient global ischemia was induced by bilateral common carotid artery occlusion and bleeding to lower the MABP to 30 to 35 mmHg, using the method originally described by Smith et al. [27] with some modifications [28, 29]. Male SOD1 Tg rats (300 to 350 mg) and their Wt littermates were anesthetized with 5%

isoflurane and maintained during surgery at a level of 2.0% isoflurane in 70% N_2O and 30% O_2 with spontaneous breathing. The rectal temperature was controlled at $37.0 \pm 0.5\,°C$ during surgery with a feedback-regulated heating pad. The femoral artery was exposed and catheterized with a PE-50 catheter to allow continuous recording of arterial blood pressure and withdrawal of blood samples for blood gas analysis. After recovery of the arterial blood pressure, the arterial blood was collected for blood gas analysis. The animals were maintained in an air-conditioned room at $20\,°C$ with free access to food and water before and after surgery. All animals were treated in accordance with Stanford University guidelines and the animal protocol approved by Stanford University's Administrative Panel on Laboratory Animal Care.

Results

SOD Activity and Superoxide Production in Wt and Tg Rats

As shown in Fig. 1 A, the average total SOD activity in the Tg rats ranged from 60.1 to 205.0 U/mg in various tissues, and 14.6 to 51.1 U/mg in their littermates. In the Wt animals, the activity was significantly greater in the striatum ($P < 0.05$), spinal cord ($P < 0.001$) and heart ($P < 0.001$) compared with that in the hippocampus. In the Tg rats, SOD activity increased approximately 4- to 6-fold compared with their Wt littermates. The activity in the hippocampal CA1 subregion was not altered after ischemia for at least 3 days in both the Tg animals and the Wt littermates (Fig. 1 B).

Superoxide Production Was Greater in Wt Animals Than in Tg Animals

In the hippocampal CA1 pyramidal neurons of the non-Tg littermates, superoxide production was shown by ethidium signals as small particles in the cytosol under normal conditions. One hour after ischemia, the hippocampal CA1 pyramidal neurons showed a marked increase in punctate cytosolic signals and diffuse cytosolic signals. In the Tg rats, small particles of ethidium were also observed in the non-ischemic CA1 subregion, but the increase in signals 1 h after ischemia was not as noticeable as in the non-Tg animals. Quantitative analysis of these signals confirmed the difference between the Tg and non-Tg rats at 1 h and revealed that the signal intensity gradually decreased until 3 days (Fig. 1 C).

Mitochondrial Release of Cytochrome C and Smac, and Subsequent Caspase-9 Cleavage

Cytochrome c and Smac immunoreactivity were evident as a single band of molecular mass of approximately 15 and 21 kDa in the cytosolic fraction of the hippocampal CA1 subregion before and after ischemia, respectively (Fig. 2). The cytosolic increase in these proteins was observed 6 h after ischemia in both the Wt and Tg groups; however, the increase in the Wt rats was greater than in the Tg ani-

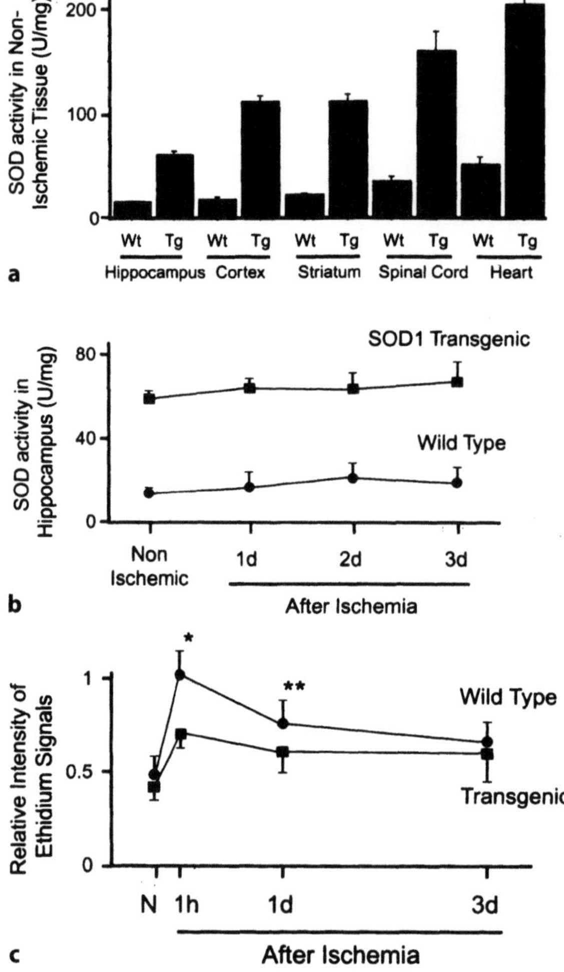

Fig. 1. SOD activity and superoxide production in Wt and SOD1 Tg rats. **a** Total SOD activity in various tissues in Wt and SOD1 Tg rats. **b** Total SOD activity in the hippocampus after cerebral ischemia. **c** Ethidium signals in rat brains after cerebral ischemia

mals. Cleaved caspase-9 was also increased in the ischemic brains, but not until 12 h of ischemia. A corresponding decrease in cytochrome c and Smac was observed in the mitochondrial fraction at the same time points. Statistical analyses ($n = 4$ each) confirmed that the cytosolic cytochrome c/Smac in the non-ischemic brain there was no significant difference between the Wt and Tg groups (OD of cytochrome c/Smac in Wt and Tg animals was $0.190 \pm 0.116/1.245 \pm 0.143$, $0.200 \pm 0.099/1.188 \pm 0.251$, respectively). However, 1 day after ischemia, these proteins were more abundant in the Wt animals than in the Tg animals (OD of cytochrome c/Smac in Wt and Tg animals was $1.435 \pm 0.383/5.927 \pm 0.781$, $0.865 \pm 0.234/$

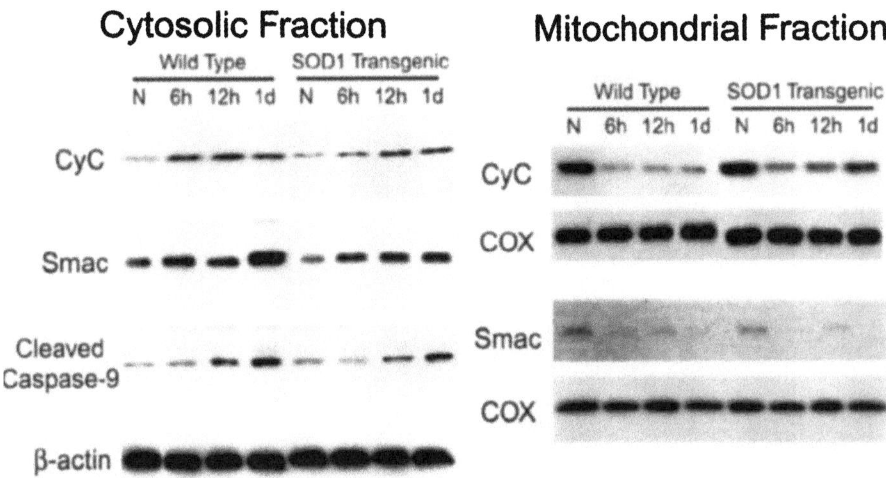

Fig. 2. Western blot analyses of cytochrome c, Smac and cleaved caspase-9 in cytosolic and mitochondrial fractions. Western blot analysis of cytosolic fraction showed that cytochrome c and Smac started to increase 6 h after ischemia and remained elevated until 1 day after ischemia, and that cleaved caspase-9 started to increase at 12 h and also remained increased at 1 day. The increase of cytochrome c and Smac was greater in the Wt animals than in the Tg animals. Correspondingly, cytochrome c and Smac both decreased at the same time in the mitochondrial fraction. The decrease in cytochrome c in the Wt rats was greater than in the Tg rats. Consistent bands of β-actin and cytochrome oxidase are also shown. Results shown are representative of three independent studies. N, non-ischemic; CyC, cytochrome c; COX, cytochrome oxidase

3.308 ± 1.005, respectively; $P = 0.0441/0.0062$). The mitochondrial cytochrome c/Smac in the non-ischemic brain was not different between the Wt and Tg groups (OD of cytochrome c/Smac in Wt and Tg animals was $4.607 \pm 0.590/0.792 \pm 0.097$, $4.525 \pm 0.473/0.725 \pm 0.176$, respectively). Yet mitochondrial cytochrome c at 1 day was less abundant in the Wt animals than in the Tg animals (OD of cytochrome c in Wt and Tg animals was 0.652 ± 0.183, 2.360 ± 0.597, respectively; $P = 0.0016$). However, there was no difference in mitochondrial Smac at 1 day between the two groups (OD of Smac in Wt and Tg animals was 0.065 ± 0.021, 0.085 ± 0.025, respectively). There was a consistent amount of β-actin in the cytosolic fraction and cytochrome oxidase IV in the mitochondrial fraction, suggesting that the amount of the loaded protein was consistent (Fig. 2). The purity of the cytosolic fraction was confirmed by the absence of cytochrome oxidase IV bands.

Overexpression of SOD1 Protected the Hippocampal CA1 Neurons From Delayed DNA-Damaged Cell Death

Delayed death of the hippocampal CA1 neurons was observed 2 to 3 days after ischemia. Most of the CA1 neurons in the Wt animals showed shrunken, triangular-shaped, condensed nuclei in H&E-stained sections; however, in a significant number of neurons, the normal features were preserved in the Tg rats. Most of the

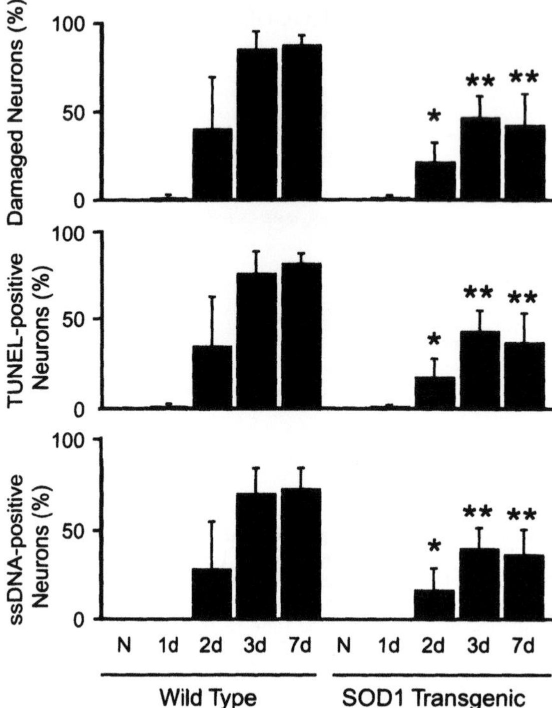

Fig. 3. Cell counting study of morphologically damaged, TUNEL-positive and ssDNA-positive cells. Cell counting analyses ($n=6$ each) show that approximately 85% of the hippocampal CA1 pyramidal neurons in the Wt rats and 45% in the Tg animals underwent delayed death, and that most of these cells were positive for TUNEL and ssDNA. The neuronal damage matured by 3 days after ischemia. There was a significant difference between the Wt and Tg groups. $*P < 0.01$, $**P < 0.001$. N, non-ischemic

morphologically damaged neurons were also positive for TUNEL and ssDNA staining. The cell counting studies ($n=6$ each) showed that the damaged neurons accounted for approximately 85% of the hippocampal CA1 pyramidal neurons in the Wt rats and 45% in the Tg rats, and that the neuronal damage matured by 3 days after ischemia (Fig. 3).

Discussion

Overexpression of SOD1 Reduced Superoxide Production in Mitochondria After Global Cerebral Ischemia

It is well established that the rat transient global ischemia model mimics the condition after transient cardiac arrest and causes selective neuronal death in vulnerable regions, such as hippocampal CA1 pyramidal cells, Purkinje cells of the cerebellum, and neurons in the third to fifth layers of the cerebral cortex [15, 25]. The vulnerability has been attributed to many factors, such as glutamate neurotoxicity, calcium, expression of cell suicide genes, activation of apoptotic proteins, mitochondrial dysfunction, and oxygen free radicals [1, 3, 28]. The oxygen free radical hypothesis seems to be especially important because superoxide production is thought to be closely associated with reperfusion injury [3, 20]. We previously developed an *in situ* imaging method for superoxide radical measurement [22]. This method is based on the selective oxidation of hydroethidine by superoxide radicals [2]. The quantitative study revealed that the intensity of these signals peaked 1 h after ischemia and gradually returned toward the normal level thereafter. The signal intensity was significantly lower in the Tg animals 1 h and 1 day after ischemia. The highest signal intensity of ethidium immediately after reperfusion is consistent with the hypothesis that superoxide is a major contributor to reperfusion injury. A significantly lower production of superoxide in the Tg rats than in the Wt rats suggests that overexpression of SOD in the cytosol can eliminate a significant amount of superoxide. Furthermore, unequal activity of the total SOD in various regions in the CNS (Fig. 1a) may help explain the vulnerability of the cells in the hippocampus and cortex. Consistent SOD activity in both the Wt and Tg animals after ischemia (Fig. 1b) suggests the stability of this enzyme after ischemia/reperfusion and that SOD can play a major role in detoxifying superoxide for at least 3 days after ischemia.

Mitochondrial Superoxide Production and the Mitochondrial Release of Cytochrome C and Smac

In the present study, our findings suggest that neuronal mitochondria were a major source of superoxide production. Because mitochondria are known as the site of superoxide production under normal or pathological conditions such as focal cerebral ischemia [24], excessive superoxide production may cause mitochondrial injury that leads to the release of proteins such as cytochrome c and Smac from their intermembrane space. Our study also demonstrated that cytochrome c and Smac are released from mitochondria to the cytosol in CA1 cells after transient global ischemia. The subcellular distribution of cytochrome c and Smac was confirmed by Western blot analyses (Fig. 2). A significant amount of mitochondrial cytochrome c and Smac was detected in the normal hippocampal CA1 subregion and was decreased 6 h to 1 day after ischemia. Correspondingly, these proteins showed a marked increase in the cytosol at the same time points. We believe that increased cytosolic cytochrome c and Smac are substantially derived from the mitochondria of CA1 neurons.

Apoptotic Cell Death in Vulnerable Hippocampal CA1 Neurons After Transient Global Ischemia

Neuronal death in the hippocampal CA1 subregion after global ischemia has been shown to occur in a delayed manner [15], and recent studies demonstrated that these neuronal deaths are caused in part by apoptosis [19, 23]. TUNEL staining has been used as a marker to detect apoptotic cells after global cerebral ischemia [23, 28], and internucleosomal DNA fragmentation has also been detected by a DNA laddering pattern [13, 28]. However, methods to detect TUNEL and DNA laddering are dependent upon detection of DNA breaks, and their specificity for apoptosis is still controversial. The method used in the present study to detect ssDNA is supposed to be more specific for apoptotic cells. Formamide-induced DNA denaturation has been reported to be specific for condensed chromatin of apoptotic cells, and the denaturation is likely caused by digestion of histones by caspases [11]. In this study, approximately 85% of the CA1 neurons in the Wt animals and 45% of the neurons in the Tg animals were morphologically damaged by 3 days after ischemia, and most of these cells were also TUNEL- and ssDNA-positive (Fig. 3). The specificity of TUNEL and ssDNA has to be further clarified. On the other hand, electron microscopic studies showed observations against apoptosis, such as early organelle swelling, disaggregation of polyribosomes, and cell and nuclear membrane breaks in CA1 neurons [8]. However, the protective effects of caspase inhibitors on delayed neuronal death in this and other studies [7] strongly suggest that the biochemical caspase cascade plays a major role in neuronal death after global ischemia.

Acknowledgements. This work was supported by NINDS contract NO1 NS82386, "Transgenic Rat for Stroke Research," by NIH grants P50 NS14543, NS25372, NS36147, NS37530 and NS38653, and by an American Heart Association Bugher Foundation Award. P.H.C. is a recipient of the Jacob Javits Neuroscience Investigator Award. We thank Dr. Xiaodong Wang for providing the antibody against Smac and Cheryl Christensen for editorial assistance.

References

1. Abe K, Aoki M, Kawagoe J, Yoshida T, Hattori A, Kogure K, Itoyama Y (1995) Ischemic delayed neuronal death. A mitochondrial hypothesis. Stroke 26:1478–1489
2. Bindokas VP, Jordan J, Lee CC, Miller RJ (1996) Superoxide production in rat hippocampal neurons: Selective imaging with hydroethidine. J Neurosci 16:1324–1336
3. Chan PH (1996) Role of oxidants in ischemic brain damage. Stroke 27:1124–1129
4. Chan PH (2001) Reactive oxygen radicals in signaling and damage in the ischemic brain. J Cereb Blood Flow Metab 21:2–14
5. Chan PH, Epstein CJ, Kinouchi H, Kamii H, Chen SF, Carlson E, Gafni J, Yang G, Reola L (1996) Neuroprotective role of CuZn-superoxide dismutase in ischemic brain damage. Adv Neurol 71:271–280
6. Chan PH, Kawase M, Murakami K, Chen SF, Li Y, Calagui B, Reola L, Carlson E, Epstein CJ (1998) Overexpression of SOD1 in transgenic rats protects vulnerable neurons against ischemic damage after global cerebral ischemia and reperfusion. J Neurosci 18:8292–8299
7. Chen J, Nagayama T, Jin K, Stetler RA, Zhu RL, Graham SH, Simon RP (1998) Induction of caspase-3-like protease may mediate delayed neuronal death in the hippocampus after transient cerebral ischemia. J Neurosci 18:4914–4928
8. Colbourne F, Sutherland GR, Auer RN (1999) Electron microscopic evidence against apoptosis as the mechanism of neuronal death in global ischemia. J Neurosci 19:4200–4210
9. Deveraux QL, Reed JC (1999) IAP family proteins – suppressors of apoptosis. Genes Dev 13:239–252
10. Du C, Fang M, Li Y, Li L, Wang X (2000) Smac, a mitochondrial protein that promotes cytochrome c-dependent caspase activation by eliminating IAP inhibition. Cell 102:33–42

11. Frankfurt OS, Krishan A (2001) Identification of apoptotic cells by formamide-induced DNA denaturation in condensed chromatin. J Histochem Cytochem 49:369–378
12. Fujimura M, Morita-Fujimura Y, Noshita N, Sugawara T, Kawase M, Chan PH (2000) The cytosolic antioxidant copper/zinc-superoxide dismutase prevents the early release of mitochondrial cytochrome c in ischemic brain after transient focal cerebral ischemia in mice. J Neurosci 20:2817–2824
13. Heron A, Pollard H, Dessi F, Moreau J, Lasbennes F, Ben-Ari Y, Charriaut-Marlangue C (1993) Regional variability in DNA fragmentation after global ischemia evidenced by combined histological and gel electrophoresis observations in the rat brain. J Neurochem 61:1973–1976
14. Kinouchi H, Epstein CJ, Mizui T, Carlson E, Chen SF, Chan PH (1991) Attenuation of focal cerebral ischemic injury in transgenic mice overexpressing CuZn superoxide dismutase. Proc Natl Acad Sci USA 88:11158–11162
15. Kirino T (1982) Delayed neuronal death in the gerbil hippocampus following ischemia. Brain Res 239:57–69
16. Kuida K, Haydar TF, Kuan CY, Gu Y, Taya C, Karasuyama H, Su MS, Rakic P, Flavell RA (1998) Reduced apoptosis and cytochrome c-mediated caspase activation in mice lacking caspase 9. Cell 94:325–337
17. Li P, Nijhawan D, Budihardjo I, Srinivasula SM, Ahmad M, Alnemri ES, Wang X (1997) Cytochrome c and dATP-dependent formation of Apaf-1/caspase-9 complex initiates an apoptotic protease cascade. Cell 91:479–489
18. Liu X, Kim CN, Yang J, Jemmerson R, Wang X (1996) Induction of apoptotic program in cell-free extracts: Requirement for dATP and cytochrome c. Cell 86:147–157
19. MacManus JP, Buchan AM, Hill IE, Rasquinha I, Preston E (1993) Global ischemia can cause DNA fragmentation indicative of apoptosis in rat brain. Neurosci Lett 164:89–92
20. McCord JM (1985) Oxygen-derived free radicals in postischemic tissue injury. N Engl J Med 312:159–163
21. Miller LK (1999) An exegesis of IAPs: Salvation and surprises from BIR motifs. Trends Cell Biol 9:323–328
22. Murakami K, Kondo T, Kawase M, Li Y, Sato S, Chen SF, Chan PH (1998) Mitochondrial susceptibility to oxidative stress exacerbates cerebral infarction that follows permanent focal cerebral ischemia in mutant mice with manganese superoxide dismutase deficiency. J Neurosci 18:205–213
23. Nitatori T, Sato N, Waguri S, Karasawa Y, Araki H, Shibanai K, Kominami E, Uchiyama Y (1995) Delayed neuronal death in the CA1 pyramidal cell layer of the gerbil hippocampus following transient ischemia is apoptosis. J Neurosci 15:1001–1011
24. Piantadosi CA, Zhang J (1996) Mitochondrial generation of reactive oxygen species after brain ischemia in the rat. Stroke 27:327–331
25. Pulsinelli WA, Brierley JB, Plum F (1982) Temporal profile of neuronal damage in a model of transient forebrain ischemia. Ann Neurol 11:491–498
26. Slee EA, Harte MT, Kluck RM, Wolf BB, Casiano CA, Newmeyer DD, Wang HG, Reed JC, Nicholson DW, Alnemri ES, Green DR, Martin SJ (1999) Ordering the cytochrome c-initiated caspase cascade: Hierarchical activation of caspases-2, -3, -6, -7, -8, and -10 in a caspase-9-dependent manner. J Cell Biol 144:281–292
27. Smith ML, Bendek G, Dahlgren N, Rosen I, Wieloch T, Siesjö BK (1984) Models for studying long-term recovery following forebrain ischemia in the rat. 2. A 2-vessel occlusion model. Acta Neurol Scand 69:385–401
28. Sugawara T, Fujimura M, Morita-Fujimura Y, Kawase M, Chan PH (1999) Mitochondrial release of cytochrome c corresponds to the selective vulnerability of hippocampal CA1 neurons in rats after transient global cerebral ischemia. J Neurosci 19:RC39
29. Sugawara T, Kawase M, Lewén A, Noshita N, Gasche Y, Fujimura M, Chan PH (2000) Effect of hypotension severity on hippocampal CA1 neurons in a rat global ischemia model. Brain Res 877:281–287
30. Verhagen AM, Ekert PG, Pakusch M, Silke J, Connolly LM, Reid GE, Moritz RL, Simpson RJ, Vaux DL (2000) Identification of DIABLO, a mammalian protein that promotes apoptosis by binding to and antagonizing IAP proteins. Cell 102:43–53
31. Yoshida H, Kong YY, Yoshida R, Elia AJ, Hakem A, Hakem R, Penninger JM, Mak TW (1998) Apaf1 is required for mitochondrial pathways of apoptosis and brain development. Cell 94:739–750

Erythropoietin: A Beneficial Approach to Neuroprotection in Stroke

H. Ehrenreich and A.-L. Sirén

Introduction

Erythropoietin (EPO) was first characterized as a hematopoietic growth factor and has been in clinical use as such for more than a decade. The expression of EPO and its receptor in brain tissue, cultured neurons and astrocytes along with its neuroprotective potential in several experimental models of stroke expands its biological role beyond hematopoiesis [1–5].

Methods

Immunohistochemistry, Western blotting, RT-PCR, cell culture, *in-vivo* and in-vitro models of hypoxia/ischemia, and ELISA were used to study mechanisms of neuroprotective effect of EPO and EPO/EPO receptor (EPOR) expression in rat and human tissues.

Results

(1) Application of EPO at the beginning of hypoxia resulted in reduction of cell death of cultured rat hippocampal neurons. (2) The neuroprotective effect of EPO is apparently via an antiapoptotic mechanism. (3) EPO exerted its neuro-protective action via restoration of hypoxia-arrested phosphorylation of extracellular signal-regulated kinases and the phosphatidyl-inositol 3-kinase pathways. (4) Hypoxia induced EPO and EPOR mRNAs in cultured neurons and astrocytes as well as in hippocampus *in vivo*. Upregulation of the EPOR protein could be observed upon hypoxia in cultured hippocampal neurons, in particular, in neuronal processes, that exhibit weak cytoplasmic EPOR staining under basal conditions. (5) In human brain tissue EPO was seen in vascular tissue and inflammatory cells acutely after stroke (<5 days), EPOR in blood vessels and neuronal and astrocytic processes within the infarcts and the peri-infarct zone reflecting perhaps an increased receptiveness of neurons to EPO upon metabolic stress. In older ischemic infarcts

Hannelore Ehrenreich and Anna-Leena Sirén, Departments of Neurology and Psychiatry, Georg-August University, and Max-Planck Institute for Experimental Medicine, Göttingen, Germany

Maturation Phenomenon in Cerebral Ischemia V
A.M. Buchan et al. (Eds.)
© Springer-Verlag Berlin Heidelberg 2004

(>18 days) and as a late response to prolonged hypoxia we observed a persistent astrocytic up-regulation of EPO production after stroke that may serve as a rapidly mobilizable source of EPO in the case of repeated hypoxic/ischemic events, i.e. constitute a correlate of acquired relative ischemic tolerance. (6) A clinical stroke trial with recombinant human EPO has been ongoing in Göttingen (The Göttingen EPO-Stroke Trial). The initial safety study demonstrated that intravenously administered EPO is able to enter the brain in acute human stroke and that EPO treatment is safe in stroke patients. The first randomized double-blind proof-of-concept study on EPO in stroke yielded a significantly better clinical outcome of EPO-treated patients.

Conclusions

EPO represents the first beneficial neuroprotective approach to stroke.

References

1. Brines ML, Ghezzi P, Keenan S, Agnello D, de Lanerolle NC, Cerami C, Itri LM, Cerami A (2000) Proc Natl Acad Sci USA 97:10526–10531
2. Lewczuk P, Hasselblatt M, Kamrowski-Kruck H, Heyer A, Unzicker C, Sirén AL, Ehrenreich H (2000) Neuroreport 11:3485–3488
3. Sakanaka M, Wen TC, Matsuda S, Masuda S, Morishita E, Nagao M, Sasaki R (1998) Proc Natl Acad Sci USA 95:4635–4640
4. Sirén A-L, Fratelli M, Brines ML, Goemans C, Casagrande S, Lewczuk P, Keenan S, Gleiter C, Pasquali C, Capobianco A et al (2001a) Proc Natl Acad Sci USA 98:4044–4049
5. Sirén A-L, Knerlich F, Poser W, Gleiter C, Brück W, Ehrenreich H (2001b) Acta Neuropathologica 101:271–276

Transgenic Mutants for the Investigation of Molecular Stroke Mechanisms

K.-A. Hossmann, R. Hata, K. Maeda, T. Trapp, and G. Mies

Key words. Focal ischemia – mouse – gene manipulation – CREB – APAF-1 – XIAP – bcl-2 – p53 – angiotensin – bradykinin

Introduction

Brain damage induced by focal interruption of blood flow can be differentiated in two pathophysiologically different categories: a hemodynamic type of injury, resulting in primary necrotic brain damage, and a molecular type of injury which leads to delayed or secondary brain injury [15]. Primary necrotic brain injury occurs when blood flow declines – and remains – below the threshold of energy failure. In anaesthetized laboratory animals, this threshold gradually increases from about 15% of control shortly after the onset of ischemia to about 30% after several hours of vascular occlusion [29].

Delayed or secondary brain damage occurs in regions called "at risk" which are not intact but which have not yet suffered tissue necrosis. In permanent focal ischemia this type of injury evolves in the peri-infarct penumbra which, by definition, is the peripheral part of the ischemic territory in which functional activity but not energy metabolism are suppressed [2]. After transient focal ischemia, the area "at risk" is the part of the ischemic territory in which energy metabolism has failed during ischemia but recovers after reversal of the vascular occlusion [26]. With increasing duration of ischemia this region expands from the central to the peripheral parts of the ischemic territory until – under conditions of irreversible energy failure – it becomes congruent with the core region of permanent vascular occlusion [49].

A hallmark of tissue "at risk" for delayed ischemic injury is the dissociation between maintained (or restored) energy metabolism and suppressed protein synthesis [29]. Another predictor of delayed cell death is the upregulation of *hsp70* mRNA which has been interpreted as evidence for the cytosolic stress response [58]. However, this response is only a marker and not a mediator of injury be-

Konstantin-Alexander Hossmann, Ryuji Hata, Keiichiro Maeda, Thorsten Trapp and Guenter Mies
Max-Planck-Institute for Neurological Research, Department of Experimental Neurology, Cologne, Germany

Correspondence to: Prof. Dr. K.-A. Hossmann, Max-Planck-Institute for Neurological Research, Department of Experimental Neurology, Gleueler Str. 50, D-50931 Cologne, Germany
Tel.: +49-221/47 26-210, Fax: +49-221/47 26-325, E-Mail: hossmann@mpin-koeln.mpg.de

Maturation Phenomenon in Cerebral Ischemia V
A.M. Buchan et al. (Eds.)
© Springer-Verlag Berlin Heidelberg 2004

cause pre-injury increase of the HSP70 protein, as induced by sublethal pre-ischemic conditioning, conveys neuroprotection to a subsequent more severe ischemic impact [58].

Obviously, precise understanding of the mechanisms of delayed ischemic brain injury is of considerable interest for the identification of molecular targets for therapeutic interventions. In the past, research into this field has been mainly accomplished by pharmacological approaches but the difficulties encountered in translating the experimental observations into clinically successful treatment protocols have raised concerns about the relevance of these interventions [20]. More recently, the availability of genetically modified laboratory animals has provided the opportunity to dissect molecular injury pathways in a much more specific way [6]. A rapidly increasing number of laboratories have embarked on this new methodology, and important data has been generated, which have helped to clarify complex molecular interactions. However, some of these data's are controversial, and the expected identification of a unique injury pathway has not yet succeeded. In the following, we review data from our own laboratory.

Material and Methods

Adult wild-type or genetically manipulated mice were submitted under halothane/ N_2O-anesthesia to either permanent or transient middle cerebral artery (MCA) occlusion. Permanent MCA occlusion was produced by intraluminal thread occlusion [25] and transient MCA occlusion by withdrawal of the intraluminal thread at 1 h after occlusion [26]. The effect of MCA occlusion and recanalization on cerebral blood flow was monitored by transcranial laser Doppler flow (LDF) recording from the ipsilateral parietal cortex. At various ischemia and recirculation times, ranging from 1 h to 3 days, animals were perfusion fixed or frozen in liquid nitrogen, and processed by one or several of the following analytical techniques.

Multiparametric Biochemical Imaging

Forty-five minutes before *in-situ* freezing mice received an intraperitoneal injection of ^3H leucine (150 µCi/animal) for autoradiographic assessment of *in vivo* cerebral protein synthesis (CPS) [27]. Adjacent cryostat sections were processed for ATP using a luciferin/luciferase reaction mix for evoking tissue bioluminescence [36]. Other consecutive cryostat sections were processed for *in-situ* hybridization of various stress and immediate-early genes. Double-strand DNA fragmentations were visualized by terminal transferase biotinylated-UTP nick-end labelling (TUNEL).

Vascular Anatomy

The cerebrovascular system was stained by intravascular infusion of black latex [44]. The peripheral branches of the anterior and middle cerebral arteries were identified on photographs taken from the cortical surface, and the anastomotic junctions, representing the outer borders of the respective vascular territories, were marked to compare the size of the supplying areas in different mouse strains.

Electron Microscopy

Brains were fixed by perfusion through the thoracic aorta with a mixture of 4% paraformaldehyde and 4% glutaraldehyde. Tissue samples were taken from various cortical and subcortical areas of the ispi- and contralateral MCA territory, and processed for standard transmission electron microscopy.

Infarct Size

The size of brain infarcts was evaluated either by histology (cresyl-violet staining or silver impregnation) or by ATP bioluminescence. Histological lesions and the areas of ATP-depletion were measured on 4 to 6 coronal slices by planimetry using appropriate edema corrections, integrated and expressed in percent of the opposite non-ischemic hemisphere.

All values are given as means ± SD.

Results

Pathophysiological Characterization of Stroke Models

Permanent intraluminal MCA thread occlusion produced large infarcts in the ipsilateral hemisphere. Shortly after vascular occlusion, the ATP-depleted tissue region was 30 to 40% smaller than the region of suppressed CPS, in accordance with the biochemical threshold concept of brain ischemia (Fig. 1). With ongoing ischemia time, the region of suppressed CPS did not change but the ATP-depleted area grew until – before the end of the first day – both regions merged. This process reflects the expansion of the infarct core into the penumbra, and probably is unrelated to apoptosis because DNA fragmentation evolves after and not before energy metabolism collapses.

After reversible thread occlusion of 1 h duration, ATP – but not CPS – promptly returned, followed after 4 to 6 h by secondary energy failure in the region of permanent CPS suppression (Fig. 2). Secondary injury was accompanied – but not preceded by TUNEL-detectable DNA fragmentation, and electron microscopy revealed both, chromatin condensations and necrotic cell shrinkage. A clear association of injury to either necrosis or apoptosis is, therefore, not possible in this model.

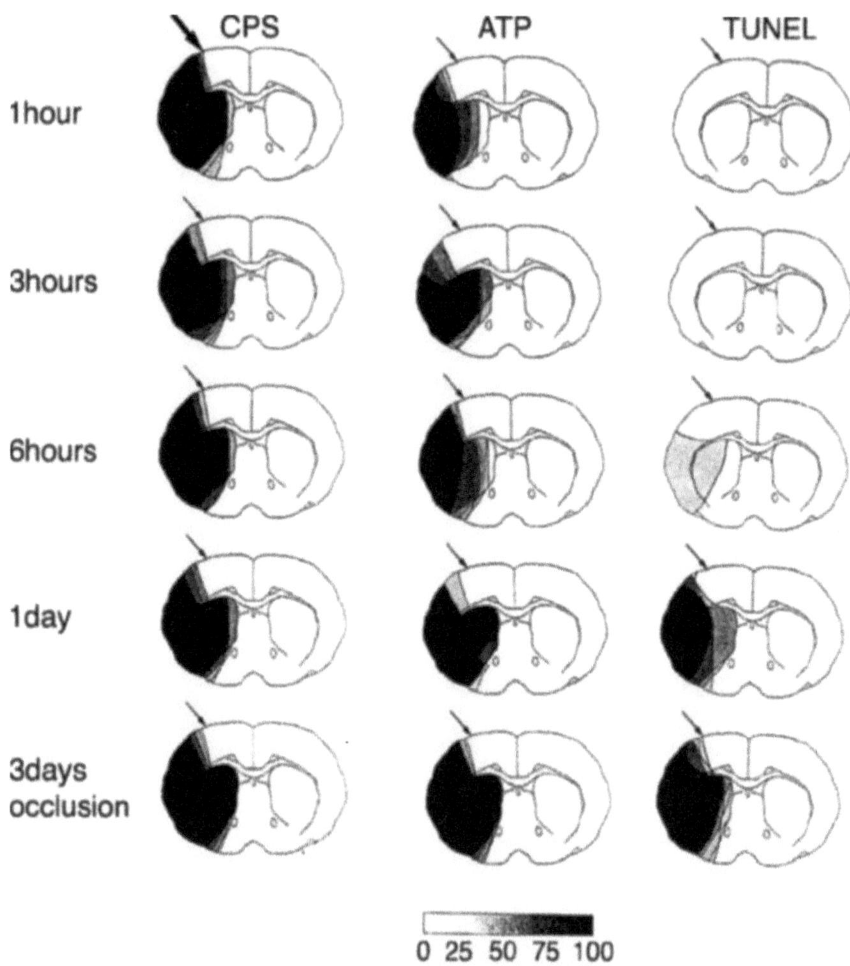

Fig. 1. Incidence maps of suppressed cerebral protein synthesis (CPS), ATP depletion and TUNEL-positive neurons on coronal sections of the mouse brain at various times after the onset of permanent middle cerebral artery occlusion. Areas of disturbed metabolism were outlined in 5 animals per time point at the level of caudate-putamen and superposed to calculate the incidence of alterations in percent of the number of animals per group. The demarcation between normal and disturbed protein synthesis in parietal cortex visible at 1 h MCA occlusion was marked by the arrows to estimate the evolution of the metabolic disturbances at later time points. Note gradual expansion of the ATP-depleted area into but not beyond the area of disturbed protein synthesis, visible after 1 h, and the delayed appearance of TUNEL within but not outside the ATP-depleted region. These findings demonstrate that early inhibition of protein synthesis heralds the final manifestation of infarction, and that DNA fragmentation occurs in the core but not in the penumbra of evolving infarcts (modified from Hata et al. [25])

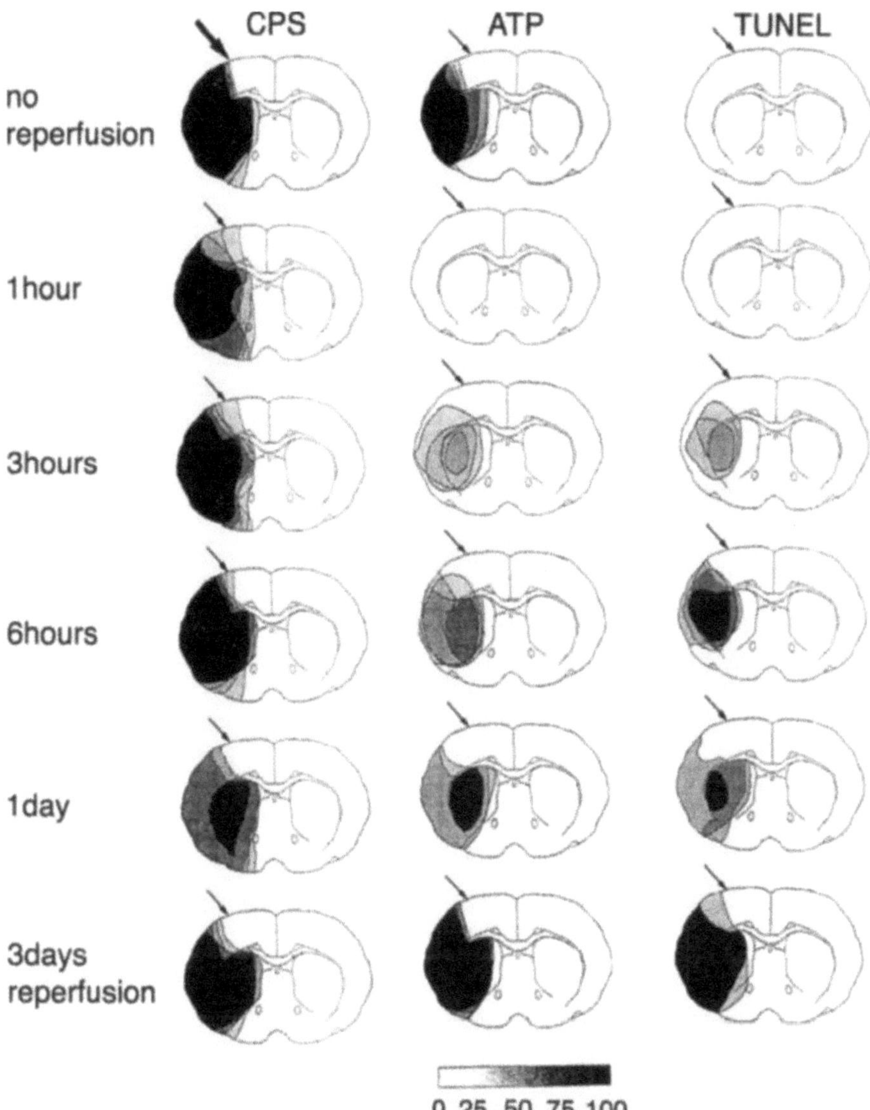

Fig. 2. Incidence maps of suppressed protein synthesis (CPS), ATP depletion and TUNEL-positive neurons on coronal sections of the mouse brain at various times after transient middle cerebral artery occlusion for 1 h. Similar presentation as in Fig. 1. Note transient recovery of ATP after the beginning of reperfusion and correlation of TUNEL with secondary energy failure (modified from Hata et al. [26])

Modulation of Infarct Size by Apoptosis-Associated Genes

The table summarizes our observations. CREB deleted animals were studied under the hypothesis that differences in *c-fos* expression might modulate apoptotic cell death. Imaging of *c-fos* expression by *in-situ* hybridization confirmed that CREB-deficient mice exhibited lower *c-fos* mRNA levels in the peri-infarct normal brain tissue [23]. However, this difference did not lead to measurable differences in infarct volume. Focal ischemia in APAF-1-deficient mice could only be investigated in heterozygotes because homozygotes are not viable. This may explain that no difference in infarct size was observed (Besselmann et al., in preparation). In contrast, overexpression of XIAP, the X-chromosome-associated inhibitor of apoptosis, led to functional improvement, a significantly smaller area of DNA fragmentation and a significant reduction in infarct size [3]. Deletion of the anti-apoptotic bcl-2 gene resulted in a gene-dose dependent enlargement of infarct size [24], but deletion of the pro-apoptotic p53 gene had the same effect, although experiments were carried out under exactly the same conditions [44]. The results of our studies therefore only partly support an apoptosis-mediated mechanism of stroke evolution.

Modulation of Infarct Size by Hemodynamically Active Genes

Compared with the effects of apoptosis-associated genes, the results of this group of experiments were more coherent. Angiotensin overexpression led to an increase in infarct size which can be readily explained by the higher vascular tone of the collateral vascular system [64]. Deletion of either angiotensinogen [42] or the angiotensin 1-receptor gene [64] improved early ischemic injury, which is also in line with the expected improvement of collateral vascular supply. This effect was only transient, but this does not contradict a hemodynamic mechanism because, due to progressing mitochondrial impairment, the threshold of ATP suppression steadily increases with increasing ischemia time [48]. An improvement of the collateral hemodynamics in the penumbra is, therefore, only temporarily able to raise the blood flow above the threshold of energy failure.

Finally, bradykinin 1-receptor deletion, which has previously been shown to aggravate ischemic heart injury, also increased the volume of brain infarction after 1 h transient middle cerebral artery occlusion (Mies et al., in preparation).

Modulation of Injury by Strain Differences of Genetic Background

In homogeneities of genetic background may modulate infarct volume in genetically modified animals. SV129 and C57 Black mice, which are most widely used for the generation of transgenic mutants, exhibit marked differences in ischemic susceptibility, both as regards the hemodynamic [18] and the molecular pathophysiology of infarct evolution [57]. We investigated the cerebral vascular anatomy in these two strains by latex infusion and measured the distance of the peripheral border of the MCA territory from the midline. At three coronal levels, 2, 4 and

Vascular anatomy

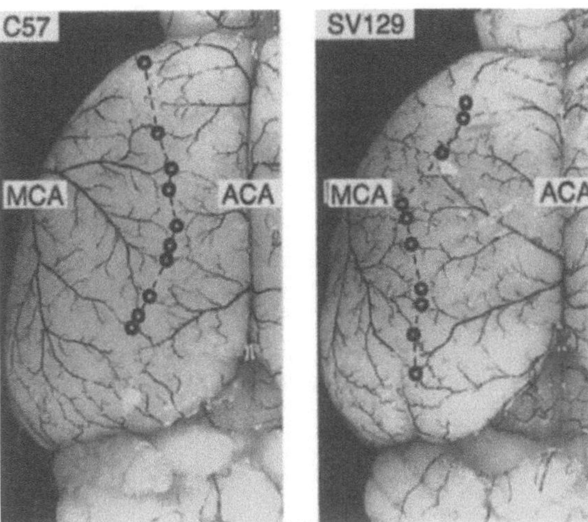

Infarct size

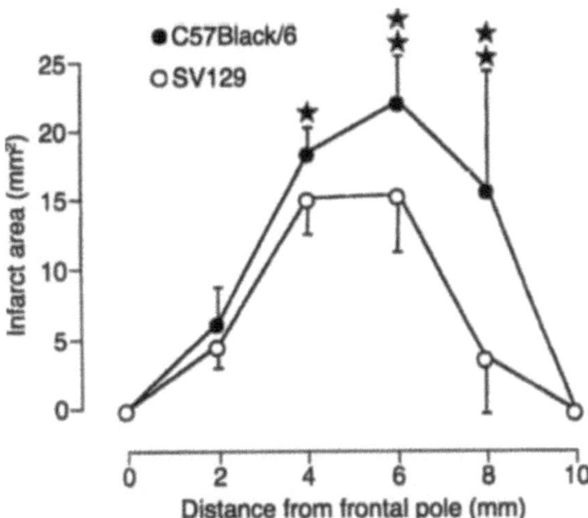

Fig. 3. Difference in mouse strain susceptibility to stroke. Above: Latex infusion of cerebral vasculature of C57 Black (left) and SV129 (right) animals. The line of anastomoses between the supplying territories of the anterior cerebral artery (ACA) and middle cerebral artery (MCA) is marked by the dotted line. The peripheral shift of the line of anastomoses in C57 Black mice reflects the larger MCA territory in this strain. Below: Comparison of infarct size in C57 Black and SV129 mice. Infarct areas were measured by planimetry on coronal slices at the indicated planes. Mean infarct volumes as calculated by integration was significantly larger in the SV129 strain (modified from Maeda et al. [44])

Table 1. Effect of transgenic mutations on ischemic brain injury

Gene manipulation	Ischemia model	Outcome criterion	Percent change
Apoptosis-associated genes			
CREB deletion	permanent MCAO	ATP depletion at 3 h	-20.9 ± 17.4 [NS]
APAF-1 deletion	1 h transient MCAO	histology at 24 h	-8.5 ± 14.9 [NS]
XIAP overexpression	1 h transient MCAO	histology at 24 h	-64.7 ± 7.4 *
bcl-2 deletion	1 h transient MCAO	histology at 24 h	$+118.8 \pm 32.4$ *
p53 deletion	1 h transient MCAO	histology at 24 h	$+210.3 \pm 155.1$ *
Hemodynamically active genes			
angiotensinogen overexpression	permanent MCAO	ATP depletion at 24 h	$+16.1 \pm 2.3$ *
angiotensinogen deletion	permanent MCAO	ATP depletion at 1 h	-37.5 ± 17.5 *
		ATP depletion at 24 h	$+3.1 \pm 21.1$ [NS]
angiotensin 1-receptor deletion	permanent MCAO	ATP depletion at 1 h	-20.3 ± 14.1 *
		ATP depletion at 24 h	-4.2 ± 11.3 [NS]
bradykinin 1-receptor deletion	1 h transient MCAO	ATP depletion at 24 h	$+71.7 \pm 6.5$ *

MCAO: middle cerebral artery occlusion. Percent change is the difference of lesion size between mutants and wildtype animals. [NS] non-significant; * $p < 0.05$

6 mm from frontal pole, the distance was significantly smaller in C57 Black mice, indicating a larger MCA supplying territory (Fig. 3). The close correspondence of the size of the MCA vascular territory with infarct volume was confirmed by occluding the MCA in the two strains. As predicted, the size of ATP-depleted and CPS-suppressed brain tissue was significantly larger in C57 than in SV129 mice, the difference amounting to about 30% [45]. Incidentally, the larger infarct size in C57 animals also documented that the previously described reduced susceptibility of this strain to glutamate toxicity [57] was not able to compensate for the difference in blood supply.

Discussion

One of the earliest topics addressed by using genetically modified animals was free radical-induced brain damage [9, 34, 37]. Free radicals are generated mainly during reoxygenation after temporary oxygen deprivation and, therefore, are of particular pathophysiological significance following transient interruption of blood flow [21, 33, 34, 37, 59, 63, 66]. In accordance with this hypothesis interventions that reduce the formation or improve the scavenging of free radicals also improve the outcome of transient ischemic brain damage. In permanent ischemia, free radical-induced injury is probably restricted to areas in which residual blood flow oscillates around the threshold of metabolic disturbance. This explains that during permanent ischemia manipulation of the radical scavenging capacity either does [19, 62] or does not [9] result in the expected modulation of brain damage.

Apoptosis, in contrast to free radicals, is thought to contribute to the evolution of injury after both permanent and transient focal ischemia [47], but our own as well as other published data are not fully in support of this hypothesis. For instance, overexpression of the anti-apoptotic gene bcl-2 led to either decrease [38, 46] or no change [4, 67], and deletion of the anti-apoptotic gene p53 either decreased [13] or increased [43] ischemic damage. There is also an inconsistency regarding the deletion of PARP which renders animals resistant to transient focal ischemia [17]. As this nuclear enzyme uses NAD for repair of DNA damage, the resulting NAD depletion – which could lead to energy failure – is thought to be the final reason for cell death [54]. However, measurements of the tissue content of NAD and ATP do not support this contention [53]. It is, therefore, questionable whether the observed beneficial effect of PARP deletion uses this mechanism for tissue protection.

Genetically modified animals have also been used for the dissection of various toxicity pathways which have been implicated to mediate ischemic or post-ischemic brain damage. In particular, evidence has been provided in support of molecular injury mediated by nitric oxide (NO) [22, 30, 31], glutamate [32, 39, 50], tissue plasminogen activator (t-PA) [51, 65] or zinc [7]. The genetic approach was particularly successful in elucidating the dual role of NO, depending on the cellular origin of generation [14]. At the vascular level NO is protective because stimulation of endothelial NOS produces vasodilation and, as a consequence, improves collateral blood supply [40]. Stimulation of neuronal or inducible NOS, in contrast, promotes ischemic injury because parenchymal increase of NO facilitates the formation of peroxynitrite [16]. Similarly, t-PA may exert a dual effect by facilitating excitotoxicity on the one hand and promoting thrombolytic reperfusion on the other [61].

Finally, experiments have been carried out with genetically modified animals to understand the role of various biologically active proteins in modulating ischemia-associated inflammation [5, 35, 60, 65], signalling [1, 10] and stress pathways [52, 55]. These studies are in general support of established hypotheses on molecular injury pathways but some negative results have been interpreted as evidence to the contrary [11, 28, 41, 56].

A major problem for the interpretation of injury outcome in gene-manipulated animals is the possibility of differences in ischemic susceptibility. The strains most commonly used for the production of such animals are SV129 for the preparation of stem cells and C57 Black for blastocyte injection. However, ligation of the middle cerebral artery in C57 Black mice results in brain infarcts which are at least 30% larger than in the SV129 strain [12, 45]. Angioarchitectural studies from our laboratory revealed that this difference is due to the larger vascular territory supplied by the MCA [44]. On the other hand, C57 Black mice are less susceptible to excitotoxic injury [57] and exhibit more robust vasoreactivity [18] which may partly counteract the angioarchitectural disadvantage of this strain.

Obviously, strain differences may interfere with the effects of targeted mutations as long as the genetic background of the transgenic animals has not become homologous. However, as full homology requires backcrossings for 12 generations which takes about two years, only few of the published studies fulfill this requirement.

Another methodological problem for the comparison of ischemic injury in transgenic and wildtype animals is the possibility of differences in body weight. Ischemic infarcts are most widely produced by intraluminal thread insertion into the MCA but to achieve consistent occlusion the tip diameter of the filament has to be adjusted to the vessel size which, in turn, varies as a function of body weight [25]. Using the same filament for MCA occlusion in differently sized animals may, therefore, produce different severities of ischemia and, in consequence, differences in stroke outcome.

Finally, the choice of the experimental stroke model may have profound influences on the effect of gene manipulation. In particular, transient and permanent focal ischemia trigger different molecular injury pathways [25, 26], and even within the same category of ischemia, the contribution of primary and secondary injury mechanisms varies, depending on the severity and duration of ischemia and/or reperfusion. Deletion or overexpression of the same gene, may, therefore, result in different outcomes in different models.

Some of these methodological problems may be overcome by using conditional mutants, but from a conceptual point of view, strain and model-dependent differences in stroke susceptibility may be conducive to the search of new molecular targets for clinical stroke therapy. Under clinical conditions, both the genetic background of patients and the pattern of flow disturbances are heterogenous, and as these heterogeneities cannot be easily assessed, only those molecular targets will be of interest for new treatment strategies which relate to an universal, model-independent mechanism of ischemic cell death. Comparing the effect of transgenic mutation in different experimental models may, therefore, help to identify such targets.

Conclusions

The use of genetically modified animals for the dissection of complex molecular injury cascades is a valuable addition to the repertoire of experimental stroke research. Important insights obtained by this approach are the differentiation between the dual role of NO for injury evolution, depending on the site of generation [14], and the elucidation of the importance of free radical scavenging systems for minimizing tissue injury after ischemia reperfusion [8]. Other investigations including the effects described here on apoptosis and collateral blood supply only partly support seemingly well-established injury pathways. It is also disappointing that to the present no common molecular target for therapeutical interventions has been identified. This may reflect the multimodality of ischemic injury but it can also be interpreted as the failure of molecular stroke research to identify key regulators of ischemic or post-ischemic tissue survival. However, whatever this regulator may be, genetically modified animals will be of considerable importance to test its relevance under various conditions of ischemic injury.

References

1. Aoki Y, Huang ZH, Thomas SS, Bhide PG, Huang I, Moskowitz MA, Reeves SA (2000) Increased susceptibility to ischemia-induced brain damage in transgenic mice overexpressing a dominant negative form of SHP2. FASEB Journal 14:1965–1973
2. Astrup J, Symon L, Siesjö BK (1981) Thresholds in cerebral ischemia – The ischemic penumbra. Stroke 12:723–725
3. Besselmann M, Föcking M, Korhonen L, Lindholm D, Hossmann K-A, Trapp T (2001) Analysis of apoptotic and survival pathways after transient cerebral ischemia in mice. Journal of Cerebral Blood Flow & Metabolism 21 (Supp 1):S9

4. Bilbao FD, Guarin E, Nef P, Vallet P, Giannakopoulos P, Dubois-Dauphin M (2000) Cell death is prevented in thalamic fields but not in injured neocortical areas after permanent focal ischaemia in mice overexpressing the anti-apoptotic protein Bcl-2. European Journal of Neuroscience 12:921–934

5. Bruce AJ, Boling W, Kindy MS, Peschon J, Kraemer PJ, Carpenter MK, Holtsberg FW, Mattson MP (1996) Altered neuronal and microglial responses to excitotoxic and ischemic brain injury in mice lacking TNF receptors. Nature Medicine 2:788–794

6. Brusa R (1999) Genetically modified mice in neuropharmacology. Pharmacological Research 39:405–419

7. Campagne MV, Thibodeaux H, van Bruggen N, Cairns B, Gerlai R, Palmer JT, Williams SP, Lowe DG (1999) Evidence for a protective role of metallothionein-1 in focal cerebral ischemia. Proceedings of the National Academy of Sciences of the United States of America 96:12870–12875

8. Chan PH (2001) Reactive oxygen radicals in signaling and damage in the ischemic brain. Journal of Cerebral Blood Flow and Metabolism 21:2–14

9. Chan PH, Kamii H, Yang GY, Gafni J, Epstein CJ, Carlson E, Reola L (1993) Brain infarction is not reduced in SOD-1 transgenic mice after a permanent focal cerebral ischemia. Neuroreport 5:293–296

10. Chen JF, Huang ZH, Ma JY, Zhu JM, Moratalla R, Standaert D, Moskowitz MA, Fink JS, Schwarzschild MA (1999) A(2a) adenosine receptor deficiency attenuates brain injury induced by transient focal ischemia in mice. Journal of Neuroscience 19:9192–9200

11. Clark WM, Rinker LG, Lessov NS, Hazel K, Hill JK, Stenzel-Poore M, Eckenstein F (2000) Lack of interleukin-6 expression is not protective against focal central nervous system ischemia. Stroke 31:1715–1720

12. Connolly ES, Winfree CJ, Stern DM, Solomon RA, Pinsky DJ (1996) Procedural and strain-related variables significantly affect outcome in a murine model of focal cerebral ischemia. Neurosurgery 38:523–531

13. Crumrine RC, Thomas AL, Morgan PF (1994) Attenuation of p53 expression protects against focal ischemic damage in transgenic mice. Journal of Cerebral Blood Flow and Metabolism 14:887–891

14. Dalkara T, Yoshida T, Irikura K, Moskowitz MA (1994) Dual role of nitric oxide in focal cerebral ischemia. Neuropharmacology 33:1447–1452

15. Dirnagl U, Iadecola C, Moskowitz MA (1999) Pathobiology of ischaemic stroke: an integrated view. Trends in Neurosciences 22:391–397

16. Eliasson MJL, Huang ZH, Ferrante RJ, Sasamata M, Molliver ME, Snyder SH, Moskowitz MA (1999) Neuronal nitric oxide synthase activation and peroxynitrite formation in ischemic stroke linked to neural damage. Journal of Neuroscience 19:5910–5918

17. Eliasson MJL, Sampei K, Mandir AS, Hurn PD, Traystman RJ, Bao J, Pieper A, Wang ZQ, Dawson TM, Snyder SH, Dawson VL (1997) Poly(ADP-ribose) polymerase gene disruption renders mice resistant to cerebral ischemia. Nature Medicine 3:1089–1095

18. Fujii M, Hara H, Meng W, Vonsattel JP, Huang ZH, Moskowitz MA (1997) Strain-related differences in susceptibility to transient forebrain ischemia in SV129 and C57Black/6 mice. Stroke 28:1805–1810

19. Fujimura M, Morita-Fujimura Y, Kawase M, Copin J-C, Calagui B, Epstein CJ, Chan PH (1999) Manganese superoxide dismutase mediates the early release of mitochondrial cytochrome c and subsequent DNA fragmentation after permanent focal cerebral ischemia in mice. Journal of Neuroscience 19:3414–3422

20. Grotta J (1995) Why do all drugs work in animals but none in stroke patients? 2. Neuroprotective therapy. Journal of Internal Medicine 237:89–94

21. Guo ZH, Kindy MS, Kruman I, Mattson MP (2000) Als-linked Cu/Zn-SOD mutation impairs cerebral synaptic glucose and glutamate transport and exacerbates ischemic brain injury. Journal of Cerebral Blood Flow and Metabolism 20:463–468

22. Hara H, Huang PL, Panahian N, Fishman MC, Moskowitz MA (1996) Reduced brain edema and infarction volume in mice lacking the neuronal isoform of nitric oxide synthase after transient MCA occlusion. Journal of Cerebral Blood Flow and Metabolism 16:605–611

23. Hata R, Gass P, Mies G, Wiessner C, Hossmann K-A (1998) Attenuated c-fos mRNA induction after middle cerebral artery occlusion in CREB knockout mice does not modulate focal ischemic injury. Journal of Cerebral Blood Flow and Metabolism 18:1325–1335

24. Hata R, Gillardon F, Michaelidis TM, Hossmann K-A (1999) Targeted disruption of the bcl-2 gene in mice exacerbates focal ischemic brain injury. Metabolic Brain Disease 14:117–124

25. Hata R, Maeda K, Hermann D, Mies G, Hossmann K-A (2000) Dynamics of regional brain metabolism and gene expression after middle cerebral artery occlusion in mice. Journal of Cerebral Blood Flow and Metabolism 20:306–315

26. Hata R, Maeda K, Hermann D, Mies G, Hossmann K-A (2000) Evolution of brain infarction after transient focal cerebral ischemia in mice. Journal of Cerebral Blood Flow and Metabolism 20:937–946
27. Hata R, Mies G, Wiessner C, Fritze K, Hesselbarth D, Brinker G, Hossmann K-A (1998) A reproducible model of middle cerebral artery occlusion in mice – hemodynamic, biochemical, and magnetic resonance imaging. Journal of Cerebral Blood Flow and Metabolism 18:367–375
28. Holschneider DP, Scremin OU, Huynh L, Chen K, Shih JC (1999) Lack of protection from ischemic injury of monoamine oxidase B-deficient mice following middle cerebral artery occlusion. Neuroscience Letters 259:161–164
29. Hossmann K-A (1994) Viability thresholds and the penumbra of focal ischemia. Annals of Neurology 36:557–565
30. Huang ZH, Huang PL, Panahian N, Dalkara T, Fishman MC, Moskowitz MA (1994) Effects of cerebral ischemia in mice deficient in neuronal nitric oxide synthase. Science 265:1883–1885
31. Iadecola C, Zhang FY, Casey R, Nagayama M, Rose ME (1997) Delayed reduction of ischemic brain injury and neurological deficits in mice lacking the inducible nitric oxide synthase gene. Journal of Neuroscience 17:9157–9164
32. Kadotani H, Namura S, Katsuura G, Terashima T, Kikuchi H (1998) Attenuation of focal cerebral infarct in mice lacking NMDA receptor subunit NR2C. Neuroreport 9:471–475
33. Keller JN, Kindy MS, Holtsberg FW, Stclair DK, Yen HC, Germeyer A, Steiner SM, Brucekeller AJ, Hutchins JB, Mattson MP (1998) Mitochondrial manganese superoxide dismutase prevents neural apoptosis and reduces ischemic brain injury: suppression of peroxynitrite production, lipid peroxidation, and mitochondrial dysfunction. Journal of Neuroscience 18:687–697
34. Kinouchi H, Epstein CJ, Mizui T, Carlson E, Chen SF, Chan PH (1991) Attenuation of focal cerebral ischemic injury in transgenic mice overexpressing CuZn superoxide dismutase. Proceedings of the National Academy of Sciences of the United States of America 88:11158–11162
35. Kitagawa K, Matsumoto M, Mabuchi T, Yagita Y, Ohtsuki T, Hori M, Yanagihara T (1998) Deficiency of intercellular adhesion molecule 1 attenuates microcirculatory disturbance and infarction size in focal cerebral ischemia. Journal of Cerebral Blood Flow and Metabolism 18:1336–1345
36. Kogure K, Alonso OF (1978) A pictorial representation of endogenous brain ATP by a bioluminescent method. Brain Research 154:273–284
37. Kondo T, Reaume AG, Huang TT, Carlson E, Murakami K, Chen SF, Hoffman EK, Scott RW, Epstein CJ, Chan PH (1997) Reduction of CuZn-superoxide dismutase activity exacerbates neuronal cell injury and edema formation after transient focal cerebral ischemia. Journal of Neuroscience 17:4180–4189
38. Lawrence MS, Ho DY, Sun GH, Steinberg GK, Sapolsky RM (1996) Overexpression of bcl-2 with herpes simplex virus vectors protects CNS neurons against neurological insults in vitro and in vivo. Journal of Neuroscience 16:486–496
39. Le D, Das SY, Wang YF, Yoshizawa T, Sasaki YF, Takasu M, Nemes A, Mendelsohn M, Dikkes P, Lipton SA, Nakanishi N (1997) Enhanced neuronal death from focal ischemia in AMPA-receptor transgenic mice. Molecular Brain Research 52:235–241
40. Lo EH, Hara H, Rogowska J, Trocha M, Pierce AR, Huang PL, Fishman MC, Wolf GL, Moskowitz MA (1996) Temporal correlation mapping analysis of the hemodynamic penumbra in mutant mice deficient in endothelial nitric oxide synthase gene expression. Stroke 27:1381–1385
41. Lukkarinen JA, Grohn OHJ, Alhonen LI, Janne J, Kauppinen RA (1999) Enhanced ornithine decarboxylase activity is associated with attenuated rate of damage evolution and reduction of infarct volume in transient middle cerebral artery occlusion in the rat. Brain Research 826:325–329
42. Maeda K, Hata R, Bader M, Walther T, Hossmann K-A (1999) Larger anastomoses in angiotensinogen-knockout mice attenuate early metabolic disturbances after middle cerebral artery occlusion. Journal of Cerebral Blood Flow and Metabolism 19:1092–1098
43. Maeda K, Hata R, Gillardon F, Hossmann K-A (2001) Aggravation of brain injury after transient focal ischemia in p53 deficient mice. Molecular Brain Research 88:54–61
44. Maeda K, Hata R, Hossmann K-A (1998) Differences in the cerebrovascular anatomy of C57Black/6 and SV129 mice. Neuroreport 9:1317–1319
45. Maeda K, Hata R, Hossmann K-A (1999) Regional metabolic disturbances and cerebrovascular anatomy after permanent middle cerebral artery occlusion in C57Black/6 and SV129 mice. Neurobiology of Disease 6:101–108
46. Martinou JC, Dubois-Dauphin M, Staple JK, Rodrigues I, Frankowski H, Missotten M, Albertini P, Talabot D, Catsicas S, Pietra C, Huarte J (1994) Overexpression of BCL-2 in transgenic mice protects neurons from naturally occurring cell death and experimental ischemia. Neuron 13:1017–1030

47. Mattson MP, Culmsee C, Yu ZF (2000) Apoptotic and antiapoptotic mechanisms in stroke. Cell and Tissue Research 301:173–187
48. Mies G, Ishimaru S, Xie Y, Seo K, Hossmann K-A (1991) Ischemic thresholds of cerebral protein synthesis and energy state following middle cerebral artery occlusion in rat. Journal of Cerebral Blood Flow and Metabolism 11:753–761
49. Mies G, Trapp T, Kilic E, Oláh L, Hata R, Hermann D, Hossmann K-A (2001) Relationship between DNA fragmentation, energy state, and protein synthesis after transient focal cerebral ischemia in mice. In: Bazan N, Ito U, Marcheselli V, Kuroiwa T, Klatzo I (eds) Maturation Phenomenon in Cerebral Ischemia IV. Springer, Berlin, Heidelberg, New York, pp 85–92
50. Morikawa E, Mori H, Kiyama Y, Mishina M, Asano T, Kirino T (1998) Attenuation of focal ischemic brain injury in mice deficient in the epsilon1 (NR2A) subunit of NMDA receptor. Journal of Neuroscience 18:9727–9732
51. Nagai N, Mol MD, Lijnen HR, Carmeliet P, Collen D (1999) Role of plasminogen system components in focal cerebral ischemic infarction. A gene targeting and gene transfer study in mice. Circulation 99:2440–2444
52. Panahian N, Yoshiura M, Maines MD (1999) Overexpression of heme oxygenase-1 is neuroprotective in a model of permanent middle cerebral artery occlusion in transgenic mice. Journal of Neurochemistry 72:1187–1203
53. Paschen W, Olah L, Mies G (2000) Effect of transient focal ischemia of mouse brain on energy state and NAD levels: No evidence that NAD depletion plays a major role in secondary disturbances of energy metabolism. Journal of Neurochemistry 75:1675–1680
54. Pieper AA, Blackshaw S, Clements EE, Brat DJ, Krug DK, White AJ, Pinto-Garcia P, Favit A, Conover JR, Snyder SH, Verma A (2000) Poly(ADP-ribosyl)ation basally activated by DNA strand breaks reflects glutamate-nitric oxide neurotransmission. Proceedings of the National Academy of Sciences of the United States of America 97:1845–1850
55. Rajdev S, Hara K, Kokubo Y, Mestril R, Dillmann W, Weinstein PR, Sharp FR (2000) Mice overexpressing rat heat shock protein 70 are protected against cerebral infarction. Annals of Neurology 47:782–791
56. Sampei K, Goto S, Alkayed NJ, Crain BJ, Korach KS, Traystman RJ, Demas GE, Nelson RJ, Hurn PD (2000) Stroke in estrogen receptor-alpha-deficient mice. Stroke 31:738–743
57. Schauwecker PE, Steward O (1997) Genetic determinants of susceptibility to excitotoxic cell death: Implications for gene targeting approaches. Proc Natl Acad Sci USA 94:4103–4108
58. Sharp FR (1998) Stress genes protect brain. Annals of Neurology 44:581–583
59. Sheng H, Bart RD, Oury TD, Pearlstein RD, Crapo JD, Warner DS (1999) Mice overexpressing extracellular superoxide dismutase have increased resistance to focal cerebral ischemia. Neuroscience 88:185–191
60. Soriano SG, Lipton SA, Wang YMF, Xiao M, Springer TA, Gutierrez-Ramos JC, Hickey PR (1996) Intercellular adhesion molecule-1-deficient mice are less susceptible to cerebral ischemia-reperfusion injury. Annals of Neurology 39:618–624
61. Tabrizi P, Wang L, Seeds N, McComb JG, Yamada S, Griffin JH, Carmeliet P, Weiss MH, Zlokovic BV (1999) Tissue plasminogen activator (tPa) deficiency exacerbates cerebrovascular fibrin deposition and brain injury in a murine stroke model. Studies in tPa-deficient mice and wild-type mice on a matched genetic background. Arteriosclerosis Thrombosis and Vascular Biology 19:2801–2806
62. Takagi Y, Mitsui A, Nishiyama A, Nozaki K, Sono H, Gon Y, Hashimoto N, Yodoi J (1999) Overexpression of thioredoxin in transgenic mice attenuates focal ischemic brain damage. Proceedings of the National Academy of Sciences of the United States of America 96:4131–4136
63. Walder CE, Green SP, Darbonne WC, Mathias J, Rae J, Dinauer MC, Curnutte JT, Thomas GR (1997) Ischemic stroke injury is reduced in mice lacking a functional NADPH oxidase. Stroke 28:2252–2258
64. Walther T, Olah L, Harms C, Maul B, Bader M, Hörtnagel H, Schultheiss H-P, Mies G (2002) Ischemic injury in experimental stroke depends on angiotensin II. FASEB Journal 16:169–176
65. Wang YMF, Tsirka SE, Strickland S, Stieg PE, Soriano SG, Lipton SA (1998) Tissue plasminogen activator (tPA) increases neuronal damage after focal cerebral ischemia in wild-type and tPA-deficient mice. Nature Medicine 4:228–231
66. Weisbrot-Lefkowitz M, Reuhl K, Perry B, Cahn PH, Inouye M, Mirochnitchenko O (1998) Overexpression of human glutathione peroxidase protects transgenic mice against focal cerebral ischemia/reperfusion damage. Molecular Brain Research 53:333–338
67. Wiessner C, Allegrini PR, Rupalla K, Sauer D, Oltersdorf T, McGregor AL, Bischoff S, Böttiger BW, Putten Hvd (1999) Neuron-specific transgene expression of Bcl-X$_L$ but not Bcl-2 genes reduced lesion size after permanent middle cerebral artery occlusion in mice. Neuroscience Letters 268:119–122

II Ischemic Infarction: Inflammation

Compartmentalization in Focal Cerebral Ischemia

G. DEL ZOPPO

Summary. During maturation of the focal cerebral ischemic injury into the fixed extracellular matrix, transit of plasma components and blood cells from the vascular compartment into the extravascular compartment, and disturbance of cell-cell relationships within the extravascular compartment, proteases identified in the target tissue following the ischemic insult which lesion major alterations in tissue barriers occur. These involve disruption of the cerebral endothelial permeability barrier and dissolution of the microvascular basal lamina (may be responsible for the opening of compartments to one another, include matrix metalloproteinases (MMPs), generators of plasmin (including urokinase, uPA) and their inhibitors, and proteases not yet identified. Within the developing lesion, proteases (their natural inhibitors) and other plasma and cell components mix allowing normally separated systems to interact. This is accentuated by tissue preparation (e.g. homogenization, sampling) during experimental focal ischemia. Hence, a central consideration in understanding the matrix-integrin responses to focal ischemia is the manner in which the compartments are disturbed. This has important implications for the appearance of injury development, the potential success of protease inhibitors put forward as interventions, and understanding the biology of protease action following ischemic injury.

Key words. Compartments – metalloproteinases – urokinase – tissue plasminogen activator (t-PA) – matrix – barrier – matrix metalloproteinase (MMP) – plasminogen activators

Introduction

The functional state of the central nervous system (CNS) depends upon the maintenance of discrete compartments which separate plasma from the cellular elements of the brain, and conducting tissues from those without known electrical properties. Tissue barriers, which preserve the separation of the circulation from the static cell compartments, are established during tissue development. For the CNS these barriers include i) the blood-brain barrier, ii) vascular matrix (e.g. basal lamina), and iii) the extracellular matrix (ECM) of the parenchymal compart-

Correspondence to: Gregory J. del Zoppo, M.D., Department of Molecular and Experimental Medicine, The Scripps Research Institute, 10550 North Torrey Pines Road, MEM 132, La Jolla, CA 92037, Tel.: 858/784-8569, Fax: 858/784-8342, E-Mail: grgdlzop@hermes.scripps.edu

Maturation Phenomenon in Cerebral Ischemia V
A.M. Buchan et al. (Eds.)
© Springer-Verlag Berlin Heidelberg 2004

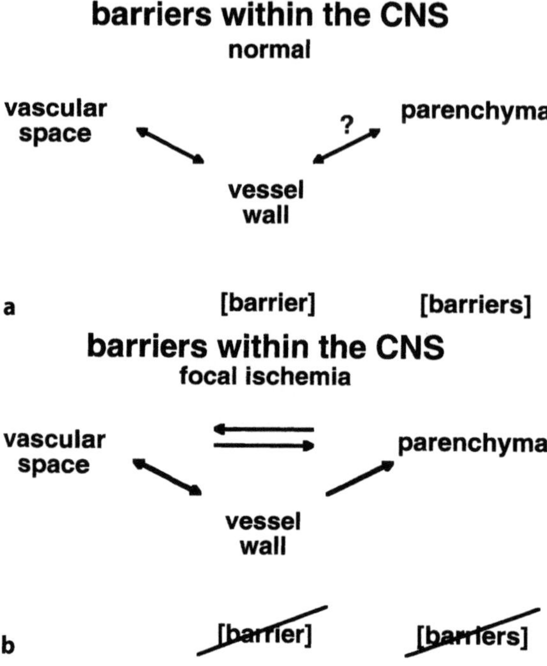

Fig. 1. Compartmentalization of cerebral microvessel unit and tissue (**a**), with disruption of compartments during focal cerebral ischemia (**b**)

ment. These latter two elements are derived from the same matrix components, the vascular ECM separates the plasma compartment from the extracellular compartment (but allows limited communication with the extracellular space under normal conditions via endothelial cell pinocytosis) and acts as a scaffold for endothelial cell and astrocyte attachment. While in the parenchymal compartment, the ECM maintains separation of cells with distinctly different properties from one another (Fig. 1). The blood-brain permeability barrier is maintained in part by inter-endothelial cell tight junctions [49, 50]. Separation (compartmentalization of the intravascular) from the extravascular space requires the cooperation of endothelial cells and astrocytes [5, 25, 33, 56]. Loss of compartmentalization occurs with disruption of the vascular matrix and cell-barriers, and has important consequences for the viability of the affected CNS tissue. Considerations of potential mechanisms of decompartmentalization during ischemia and their consequences for injury maturation are the subject of this presentation.

Formation of Matrix Barriers

Cerebral microvessels are structurally unique. Capillaries in the CNS consist of endothelium, the basal lamina/ECM, and the circumferential astrocyte end-feet (the *glia limitans*). Astrocyte end-feet and pericytes participate in cerebral microvessels where the adventitia exists in non-cerebral vessels. The basal lamina/ECM separates the specialized endothelium from the astrocyte attachments (the *glia limitans*), and their connections to neurons. Generated by endothelial cells and astrocytes in concert during development, the basal lamina forms a biologically active connection between these two cell constituents of the vessel wall. Organotypic tissue cultures (murine spinal cord/pia-arachnoid explants) have shown that intact basal lamina requires the juxtaposition of microvascular endothelium and astrocytes [5, 33]. In noncapillary microvessels, individual smooth muscle cells are encased in ECM, which is continuous with the basal lamina [48]. Microvascular endothelial cells and astrocytes play reciprocal roles in the generation of matrix proteins. In culture, astrocytes secrete laminin, fibronectin, and chondroitin sulfate proteoglycan, while collagens stimulate astrocyte-induced endothelial cell maturation [4, 56, 64]. Conversely, endothelial cell-derived ECM components stimulate astrocyte growth and function (e.g., glutamine synthetase activity) [32, 43]. These developmental interrelationships highlight the close functional association of endothelial cells and astrocytes. The bloodbrain barrier also relies upon the interdependence of endothelial cells and astrocytes, as elegantly shown in chick-quail adrenal vascular tissue/brain tissue xenograft and fetal-adult hippocampal/neocortex allograft preparations [25]. Astrocytes promote many microvascular blood-brain barrier properties including endothelial cell tight junctions, Evan's blue exclusion, and HT7/neurothelin generation [36, 49, 55]. Soluble factor(s) generated by astrocytes are necessary to maintain endothelial bloodbrain barrier characteristics including the induction of tight-junctions, transendothelial resistance, and glucose/amino acid transport polarity [13, 25, 41]. Disruption of both barriers, as during ischemic stroke, contributes to edema formation and hemorrhagic transformation [19].

Intact matrix in the parenchyma is required for cell viability. Glycosaminoglycans, heparan sulfate proteoglycans (HSP), and chondroitin sulfate are found in the adult brain [38, 39]. Cat-301 proteoglycan is an extracellular matrix component expressed by subsets of neurons which coat the perikaryon, and the proximal portions of dendrites and axonal segments [7]. HSP participates in neuronal viability [7]. It is likely that processes which disrupt the vascular matrix may also affect the extravascular matrix which separates cells within the neuropil. Little is yet known about the interplay of these processes during ischemia and the fate of the matrix components involved.

Barrier Disruption

Focal cerebral ischemia produces well-described alterations in the endothelial permeability barrier, in the microvascular basal lamina, and in astrocyte, microglia, and neuron structure and function [14, 15, 20, 50]. The two barriers represent defense systems separating events within the vasculature from the surrounding tissue. Loss of compartmentalization means that cellular localization and function may also be lost. Hamann et al. described significant loss of basal lamina/ECM components, including laminin-1, collagen type IV, and fibronectin of cellular origin within 24 h following middle cerebral occlusion (MCA:O) [20]. A significant loss in immunoreactivity of endothelial cell and astrocyte integrin $\alpha_1\beta_1$, and integrin $\alpha_6\beta_4$ on astrocyte end-feet of cerebral microvessels occurs in within 1–2 h following MCA:O the ischemic core [57, 61]. In the same time frame, leakage of plasma components into the ischemic core is detectable in many model systems. The consequences of the loss of vascular barrier for cell-cell interactions in the extravascular compartment are not well-defined. Based upon the role of ECM components on neuron and astrocyte to function [7, 32, 43], it is hypothesized that within the extravascular parenchyma disruption of the intercellular matrix network may alter cell viability.

Disruption of the vascular-extravascular barriers, posed by the inter-endothelial cell tight junctions and the microvascular ECM, during focal ischemia exposes the parenchyma to the vascular space, as well as components of the vessel wall usually isolated from the cells of the neuropil (Fig. 1). These events during ischemia, in addition to the potential changes intercellular matrix of the parenchyma, result in degradation of the compartment barriers, extension of the plasma space to include the ischemic core, and loss of cellular localization.

Proteases of several families are capable of altering such matrix barriers, including the matrixins, the plasminogen activator family, serine proteases, and cysteine proteases. It is hypothesized that the timely generation and secretion of members of these families contribute to vascular barrier degradation.

Resolution

Defining the nature of the compartments in functional terms is quite difficult. The topographical changes occur in all dimensions. They involve a number of considerations: i) the communication between the plasma compartment, and the vessel wall and abluminal tissue, ii) the differences in the resolution of imaging studies and the molecular events involved with barrier disruption, and iii) the detection of alterations in the apparently static components of the ECM, which are biologically active. Hence, events at the molecular level must be examined in detail, mindful of the processes which maintain separation of the cellular, and the vascular-extravascular compartments in the normal state.

Ischemic injury and the consequent loss of compartmentalization allow mixing of cells and cell components which are normally separate, including cytoplasmic and membrane constituents. The consequences of these unusual interactions may contribute to tissue injury. Distinguishing the effects of matrix dissolution from ischemia on cell viability is the central challenge. Furthermore, in the experimental setting enzyme systems which normally may be separate from their substrates/ligands mix, thereby providing a false indication of active participation in the ischemic state. Conversely, the relative absence of active metalloproteinase (MMP)-2 in most experimental preparations in which the upregulation of this matrixin has been documented following MCA:O implies either the production of very small quantities of the active enzyme in specially localized regions, rapid turnover of the active form, and/or separation of the active and inactive forms. Furthermore, polymorphonuclear (PMN) leukocytes transit across barriers with the generation of local defects in the vascular ECM, allowing components of the vascular space to enter the parenchyma. These observations suggest dynamic and limited exposure of one compartment to another by protease generation. The assumption is made that the presence of proteases during ischemia is linked to their active participation in ECM degradation. Their activation and their activity in this setting are unproven.

Matrix Alterations Following MCA:O

Most observations concerned with loss of compartmentalization have focused on the blood-brain barrier and the fate of the microvascular ECM. Focal degradation of the basal lamina with loss of laminin-1 and -5, collagen type IV, and fibronectin of cellular origin occurs shortly after MCA:O and proceeds most extensively within the ischemic core [20]. This loss is associated with evidence of hemorrhagic transformation [20]. It has been proposed that the microvascular ECM is a barrier to the movement of cells from the vascular lumen into the tissue, as during hemorrhage and leukocyte transit. The exposure of the parenchymal tissue to the vascular space during the ischemic insult may involve the generation of specific active proteases.

Metalloproteinases Generation During Focal Cerebral Ischemia

The distribution and compartmentalization of specific MMPs, their activation systems and relevant ligands/substrates are likely to be important to their participation in cerebral vascular injury, although little is known about their relationships in the CNS. MMPs and plasminogen activators (PAs) belong to protease families responsible for degradation of many vascular ECM components. Among them, pro-MMP-2, which in active form can degrade type IV collagen and laminins, is significantly upregulated in the ischemic core of the striatum within 1 h after MCA:O in the non-human primate [20, 23, 24, 52] (Fig. 2), but later in rodent models of MCA:O [2, 17, 28, 52, 53].

Pro-MMP-2 requires proteolytic digestion of the propeptide domain for activation by membrane-type MMPs (MT-MMPs) [54] and u-PA [40]. It has been recently reported that in addition to the ability to activate pro-MMP-2 and pro-MMP-13, MT-MMPs can also degrade a number of extracellular matrix proteins (e.g., fibronectin, collagen IV, and laminin) [31]. Pro-MMP-2 expression in the normal brain has been described variously as absent [35, 44] or confined to microvessels, leukocytes, microglia, or astrocytes [3, 9, 37]. The significant increase in latent MMP-2 by 2 h following MCA:O is significantly correlated with the density of injured neurons and the region of ischemic injury in the primate basal ganglia [23, 58] (Fig. 3). Furthermore, MMP-2 transcripts appear on select microvessels and on nonvascular cells in the same interval within the ischemic core. These are distributed in a heterogeneous pattern in the ischemic core [1]. The cell association of active MMP-2 and the precise localization of the antigen have not been described in relation to the time course of pro-MMP-2 activity generation following MCA:O, although evidence for rapid pro-MMP-2 synthesis supports the search for these sources [23, 51]. Rosenberg et al. more recently reported the appearance of

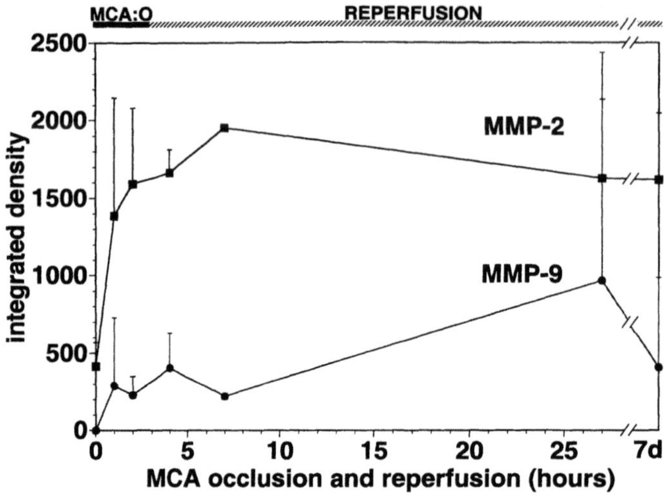

Fig. 2. Appearance of pro-MMP-2 and pro-MMP-9 following MCA:O. From [23]

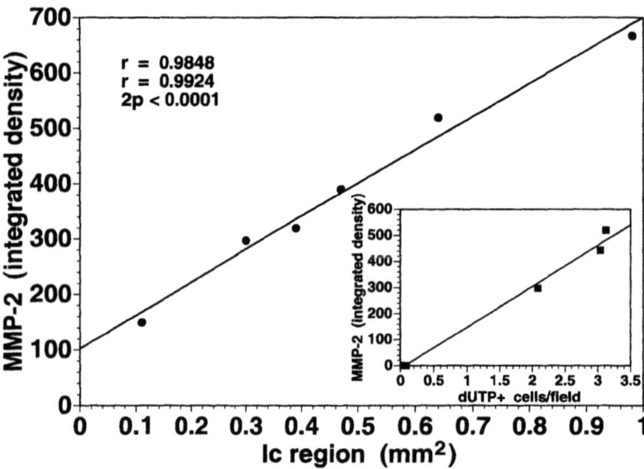

Fig. 3. Correlation of pro-MMP-2 with evidence of neuron injury at 2 h following MCA:O. Neuron injury was determined by evidence of DNA scission and marker studies. From [23, 46]

edema and pro-MMP "activity", identified by gel zymography in the rat, as early as 3 h after MCA:O. However, the presence and sites of pro-MMP-2 activation were not identified [52]. MMP-2 immunoreactivity was assigned to perivascular astrocytes in Sprague-Dawley rats [42, 51]. But, a unique relationship between microvascular permeability and active MMP-2 cannot be definitely made because other causes of vascular permeability (including VEGF and thrombin) also coincide with albumin and IgG extravasation in all models [1, 45, 47, 59, 60].

Plasminogen Activator Family

The members of the plasminogen activator family include tissue plasminogen activator (t-PA), urokinase (u-PA), the inhibitor plasminogen activator inhibitor (PAI)-1, and the urokinase receptor (u-PAR). u-PA and t-PA activate plasminogen to plasmin, which can degrade laminin, collagen IV, fibronectin, and myelin-basic protein directly or indirectly by activation of pro-MMP-2 and pro-MMP-9 by plasmin [27, 40]. We have demonstrated that u-PA and PAI-1, but not t-PA, are rapidly generated following MCA:O in ischemic primate brain tissue [24] (Fig. 4). While t-PA antigen and activity were not substantially altered within the ischemic core even up to 7 days after MCA:O in the primate, a transient decrease in parenchymal t-PA activity was due to its binding with the inhibitor PAI-1 in tissue homogenates [24]. This indicates, in contrast to earlier reports [26], that endogenous brain tissue t-PA production is not increased during ischemia.

PAI-1 antigen and mRNA expression increased in the ischemic core within 1–2 h after MCA:O, and remained elevated until 7 days. PAI-1 antigen appeared simultaneously in a concentration-equivalent fashion in both the parenchyma and the plasma. PAI-1 appears to derive from the microvascular endothelium [24]. However, the role of PAI-1 in modulating ischemic tissue injury is not clear from that and other studies [8, 24]. PAI-1, by blocking t-PA or u-PA in plasma, could alter vascular fibrinolytic activity, but its roles in the parenchymal compartment are yet to be defined.

The specific cellular receptor for u-PA, u-PAR, serves to localize and concentrate u-PA in many cell systems [6, 22]. u-PAR has been detected on endothelial cells, neurons, and microglia *in vitro*, where its expression can be stimulated by hypoxia [18, 21, 63]. u-PA is secreted from cells as the inactive single chain u-PA (scu-PA), whose activation by plasmin to the two-chain form is accelerated when

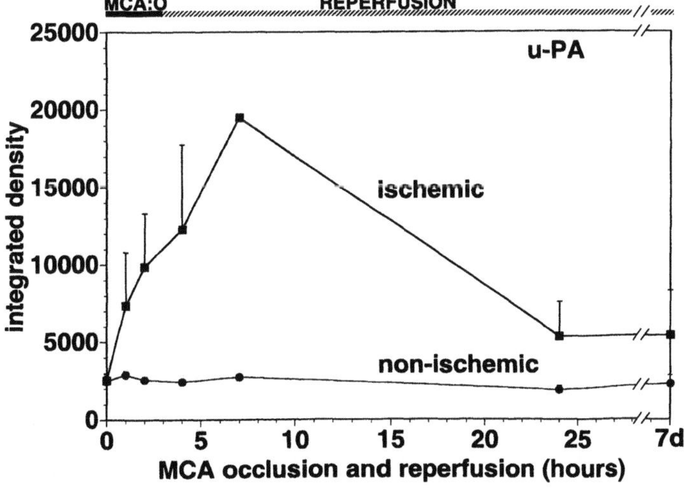

Fig. 4. Appearance of u-PA activity following MCA:O. From [24]

u-PA is bound to its specific, high-affinity cell surface receptor u-PAR [8]. u-PA antigen and activity increased significantly 1–2 h after MCA:O in the ischemic core in the non-human primate, in concert with pro-MMP-2 and PAI-1. Its pattern of appearance within the parenchyma of the basal ganglia in the acute phase of focal ischemia suggests generation from the extravascular compartment. Other investigators have described increased u-PA activity significantly later (at 12 to 24 h, and up to 5 days) in the rat ischemic hemisphere, and decreased t-PA activity at 1, 12, and 24 h [2, 53]. The precise sources of these activities have not been described. In all reports so far, homogenized tissues were used for the gel zymographic and antigen analyses. Furthermore, in human post-mortem materials immunoreactive u-PA, u-PAR, and PAI-1 were over-expressed in endothelial cells, reactive astrocytes, and microglia around, but not within the infarct [10]. The inconsistency among those reports most probably reflects the differences in tissues, the resolution of the detection techniques, and other factors.

Microvascular Remodeling and Decompartmentalization

The simultaneous appearance of pro-MMP-2 and u-PA suggests that these enzymes may also play roles in angiogenesis. Vascular endothelial growth factor (VEGF) has been reported to increase the expression of MMP-2, u-PA, and u-PAR in endothelial cells in culture [29, 34, 65]. Furthermore, volocalization of u-PA and u-PAR on the endothelium accompanies new vessel formation [30]. Abumiya et al. demonstrated that VEGF and integrin $\alpha_v\beta_3$ are rapidly and coordinately upregulated within activated microvessels (mostly precapillary arterioles) in the ischemic core of the stricken basal ganglia following MCA:O [1].

Stimuli for expression of VEGF and integrin $\alpha_v\beta_3$ early following MCA:O have not been identified. However, there is evidence that at least two ligands for integrin $\alpha_v\beta_3$ appear in ischemic tissue, fibrin(ogen) and osteopontin [12, 45, 62]. Fibrin, in the intravascular compartment, appeared simultaneously with integrin $\alpha_v\beta_3$ expression in non-capillary microvessels [45]. This somewhat unusual presentation of fibrin(ogen) as a ligand for integrin $\alpha_v\beta_3$ could be explained by increased permeability of select non-capillary microvessels with loss of compartment separation. The co-expression of VEGF and integrin $\alpha_v\beta_3$ implies a possible mechanism for fibrin generation [1]. Other proteases, including thrombin and leukocyte-derived MMP-2 and MMP-9, may also alter microvascular permeability. In contrast to pro-MMP-2 and u-PA following ischemia onset, VEGF and integrin $\alpha_v\beta_3$ are strictly confined to select microvessels. Nonetheless, the presence of proteases together with known signals and receptors of angiogenesis in other vascular beds support the potential for neovascularization. Extension of the vascular compartment into the surrounding tissue at the time because of increased vascular permeability when cellular barriers are disrupted underscores the proper assignment of specific enzymes to their cellular sources.

The immediate loss of vascular integrin antigens [20], vascular matrix integrity [61], and normal neuron morphometry and function [11, 14, 15, 16] very early following MCA:O is consistent with the hypothesis that proteolytic degradation of matrix during focal cerebral ischemia may contribute to neuronal injury.

Conclusion

Experimental evidence to date from several model systems suggests that protease generation is confined to specific compartments. During focal cerebral ischemia, constitutive generation of pro-MMP-2 appears within the extravascular compartment (confined to the ischemic core), but does not produce an increase in plasma pro-MMP-2 [23]. One recent report suggests that among stroke patients there is a significant increase in plasma pro-MMP-9 in association with parenchymal hemorrhage [23]. That observation is consistent with the association of pro-MMP-9 with hemorrhagic transformation in the non-human primate [23]. u-PA activity and antigen increase simultaneously and in topographical concert with pro-MMP-9 appearance in the extravascular tissue. In contrast, PAI-1 appears in both the plasma and the ischemic tissue, suggesting that the endothelium is the origin of PAI-1 synthesis [24]. The fate of the receptor for urokinase, u-PAR, is currently under study. While these compartment associations are compatible for the most part with their origins in the extravascular cell populations, *in vitro* data has not supported a single cell population for each element. Furthermore, the appearance of t-PA/PAI-1 complexes by 2 h following MCA:O in the non-human primate ischemic striatum [24], implies either i) the association of the two elements in the extracellular space, or ii) their association within plasma which communicates with the extravascular compartment during edema formation (Fig. 5).

Technical considerations are important for interpreting these results. For instance, the association of free t-PA and PAI-1 at a discrete time point may also result from homogenization of the ischemic tissues in sample preparation for gel zymography or western blots. More important may be the maturation of the ischemic lesion to infarction, which may perforce involve destruction of the cell-cell barriers and redistribution of cellular contents bringing together the proteases and their inhibitors contained within individual cell (e.g. astrocytes, endothelial

compartments of expression

protease	site of expression				
	plasma	vessel wall		tissue	
		endo	*smc*	*astro*	*Ic*	*Ip*
pro-MMP-2	+	+	±	±	+
t-PA		±			±
PAI-1	+ ←——————————→				+
u-PA		+	±	±	+

Fig. 5. Compartments of protease expression following MCA:O. *endo* = endothelial cell, *smc* = smooth muscle cell, *astro* = astrocyte, *Ic* = ischemic core, *Ip* = tissue peripheral to ischemic core

cells, microglia) compartments. However, the appearance of pro-MMP-2 and u-PA indicates active synthesis of these proteases directly following MCA:O. Furthermore, the appearance of PAI-1 in the same time frame implies the active participation of endothelial cells in the response to ischemia. This synthesis also supports active processes involving enzyme systems known to degrade matrix.

The mixture of vascular and extravascular compartments has obvious implications for the delivery of inhibitors. It is an open question whether the integrity of these compartments can be preserved. While decompartmentalization appears to be a consequence of ongoing ischemia, and may contribute to the maturation of the ischemic lesion, it may also represent responses to ischemia to i) minimize injury or ii) promote recovery from injury. In this context, the elements of angiogenesis observed early following MCA:O in the non-human primate may have some relevance. Nonetheless, the appearance of secreted proteases (e.g. MMPs), and their cell surface expression offer opportunity for carefully testing modalities which may inhibit matrix degradation. While vascular matrix integrity presents an obvious accessible marker of tissue injury, matrix constituents also form barriers separating glia and neurons within the extravascular compartment, but below the resolution of light microscopic systems. It is very likely that proteases generated during the evolution of the ischemic injury also affect those cell-cell separations. In this way, the loss of vascular matrix integrity and other compartment delimiting structures may contribute to neuronal injury following ischemia. However, it cannot be excluded that methodologies used to explore these processes may artificially bring together cell components that are usually separated, even during maturation of the ischemic lesion.

Acknowledgements. This manuscript is supported by grants NS26945 and NS38710 of the National Institutes of Health.

References

1. Abumiya T, Lucero J, Heo JH, Tagaya M, Koziol JA, Copeland BR, del Zoppo GJ (1999) Activated microvessels express vascular endothelial growth factor and integrin $\alpha_v\beta_3$ during focal cerebral ischemia. J Cereb Blood Flow Metab 19:1038–1050
2. Ahn MY, Zhang ZG, Tsang W, Chopp M (1999) Endogeneous plasminogen activator expression after embolic focal cerebral ischemia in mice. Brain Res 837:169–176
3. Anthony DC, Ferguson B, Matyzak MK, Miller KM, Esiri MM, Perry VH (1997) Differential matrix metalloproteinase expression in cases of multiple sclerosis and stroke. Neuropathol Appl Neurobiol 23:406–415
4. Ard MD, Faissner A (1991) Components of astrocytic extracellular matrix are regulated by contact with axons. Ann NY Acad Sci 633:566–569
5. Bernstein JJ, Getz R, Jefferson M, Kelemen M (1985) Astrocytes secrete basal lamina after hemisection of rat spinal cord. Brain Res 327:135–141
6. Blasi F (1999) Proteolysis, cell adhesion, chemotaxis, and invasiveness are regulated by the u-PA-u-PAR-PAI-1 system. Thrombosis & Haemostasis 82:298–304
7. Carlson SS, Hockfield S (1996) Central nervous system. In: Extracellular Matrix, Volume 1, pp 1–23
8. Conese M, Biasi F (1995) Urokinase/urokinase receptor system: Internalization/degradation of urokinase-serpin complexes: mechanism and regulation. Biol Chem Hoppe Seyler 376:143–155

9. Cuzner ML, Gveric D, Strand C, Loughlin AJ, Paemen L, Opdenakker G, Newcombe J (1996) The expression of tissue-type plasminogen activator, matrix metalloproteinases and endogenous inhibitors in the central nervous system in multiple sclerosis: comparison of stages in lesion evolution. J Neuropathol Exp Neurol 55:1194–1204

10. Dietzmann K, von Bossanyi P, Krause D, Wittig H, Mawrin C, Kirches E (2000) Expression of the plasminogen activator system and the inhibitors PAI-1 and PAI-2 in posttraumatic lesions of the CNS and brain injuries following dramatic circulatory arrests: an immunohistochemical study. Pathol Res Pract 196:15–21

11. Eke A, Conger KA, Anderson M, Garcia JH (1990) Histologic assessment of neurons in rat models of cerebral ischemia. Stroke 21:299–304

12. Ellison JA, Velier JJ, Spera P, Johak ZL, Wang X, Barone FC, Feuerstein GZ (1998) Osteopontin and its integrin receptor $a_v\beta_3$ are upregulated during formation of the glial scar after focal stroke. Stroke 29:1698–1706

13. Estrada C, Bready JV, Berliner JA, Pardridge WM, Cancilla PA (1990) Astrocyte growth stimulation by a soluble factor produced by cerebral endothelial cells in vitro. J Neuropathol Exp Neurol 49:539–549

14. Garcia JH, Kamijyo Y (1974) Cerebral infarction. Evolution of histopathological changes after occlusion of a middle cerebral artery in primates. J Neuropathol Exp Neurol 33:409–421

15. Garcia JH, Mitchem HL, Briggs L, Morawetz R, Hudetz AG, Hazelrig JB, Halsey JH Jr, Conger KA (1983) Transient focal ischemia in subhuman primates: neuronal injury as a function of local cerebral blood flow. J Neuropathol Exp Neurol 42:44–60

16. Garcia JH, Yoshida Y, Chen H, Li Y, Zhang ZG, Liam J, Chen S, Chopp M (1993) Progression from ischemic injury to infarct following middle cerebral artery occlusion in the rat. Am J Pathol 142:623–635

17. Gasche Y, Fujimura M, Morita-Fujimura Y, Copin JC, Kawase M, Massengale J, Chan PH (1999) Early appearance of activated matrix metalloproteinase-9 after focal cerebral ischemia in mice: a possible role in blood-brain barrier dysfunction. J Cereb Blood Flow Metab 19:1020–1028

18. Graham CH, Forsdike J, Fitzgerald CJ, Macdonald-Goodfellow S (1999) Hypoxia-mediated stimulation of carcinoma cell invasiveness via upregulation of urokinase receptor expression. Int J Cancer 80:617–623

19. Hamann GF, Okada Y, del Zoppo GJ (1996) Hemorrhagic transformation and microvascular integrity during focal cerebral ischemia/reperfusion. J Cereb Blood Flow Metab 16:1373–1378

20. Hamann GF, Okada Y, Fitridge R, del Zoppo GJ (1995) Microvascular basal lamina antigens disappear during cerebral ischemia and reperfusion. Stroke 26:2120–2126

21. Hayden SM, Seeds NW (1996) Modulated expression of plasminogen activator system components in cultured cells from dissociated mouse dorsal root ganglia. J Neurosci 16:2307–2317

22. Hendreasen PA, Kjoller L, Christensen L, Duffy MJ (1997) The urokinase-type plasminogen activator system in cancer metastasis. Int J Cancer 72:1–22

23. Heo JH, Lucero J, Abumiya T, Koziol JA, Copeland BR, del Zoppo GJ (1999) Matrix metalloproteinases increase very early during experimental focal cerebral ischemia. J Cereb Blood Flow Metab 19:624–633

24. Hosomi N, Lucero J, Heo JH, Koziol J, Copeland BR, del Zoppo GJ (2001) Rapid differential endogenous plasminogen activator expression after acute middle cerebral artery occlusion. Stroke 32:1341–1348

25. Hurwitz AA, Berman JW, Rashbaum WK, Lyman WD (1993) Human fetal astrocytes induce the expression of blood-brain barrier specific proteins by autologous endothelial cells. Brain Res 625:238–243

26. Ito T, Takenaka K, Sakai H, Yoshimura S, Hayashi K, Noda S, Sakai N (2000) Elevation of mRNA levels of tissue-type plasminogen activator and urokinase-type plasminogen activator in hippocampus and cerebral cortex following middle cerebral artery occlusion in rats. Neurol Res 22:413–419

27. Keski-Oja J, Lohi J, Tuuttila A, Tryggvason K, Vartio T (1992) Proteolytic processing of the 72000-Da type IV collagenase by urokinase plasminogen activator. Exp Cell Res 202:471–476

28. Kimura H, Weisz A, Ogura T, Hitomi Y, Kurashima Y, Hashimoto K, D'Acquisto F, Makuuchi M, Esumi H (2001) Identification of hypoxia-inducible factor 1 ancillary sequence and its function in vascular endothelial growth factor gene induction by hypoxia and nitric oxide. J Biol Chem 276(3):2292–2298

29. Koolwijk P, van Erck MG, de Vree WJ, Vermeer MA, Weich HA, Hanemaaijer R, van Hinsbergh VW (1996) Cooperative effect of TNFa, bFGF, and VEGF on the formation of tubular structures of human microvascular endothelial cells in a fibrin matrix. Role of urokinase activity. J Cell Biol 132:1177–1188

30. Korff T, Kimmina S, Martiny-Baron G, Augustin HG (2001) Blood vessel maturation in a 3-dimensional spheroidal coculture model: direct contact with smooth muscle cells regulates endothelial cell quiescence and abrogates VEGF responsiveness. FASEB J 15:447–457
31. Koshikawa N, Giannelli G, Cirulli V, Miyazaki K, Quaranta V (2000) Role of cell surface metalloprotease MT1-MMP in epithelial cell migration over laminin-5. J Cell Biol 148:615–624
32. Kozlova M, Kentroti S, Vernadakis A (1993) Influence of culture substrata on the differentiation of advanced passage glial cells in cultures from aged mouse cerebral hemispheres. Int J Dev Neurosci 11:513–519
33. Kusaka H, Hirano A, Bornstein MB, Raine CS (1985) Basal lamina formation by astrocytes in organotypic cultures of mouse spinal cord tissue. J Neuropathol Exp Neurol 44:295–303
34. Lamoreaux WJ, Fitzgerald MC, Reiner A, Hasty KA, Charles ST (1998) Vascular endothelial growth factor increases release of gelatinase A and decrease release of tissue inhibitor of metalloproteinases by microvascular endothelial cells. Microvasc Res 55:29–42
35. Lim GP, Backstrom JR, Cullen MJ, Miller CA, Atkinson RD, Tökes ZA (1996) Matrix metalloproteinases in the neocortex and spinal cord of amyotrophic lateral sclerosis patients. J Neurochem 67:251–259
36. Lobrinus JA, Juillerat-Jeanneret L, Darekar P, Schlosshauer B, Janzer RC (1992) Induction of the blood-brain barrier specific HT7 and neurothelin epitopes in endothelial cells of the chick chorioallantoic vessels by a soluble factor derived from astrocytes. Dev Brain Res 70:207–211
37. Maeda A, Sobel RA (1996) Matrix metalloproteinases in the normal human central nervous system, microglial nodules, and multiple sclerosis lesions. J Neuropathol Exp Neurol 55:300–309
38. Margolis RK, Margolis RU (1993) Nervous tissue proteoglycans. Experientia 49:429–446
39. Margolis RU, Margolis RK, Chang LBPC (1975) Glycosaminoglycans of brain during development. Biochemistry 14:85–88
40. Mazzieri R, Masiero L, Zanetta L, Monea S, Onisto M, Garbisa S, Mignatti P (1997) Control of type IV collagenase activity by components of the urokinase-plasmin system: a regulatory mechanism with cell-bound reactants. EMBO J 16:2319–2332
41. Minakawa T, Bready J, Berliner J, Fisher M, Cancilla PA (1991) In vitro interaction of astrocytes and pericyte with capillary-like structures of brain microvessel endothelium. Lab Invest 65:32–40
42. Mun-Bryce S, Lukes A, Wallace J, Lukes-Marx M, Rosenberg GA (2002) Stromelysin-1 and gelatinase A are upregulated before TNF-alpha in LPS-stimulated neuroflammation. Brain Res 933:42–49
43. Nagano N, Aoyagi M, Hirakawa K (1993) Extracellular matrix modulates the proliferation of rat astrocytes in serum-free culture. GLIA 8:71–76
44. Nakagawa T, Kubota T, Kabuto M, Sato K, Kawano H, Hayakawa T, Okada Y (1994) Production of matrix metalloproteinases and tissue inhibitor of metalloproteinases-1 by human brain tumor. J Neurosurg 81:69–77
45. Okada Y, Copeland BR, Fitridge R, Koziol JA, del Zoppo GJ (1994) Fibrin contributes to microvascular obstructions and parenchymal changes during early focal cerebral ischemia and reperfusion. Stroke 25:1847–1854
46. Okada Y, Copeland BR, Hamann GF, Koziol JA, Cheresh DR, del Zoppo GJ (1996) Integrin $a_v\beta_3$ is expressed in selected microvessels following focal cerebral ischemia. Am J Pathol 149:37–44
47. Olesen SP (1986) Rapid increase in blood-brain barrier permeability during severe hypoxia and metabolic inhibition. Brain Res 368:24–29
48. Peters H, Palay SL, Webster HD (1991) The Fine Structure of the Nervous System. Neurons and their Supporting Cells 3:344–355
49. Risau W, Hallmann R, Albrecht U, Henke-Fahle S (1986) Brain astrocytes induce the expression of an early cell surface marker for blood-brain barrier specific endothelium. EMBO J 5:3179–3183
50. Risau W, Wolburg H (1990) Development of the blood-brain barrier. TINS 13:174–178
51. Rosenberg GA, Cunningham LA, Wallace J, Alexander S, Estrada EY, Grossetete M, Razhagi A, Miller K, Gearing A (2001) Immunohistochemistry of matrix metalloproteinases in reperfusion injury to rat brain: activation of MMP-9 linked to stromelysin-1 and microglia in cell cultures. Brain Res 893:104–112
52. Rosenberg GA, Estrada EY, Dencoff JE (1998) Matrix metalloproteinases and TIMPs are associated with blood-brain barrier opening after reperfusion in rat brain. Stroke 29:2189–2195
53. Rosenberg GA, Navratil M, Barone F, Feuerstein G (1996) Proteolytic cascade enzymes increase in focal cerebral ischemia in rat. J Cereb Blood Flow Metab 16:360–366
54. Sato H, Takino T, Akada Y, Cao J, Shinagawa A, Yamamoto E, Seiki M (1994) A matrix metalloproteinase expressed on the surface of invasive tumor cells. Nature 370(7):61–65

55. Schlosshauer B, Herzog KH (1990) Neurothelin: an inducible cell surface glycoprotein of blood-brain barrier specific endothelial cells and distinct neurons. J Cell Biol 110:1261–1274
56. Tagami M, Yamagata K, Fujino H, Kubota A, Nara Y, Yamori Y (1992) Morphological differentiation of endothelial cells co-cultured with astrocytes on type-I or type-IV collagen. Cell Tissue Res 268:225–232
57. Tagaya M, Haring H-P, Stuiver I, Wagner S, Abumiya T, Lucero J, Lee P, Copeland BR, Seiffert D, del Zoppo GJ (2001) Rapid loss of microvascular integrin expression during focal brain ischemia reflects neuron injury. J Cereb Blood Flow Metabol 21:835–846
58. Tagaya M, Liu K-F, Copeland B, Seiffert D, Engler R, Garcia JH, del Zoppo GJ (1997) DNA scission after focal brain ischemia: temporal differences in two species. Stroke 28:1245–1254
59. Thomas WS, Mori E, Copeland BR, Yu J-Q, Morrissey JH, del Zoppo GJ (1993) Tissue factor contributes to microvascular defects following cerebral ischemia. Stroke 24:847–853
60. Todd NV, Picozzi P, Crockard HA, Russell RWR (1986) Duration of ischemia influences the development and resolution of ischemic brain edema. Stroke 17:466–471
61. Wagner S, Tagaya M, Koziol JA, Quaranta V, del Zoppo GJ (1997) Rapid disruption of an astrocyte interaction with the extracellular matrix mediated by integrin $\alpha_6\beta_4$ during focal cerebral ischemia/reperfusion. Stroke 28:858–865
62. Wang X, Louden C, Yue TL, Ellison JA, Barone FC, Solleveld HA, Feuerstein GZ (1998) Delayed expression of osteopontin after focal stroke in the rat. J Neurosci 18:2075–2083
63. Washington RA, Becher B, Balabanov R, Antel J, Dore-Duffy P (1996) Expression of the activation marker urokinase plasminogen-activator receptor in cultured human central nervous system microglia. J Neurosci Res 45:392–399
64. Webersinke G, Bauer H, Amberger A, Zach O, Bauer HC (1992) Comparison of gene expressionextracellular matrix molecules in brain microvascular endothelial cells and astrocytes. Biochem Biophys Res Commun 189:877–879
65. Zucker S, Mirza H, Conner CE, Lorenz AF, Drews MH, Bahou WF, Jesty J (1998) Vascular endothelial growth factor induces tissue factor and matrix metalloproteinase production in endothelial cells: conversion of prothrombin to thrombin results in progelatinase A activation and cell proliferation. Int J Cancer 75:780–786

Modulation of the Post-Ischemic Immune Response Improves Outcome in Focal Cerebral Ischemia: A Role for Lymphocytes in Stroke?

KYRA J. BECKER*, DARIN L. KINDRICK*, JOHN M. HALLENBECK**, RICHARD M. McCARRON†, and ROBERT K. WINN*

Key words. Stroke – Inflammation – Tolerance – Brain – Cytokines

Inflammation appears to contribute to ischemic brain injury [5, 16]. The factors that incite the inflammatory response are not clear, but break down of the blood-brain barrier (BBB) exposes the brain to the peripheral circulation and thus the systemic immune system. Following stroke, antigens found either primarily or exclusively in the central nervous system (CNS), like neuron-specific enolase (NSE), glial fibrillary acidic protein (GFAP), myelin basic protein (MBP), S-100 protein, and creatine phosphokinase (CPK)-BB, can be found within the plasma [8, 13, 14, 43, 57]. Furthermore, the plasma levels of these antigens correlate to eventual infarct size and neurological outcome [8, 13, 14, 43, 57]. This systemic encounter with CNS antigens may explain why there is an increase in the number of circulating lymphocytes and immunoglobulins that react with CNS antigens following stroke [1, 28, 60].

The mucosal immune system processes antigens in a fundamentally different manner than the systemic immune system, and mucosal exposure to an antigen results in immunologic tolerance to that antigen [30]. Mucosal tolerance can be induced by oral administration or inhalation of antigen. In rodents, administration of the antigen by the nasal route is particularly effective in inducing tolerance [36–38]. The nature of the acquired tolerance depends upon the amount of antigen delivered. Clonal deletion of antigen-reactive T cells occurs after a single high dose exposure to antigen [10, 55], while active tolerance occurs after repetitive low dose feeding of antigen [9, 11, 34, 42]. Upon restimulation with the appropriate antigen, T cells in those animals tolerized with a low dose regimen secrete cytokines such as TGF-β1, IL-4, and IL-10, which suppress cell-mediated (TH1) immune responses [9, 27, 34, 42]. These cytokines help drive the immune response towards a humoral or TH2-type response [20, 35, 51]. In tolerized animals, T cells are activated in an antigen-specific fashion, but the effects of activation, i.e. secretion of immunomodulatory cytokines, have antigen-independent effects. Thus, suppression of cell-mediated immune responses will occur anywhere antigen is present, regardless of

Kyra J. Becker*, Darin L. Kindrick*, John M. Hallenbeck**, Richard M. McCarron†, Robert K. Winn*
*University of Washington School of Medicine, Seattle Washington, **National Institute of Neurological Disorders and Stroke, National Institutes of Health, Bethesda Maryland, †Resuscitative Medicine, Naval Medical Research Center, Silver Spring Maryland

Address all inquiries to: Kyra J. Becker, MD, Box 359775, Harborview Medical Center, 325 Ninth Avenue, Seattle, WA 98104-2499, Tel.: 206-731-3251, Fax: 206-731-8787, E-Mail: kjb@u.washington.edu

Maturation Phenomenon in Cerebral Ischemia V
A.M. Buchan et al. (Eds.)
© Springer-Verlag Berlin Heidelberg 2004

whether or not the antigen prompted the initial inflammatory response. This phenomenon, referred to as *bystander suppression* [41, 52], leads to relatively organ-specific immunosuppression [46].

We previously showed that infarct size could be decreased following transient focal cerebral ischemia by creating immunologic tolerance to the CNS antigen MBP through *oral* administration of MBP [4]. The protection associated with *oral tolerance* in this model of stroke seemed to be related to brain-specific immunosuppression, a phenomenon which capitalizes upon breakdown of the BBB and appears to be, in part, related to local TGF-β1 production by lymphocytes [4]. In the current study, we sought to replicate our prior findings by inducing mucosal tolerance through nasal instillation of antigen. We also sought to better define the mechanism of neuroprotection afforded by immunologic tolerance through assessment of antigen-specific cytokine production by lymphocytes.

Methods and Materials

Animals. Experiments were approved by the Institution's Animal Care and Use Committee. Male Lewis rats (250–300 g) were used for all studies.

Induction of Tolerance. Animals were tolerized to bovine MBP or OVA (Sigma) through nasal installation of the antigen on five separate occasions over a period of 2 weeks. Following brief exposure to anesthesia (halothane 3%), 100 µg of antigen suspended in 40λ PBS was instilled into each nare (for a total of 200 µg Ag/ 80λ of PBS). In initial experiments, animals were sacrificed 24 h after MCAO for determination of infarct size (MBP, n = 12; OVA, n = 14); in subsequent experiments, animals tolerized to MBP (n = 13) and animals that had not been tolerized (n = 5) were sacrificed 48 h after MCAO and lymphocytes isolated from their brains.

Stroke. Anesthesia was induced with 5% halothane and maintained with 1.5% halothane. Following a mid-line neck incision, the right common carotid, internal carotid, and pterygopalatine arteries were ligated. A monofilament suture (4.0) was inserted into the common carotid artery and advanced into the internal carotid artery approximately 20 mm [61]. Animals were maintained at normothermia during the surgery and allowed to spontaneously thermoregulate thereafter. Reperfusion was performed 3 h after the onset of ischemia. Temperature was monitored at 1, 2, 3, 4, 5, 6, and 24 h after ischemia. Body weight was assessed prior to surgery and 24 h after surgery.

Determination of Infarct Volume. At the time of sacrifice, brains were removed and placed in chilled PBS (20°C) for 10 minutes. The brains were then sectioned at 2 mm intervals and incubated in 2% 2,3,5-triphenyltetrazolium chloride (TTC) (Sigma) at 38°C for 15 minutes [6]. Brain sections (bregma +2.40, +1.00,

–0.40, –1.80, –3.20, –4.40) were scanned and digitized; infarct volume, corrected for the presence of edema [50], was determined using the MetaMorph® Image System V4.1.1 (Universal Imaging Corp.).

Lymphocyte Isolation. Lymphocytes were isolated from the brain using previously described methods [32]. Briefly, brains were removed and gently pressed through a 70 μ mesh into a solution containing HBSS (Bio-Whittaker), 10 μl/ml DNAse (Sigma), and 0.1% collagenase (Boehringer-Mannheim). The cell suspension was layered over Ficoll-Paque™ Plus (Amersham-Pharmacia Biotech) and centrifuged at 1500 rpm for 30 minutes. Erythrocytes were removed with Ack lysing reagent (Bio-Whittaker).

ELISPOT Assay. Plates were prepared by incubating with the capture antibody (TGF-β1; R&D Systems) at 4 °C overnight. The plates were washed and blocked with RPMI with 5% FCS at 37 °C in 5% CO_2 for 1 h. Isolated cells (1×10^5 in 100 μl) were incubated with MBP (25 μg/ml) or media alone at 37 °C in 5% CO_2. (Media consisted of RPMI-1640 supplemented with 10% fetal calf serum, 2-mercaptoethanol, sodium pyruvate, non-essential amino acids, L-glutamine, penicillin-streptomycin, and HEPES.) All cell culture reagents were purchased from Bio-Whittaker. After 48 h, plates were washed and incubated with the appropriate biotinylated secondary (R&D Systems antibody) for 2 h at 37 °C in 5% CO_2. After washing, streptavidin alkaline phosphatase conjugate (1:2000, Calbiochem) was added to the wells and the plates incubated for 1 h at room temperature. After a final wash, the plates were developed using the Vector alkaline phosphatase substrate kit (BCIP/NBT; Vector Laboratories). All experiments were done in triplicate. The capture antibody was omitted for a negative control. Spots were counted under a dissecting microscope by 2 blinded investigators. Data are presented as the number of spots per 1×10^5 cells and represent the average of the 2 investigators values.

Statistics. Statistics were performed using the student t-test. Animal data within groups was compared using the paired t-test. All data are presented as mean ± standard deviation. Significance was set at $p < 0.05$.

Results

Mucosal Tolerance. All animals undergoing MCAO developed hyperthermia. The maximal recorded temperatures in both groups occurred shortly (2–3 h) after reperfusion. The elevation in temperature of MBP and OVA tolerized animals paralleled each other until reperfusion; after reperfusion the temperatures of the MBP tolerized animals began to plateau while the temperatures of OVA tolerized animals continued to rise (Fig. 1). There was a significant difference in temperature between MBP and OVA tolerized animals by 5 h (38.8 ± 0.76 °C *vs.* 39.4 ± 0.45 °C; p = 0.02) that persisted at 6 h (38.8 ± 0.65 °C vs. 39.4 ± 0.47 °C; p = 0.01) and 24 h (37.3 ± 0.62 °C vs.

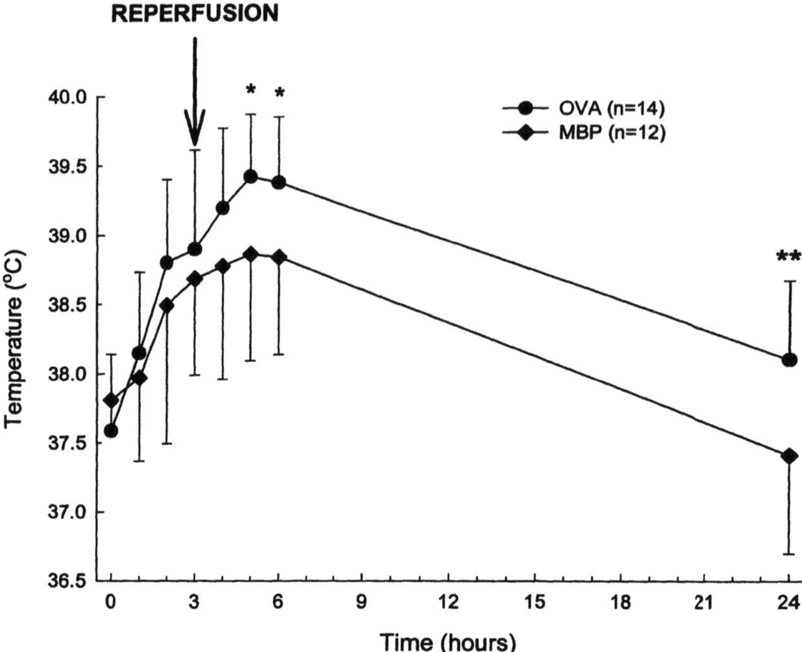

Fig. 1. Mucosal tolerance. Temperatures of MBP tolerized animals begin to plateau after reperfusion and are significantly less than the temperatures of OVA tolerized animals at 5, 6, and 24 h after ischemia ($^*p < 0.05$; $^{**}p < 0.01$).

$38.0 \pm 0.46\,^\circ$C; $p < 0.01$). There was a non-significant trend for higher temperatures at 24 h than at baseline in both OVA ($38.0 \pm 0.46\,^\circ$C $vs.$ $37.6 \pm 0.55\,^\circ$C; $p = 0.06$) and MBP tolerized animals ($37.3 \pm 0.62\,^\circ$C vs. $37.9 \pm 0.17\,^\circ$C; $p = 0.10$). The decrease in body weight at 24 h was similar in both groups. Infarct volumes 24 h after MCAO in MBP tolerized animals was significantly less than that in OVA tolerized animals (79.5 ± 48.3 mm^3 vs. 148 ± 61.6 mm^3; $p < 0.01$) (Fig. 2).

ELISPOT. At 48 h after ischemia, the total number of lymphocytes isolated from the ischemic hemisphere in MBP tolerized animals was less than in non-tolerized animals (4.1×10^6 vs. 8.1×10^6; $p = 0.01$) (Fig. 3); the number of these cells secreting TGF-β1 in response to MBP (per 1×10^5 total lymphocytes) was significantly greater in animals tolerized to MBP (1.01 ± 1.81 vs. 0.12 ± 0.25; $p = 0.02$) (Fig. 4).

Discussion

We previously showed that induction of immunologic tolerance to the CNS antigen MBP through *oral* administration of MBP could decrease infarct size following transient focal cerebral ischemia [4]. In the current study, we replicated those re-

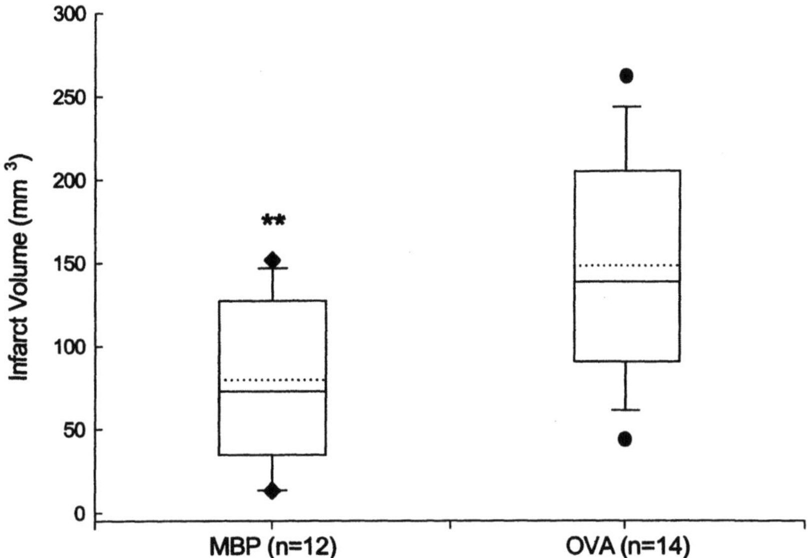

Fig. 2. Mucosal tolerance. Infarct volume is significantly less in MBP tolerized animals (**p < 0.01).

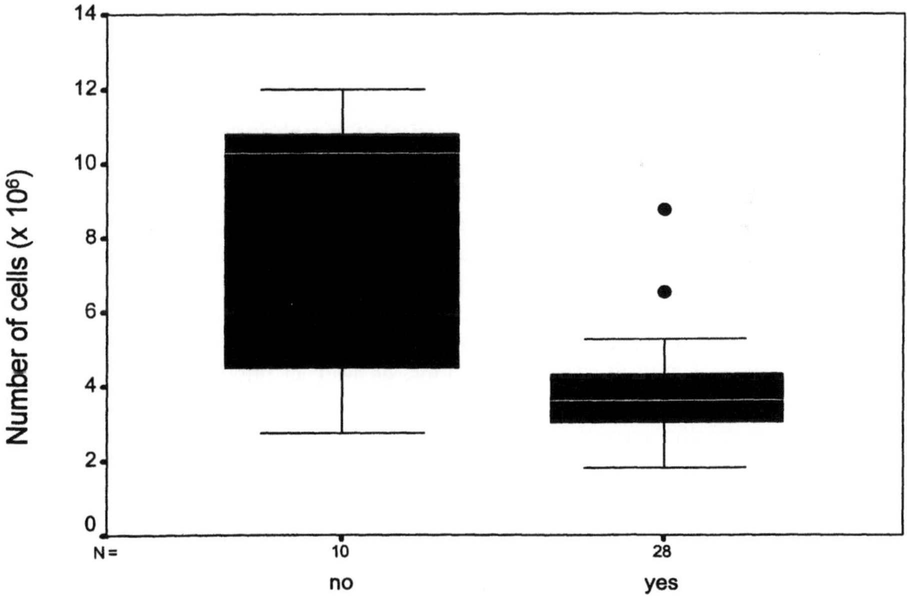

MBP tolerized

Fig. 3. Fewer lymphocytes are found within the ischemic hemispheres of animals tolerized to MBP 48 h after MCAO (*p < 0.05).

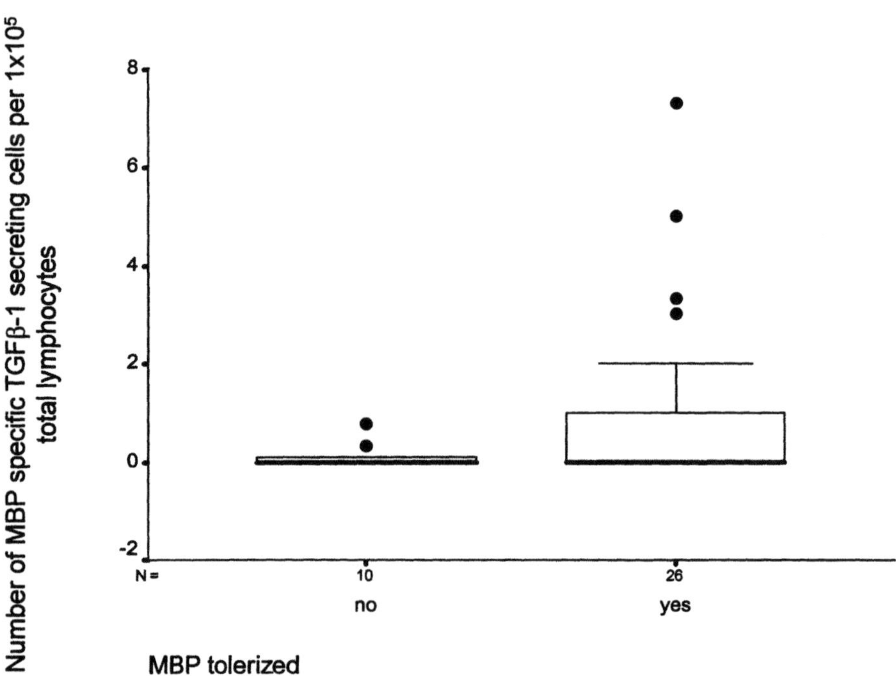

Fig. 4. TGF-β1 secretion by lymphocytes isolated from the ischemic hemispheres of MBP tolerized and non-tolerized animals 48 h after MCAO (*$p < 0.05$).

sults using an alternative and more potent method of inducing tolerance. *nasal* installation of MBP [36–38]. We also showed that lymphocytes isolated from the ischemic hemisphere of animals tolerized with MBP were more likely to secrete the immunomodulatory cytokine TGF-β1 when cultured with MBP than lymphocytes from the brains of non-tolerized animals. Taken together, these results show that manipulation of the immune system can attenuate ischemic brain injury.

Immunologic tolerance is mediated by T cells [22, 26, 30, 31, 42, 44, 47]. In animals tolerized through repetitive low dose mucosal exposure to antigen, re-encounter with antigen tends to result in a TH2-like immune response with secretion of IL-4, IL-10, and TGF-β1. These cytokines, as well as others, promote humoral immune responses and prevent cytotoxic or cell-mediated immune responses. Thus, antigen-specific activation of cells leads to production of antigen non-specific cytokines that have the potential to modulate any ongoing inflammatory response, irrespective of the trigger for that inflammatory response. This phenomenon is referred to as "by-stander suppression" [41, 52] and has been capitalized upon for the treatment of several different autoimmune diseases [54].

While our model utilizes the CNS antigen MBP to manipulate the immune system, it does not imply that MBP incites a deleterious inflammatory response following stroke. We have chosen to use MBP as the antigen with which to initiate the TH2-like response in brain after stroke, since MBP is released into the circulation following breakdown of the BBB and there is extensive literature regarding induction of tolerance to MBP. In the current study, body temperature in animals tolerized to MBP plateaued following reperfusion (Fig. 1), suggesting that immunomodulation began when the CNS antigens entered the systemic circulation and tolerized lymphocytes were given access to the ischemic brain, since these lymphocytes secrete cytokines like TGF-β1 when they encounter MBP and thereby suppress the inflammatory response [7, 19, 24, 39, 45, 56]. In addition to leukocytes, neurons and astrocytes also display receptors for TGF-β1 [53] and TGF-β1 may also have neuroprotective properties that are independent of its immunomodulatory properties [18, 29, 35]. In fact, administration of exogenous TGF-β1 also decreases infarct size [25, 29], but because of the potential detrimental effects of prolonged or excessive exposure of both CNS [58, 59] and non-CNS tissue [3, 12, 33] to TGF-β1, however, its direct use as a therapeutic agent may be limited.

In animals with immunologic tolerance, antigen-specific lymphocytes tend to traffic to the site of highest antigen expression. Thus, mucosal tolerance to CNS antigens can be viewed as a way to induce a relatively organ- (and lesion-) specific immunomodulatory response at the time it is needed, i.e. at the time of BBB disruption or brain injury. Mucosal tolerance to brain antigens could therefore be induced in persons at risk for cerebrovascular events (i.e. patients with transient ischemic attacks, carotid artery disease and those undergoing coronary artery bypass grafting), somewhat like a "vaccination". In fact, a recent paper describing a "vaccine" for stroke utilized an oral route for administration of a viral vector engineered to express a portion of the N-methyl-D-aspartate (NMDA) receptor [17]. The authors hypothesized that the resulting humoral immune response led to production of antibodies that blocked NMDA receptor activation following stroke. The possibility that the benefit observed in this model was related to mucosal tolerance was not addressed, and it remains possible that inhalation of the NMDA peptide itself would have produced similar benefit.

Changes in leukocyte function and leukocyte trafficking occur within hours after mucosal exposure to an antigen [23], so there is also potential to utilize immunologic tolerance for the treatment of stroke soon after it occurs. While the inflammatory response begins shortly after stroke onset (neutrophils are evident in the ischemic vasculature within minutes to hours of stroke) [15, 21], it persists for days and becomes a predominantly mononuclear and lymphocyte-mediated process with the progression of time [2, 48, 49]. Thus, interventions that modulate the inflammatory response could be given even after stroke onset with potential benefit.

In our prior study, we documented the effect of the tolerizing regimen on cell-mediated immunity [4]. We did not directly assess the effect of the tolerizing regimen on MBP-mediated immune responses in this study, but the decrease in infarct size among animals directly tolerized to MBP was nearly identical in both. Detailed immunohistochemistry would help to verify that a TH2 deviation in the CNS immune response was responsible for the observed neuroprotection, although the ELISPOT data suggest a role for TGF-β1. Because mucosal administration of antigen can occasionally lead to sensitivity to that antigen, long-term studies will need to be done to insure that induction of tolerance to CNS is safe [40].

Conclusions

Immunologic tolerance to CNS-specific antigens induced by mucosal administration of antigen reduces infarct size in a rodent model of transient focal cerebral ischemia. Active immunologic tolerance to brain antigens, by virtue of *bystander suppression*, could be exploited for the expectant treatment of stroke and other forms of brain injury.

References

1. Alvord EC, Jr., Hsu PC, Thron R (1974) Leukocyte sensitivity to brain fractions in neurological diseases. Arch Neurol 30:296–299
2. Barone FC, Hillegass LM, Tzimas MN, et al (1995) Time-related changes in myeloperoxidase activity and leukotriene B4 receptor binding reflect leukocyte influx in cerebral focal stroke. Mol Chem Neuropathol 24:13–30
3. Baughman RP, Lower EE, Miller MA, et al (1999) Overexpression of transforming growth factor-alpha and epidermal growth factor-receptor in idiopathic pulmonary fibrosis. Sarcoidosis Vasc Diffuse Lung Dis 16:57–61
4. Becker KJ, McCarron RM, Ruetzler C, et al (1997) Immunologic tolerance to myelin basic protein decreases stroke size after transient focal cerebral ischemia. Proc Natl Acad Sci USA 94:10873–10878
5. Becker KJ (2001) Targeting the central nervous system inflammatory response in ischemic stroke. Curr Opin Neurol 14:349–353
6. Bederson JB, Pitts LH, Germano SM, et al (1986) Evaluation of 2,3,5-triphenyltetrazolium chloride as a stain for detection and quantification of experimental cerebral infarction in rats. Stroke 17:1304–1308
7. Bridoux F, Badou A, Saoudi A, et al (1997) Transforming growth factor beta (TGF-beta)-dependent inhibition of T helper cell 2 (Th2)-induced autoimmunity by self-major histocompatibility complex (MHC) class II-specific, regulatory CD4(+) T cell lines. J Exp Med 185:1769–1775
8. Butterworth RJ, Wassif WS, Sherwood RA, et al (1996) Serum neuron-specific enolase, carnosinase, and their ratio in acute stroke. An enzymatic test for predicting outcome? Stroke 27:2064–2068
9. Chen Y, Kuchroo VK, Inobe J, et al (1994) Regulatory T cell clones induced by oral tolerance: suppression of autoimmune encephalomyelitis. Science 265:1237–1240
10. Chen Y, Inobe J, Marks R, et al (1995) Peripheral deletion of antigen-reactive T cells in oral tolerance [published erratum appears in Nature 1995 Sep 21;377(6546):257]. Nature 376:177–180
11. Chen Y, Inobe J, Kuchroo VK, et al (1996) Oral tolerance in myelin basic protein T-cell receptor transgenic mice: suppression of autoimmune encephalomyelitis and dose-dependent induction of regulatory cells. Proc Natl Acad Sci USA 93:388–391
12. Clouthier DE, Comerford SA, Hammer RE (1997) Hepatic fibrosis, glomerulosclerosis, and a lipodystrophy-like syndrome in PEPCK-TGF-beta1 transgenic mice. J Clin Invest 100:2697–2713
13. Cunningham RT, Young IS, Winder J, et al (1991) Serum neurone specific enolase (NSE) levels as an indicator of neuronal damage in patients with cerebral infarction. Eur J Clin Invest 21:497–500
14. Cunningham RT, Watt M, Winder J, et al (1996) Serum neurone-specific enolase as an indicator of stroke volume. Eur J Clin Invest 26:298–303
15. del Zoppo GJ, Schmid-Schonbein GW, Mori E, et al (1991) Polymorphonuclear leukocytes occlude capillaries following middle cerebral artery occlusion and reperfusion in baboons. Stroke 22:1276–1283
16. del Zoppo GJ, Becker KJ, Hallenbeck JM (2001) Inflammation after stroke: is it harmful? Arch Neurol 58:669–672
17. During MJ, Symes CW, Lawlor PA, et al (2000) An oral vaccine against NMDAR1 with efficacy in experimental stroke and epilepsy. Science 287:1453–1460
18. Flanders KC, Ren RF, Lippa CF (1998) Transforming growth factor-betas in neurodegenerative disease. Prog Neurobiol 54:71–85
19. Fontana A, Constam DB, Frei K, et al (1992) Modulation of the immune response by transforming growth factor beta. Int Arch Allergy Immunol 99:1–7

20. Fox CJ, Danska JS (1997) IL-4 expression at the onset of islet inflammation predicts nondestructive insulinitis in nonobese diabetic mice. J Immunol 158:2414–2424
21. Garcia JH, Liu KF, Yoshida Y, et al (1994) Influx of leukocytes and platelets in an evolving brain infarct (Wistar rat). Am J Pathol 144:188–199
22. Garside P, Steel M, Liew FY, Mowat AM (1995) CD4+ but not CD8+ T cells are required for the induction of oral tolerance. Int Immunol 7:501–504
23. Gonnella PA, Chen Y, Inobe J, et al (1998) In situ immune response in gut-associated lymphoid tissue (GALT) following oral antigen in TCR-transgenic mice. J Immunol 160:4708–4718
24. Gordon LB, Nolan SC, Ksander BR, et al (1998) Normal cerebrospinal fluid suppresses the in vitro development of cytotoxic T cells: role of the brain microenvironment in CNS immune regulation. J Neuroimmunol 88:77–84
25. Gross CE, Bednar MM, Howard DB, Sporn MB (1993) Transforming growth factor-beta 1 reduces infarct size after experimental cerebral ischemia in a rabbit model. Stroke 24:558–562
26. Gutgemann I, Fahrer AM, Altman JD, et al (1998) Induction of rapid T cell activation and tolerance by systemic presentation of an orally administered antigen. Immunity 8:667–673
27. Haneda K, Sano K, Tamura G, et al (1997) TGF-β induced by oral tolerance ameliorates experimental tracheal eosinophilia. J Immunol 159:4484–4490
28. Hashim GA, Brewen M (1985) Myelin basic protein-responsive blood T lymphocytes in patients with multiple sclerosis. J Neurosci Res 13:349–355
29. Henrich-Noack P, Prehn JH, Krieglstein J (1996) TGF-β1 protects hippocampal neurons against degeneration caused by transient global ischemia. Dose-response relationship and potential neuroprotective mechanisms. Stroke 27:1609–1614
30. Hoyne GF, Lamb JR (1997) Regulation of T cell function in mucosal tolerance. Immunol Cell Biol 75:197–201
31. Husby S, Mestecky J, Moldoveanu Z, et al (1994) Oral tolerance in humans. T cell but not B cell tolerance after antigen feeding. J Immunol 152:4663–4670
32. Irani DN, Griffin DE (1991) Isolation of brain parenchymal lymphocytes for flow cytometric analysis. Application to acute viral encephalitis. J Immunol Methods 139:223–231
33. Kanzler S, Lohse AW, Keil A, et al (1999) TGF-beta1 in liver fibrosis: an inducible transgenic mouse model to study liver fibrogenesis. Am J Physiol 276:G1059–1068
34. Khoury SJ, Hancock WW, Weiner HL (1992) Oral tolerance to myelin basic protein and natural recovery from experimental autoimmune encephalomyelitis are associated with downregulation of inflammatory cytokines and differential upregulation of transforming growth factor beta, interleukin 4, and prostaglandin E expression in the brain. J Exp Med 176:1355–1364
35. King C, Davies J, Mueller R, et al (1998) TGF-β1 alters APC preference, polarizing islet antigen responses toward a Th2 phenotype. Immunity 8:601–613
36. Kuper CF, Koornstra PJ, Hameleers DM, et al (1992) The role of nasopharyngeal lymphoid tissue. Immunol Today 13:219–224
37. Li HL, Liu JQ, Bai XF, et al (1998) Dose-dependent mechanisms relate to nasal tolerance induction and protection against experimental autoimmune encephalomyelitis in Lewis rats. Immunology 94:431–437
38. Ma CG, Zhang GX, Xiao BG, et al (1995) Suppression of experimental autoimmune myasthenia gravis by nasal administration of acetylcholine receptor. J Neuroimmunol 58:51–60
39. Malygin AM, Meri S, Timonen T (1993) Regulation of natural killer cell activity by transforming growth factor-beta and prostaglandin E2. Scand J Immunol 37:71–76
40. McFarland HF (1996) Complexities in the treatment of autoimmune disease. Science 274:2037–2038
41. Miller A, Lider O, Weiner HL (1991) Antigen-driven bystander suppression after oral administration of antigens. J Exp Med 174:791–798
42. Miller A, Lider O, Roberts A, et al (1992) Suppressor T cells generated by oral tolerization to myelin basic protein suppress both in vitro and in vivo immune responses by the release of transforming growth factor beta after antigen-specific triggering. Pro Natl. Acad Sci USA 89:421–425
43. Missler U, Wiesmann M, Friedrich C, Kaps M (1997) S-100 protein and neuron-specific enolase concentrations in blood as indicators of infarction volume and prognosis in acute ischemic stroke [see comments]. Stroke 28:1956–1960
44. Mondino A, Khoruts A, Jenkins MK (1996) The anatomy of T-cell activation and tolerance. Proc Natl Acad Sci USA 93:2245–2252
45. Rook AH, Kehrl JH, Wakefield LM, et al (1986) Effects of transforming growth factor beta on the functions of natural killer cells: depressed cytolytic activity and blunting of interferon responsiveness. J Immunol 136:3916–3920

46. Santambrogio L, Hochwald GM, Saxena B, et al (1993) Studies on the mechanisms by which transforming growth factor beta (TGF-beta) protects against allergic encephalomyelitis. J Immunol 151:1116–1127
47. Santos LM, al-Sabbagh A, Londono A, Weiner HL (1994) Oral tolerance to myelin basic protein induces regulatory TGF-beta-secreting T cells in Peyer's patches of SJL mice. Cell Immunol 157:439–447
48. Schroeter M, Jander S, Witte OW, Stoll G (1994) Local immune responses in the rat cerebral cortex after middle cerebral artery occlusion. J Neuroimmunol 55:195–203
49. Schroeter M, Jander S, Huitinga I, et al (1997) Phagocytic response in photochemically induced infarction of rat cerebral cortex. The role of resident microglia. Stroke 28:382–386
50. Swanson RA, Morton MT, Tsao-Wu G, et al (1990) A semiautomated method for measuring brain infarct volume. J Cereb Blood Flow Metab 10:290–293
51. Takeuchi M, Alard P, Streilein JW (1998) TGF-beta promotes immune deviation by altering accessory signals of antigen-presenting cells. J Immunol 160:1589–1597
52. Teng YT, Gorczynski RM, Hozumi N (1998) The function of TGF-beta-mediated innocent bystander suppression associated with physiological self-tolerance in vivo. Cell Immunol 190:51–60
53. Vivien D, Bernaudin M, Buisson A, et al (1998) Evidence of type I and type II transforming growth factor-beta receptors in central nervous tissues: changes induced by focal cerebral ischemia. J Neurochem 70:2296–2304
54. Wardrop RM, 3rd, Whitacre CC (1999) Oral tolerance in the treatment of inflammatory autoimmune diseases. Inflamm Res 48:106–119
55. Whitacre CC, Gienapp IE, Orosz CG, Bitar DM (1991) Oral tolerance in experimental autoimmune encephalomyelitis. III. Evidence for clonal anergy. J Immunol 147:2155–2263
56. Winkler MK, Beveniste EN (1998) Transforming growth factor-inhibition of cytokine-induced vascular cell adhesion molecule-1 expression in human astrocytes. Glia 22:171–179
57. Wunderlich MT, Ebert AD, Kratz T, et al (1999) Early neurobehavioral outcome after stroke is related to release of neurobiochemical markers of brain damage. Stroke 30:1190–1195
58. Wyss-Coray T, Lin C, Sanan DA, et al (2000) Chronic overproduction of transforming growth factor-beta1 by astrocytes promotes Alzheimer's disease-like microvascular degeneration in transgenic mice. Am J Pathol 156:139–150
59. Wyss-Coray T, Lin C, von Euw D, et al (2000) Alzheimer's disease-like cerebrovascular pathology in transforming growth factor-beta 1 transgenic mice and functional metabolic correlates [In Process Citation]. Ann NY Acad Sci 903:317–323
60. Youngchaiyud U, Coates AS, Whittingham S, Mackay IR (1974) Cellular-immune response to myelin protein: absence in multiple sclerosis and presence in cerebrovascular accidents. Aust N Z J Med 4:535–538
61. Zea-Longa E, Weinstein PR, Carlson S, Cummins R (1989) Reversible middle cerebral artery occlusion without craniectomy in rats. Stroke 20:84–91

Tumor Necrosis Factor-Alpha (TNF-α) and Ceramide Induce Tolerance to Ischemic and Hypoxic Insults to Brain and Brain Cells Associated with Changes in NFκB Function

I. GINIS and J. HALLENBECK

Summary. *In-vivo* and *in-vitro* studies by our group have shown that TNF-α and ceramide are involved in the regulation of tolerance to hypoxia and ischemia of brain. Preconditioning with these signaling molecules alters the function of NFκB. Although the nuclear translocation and DNA binding of the NFκB heterodimer, p50 and p65/RelA, are unaffected, p65 binding to the coactivator, p300/CBP, becomes disrupted.

Key words. Astrocytes – tolerance – NFκB – signaling – cell culture

Introduction

The study of tolerance to brain ischemia holds out the promise that molecular mechanisms can be identified, mechanisms that hold in check the network of pathophysiologic processes that give rise to progressive brain damage in the early hours of a stroke. A number of mechanisms have been implicated in ischemic tolerance [2], but the search for "master switch" targets for cytoprotection continues. Work in our laboratory with animal stroke models has implicated the stress-related cytokine, TNF-α, as an upstream signaling molecule in tolerance to brain ischemia [10, 13]. In addition, the signaling that regulates tolerance has been examined in primary cultures of astrocytes and neurons from the cortices of 2–3 day-old Sprague Dawley rats. This work has further supported a role for TNF-α in hypoxic tolerance and has identified ceramide as a downstream mediator of the state [6, 9].

More recent work *in vivo* has explored the role of ceramide in tolerance and brain ischemia in an effort to validate the results derived from primary cultures. One approach was to measure brain and plasma ceramide after intravenous injection of a tolerizing dose of lipopolysaccharide (LPS) (0.9 mg/kg) in spontaneously hypertensive rats (SHR) to test the hypothesis that ceramide contributes to tolerance in this model [20]. LPS in this model induces expression of TNF-α and leads to TNF receptor-mediated intracellular signaling that could increase ceramide lev-

*Correspondence to: John Hallenbeck, MD, Stroke Branch, National Institute of Neurological Disorders and Stroke, National Institutes of Health, Building 36, Room 4A03, MSC 4128, Convent Drive, Bethesda, Maryland 20892-4128, USA, Tel.: (301) 496-6231, Fax: (301) 402-2769, E-Mail: hallenbj@ninds@nih.gov

Maturation Phenomenon in Cerebral Ischemia V
A. M. Buchan et al. (Eds.)
© Springer-Verlag Berlin Heidelberg 2004

els. Ceramide measured by reverse phase HPLC significantly increased to 8.32 ± 1.14 pmol/µl plasma 24 h after LPS as compared to 2.65 ± 0.62 pmol/µl plasma in saline controls. In brain, ceramide levels significantly increased from 12 to 48 h after LPS compared to saline controls. The peak level at 48 h post LPS was 182% above the baseline value of 2.7 ± 0.2 pmol/nmol lipid phosphate. These results indicate that LPS preconditioning leads to elevation of ceramide in brain and plasma and, in conjunction with previous work, suggests that ceramide plays a role in LPS-induced tolerance to brain ischemic injury *in vivo*. The capacity of ceramide to protect immature rat brain from standardized hypoxia-ischemia (HI) was tested in postnatal day 7 Sprague-Dawley rats [3]. Thirty minutes after the HI, an optimal dose of 150 µg/kg C_2-ceramide was introduced into the lateral ventricle. C2-ceramide reduced HI-induced brain damage by 45 to 65% compared with controls. The cytoprotection was associated with augmented Bcl-2 and Bcl-xl protein levels in brain and decreased numbers of TUNEL-positive cells. These results supported a protective role for ceramide in neonatal HI and implicated anti-apoptotic proteins in this effect. C_2-ceramide was also infused intraventricularly 1 h before middle cerebral artery occlusion (MCAO) in 13-16 week spontaneously hypertensive rats (SHR). In another series of 13–16 week SHR, C_8-ceramide was injected intravenously 48 h or 24 h before MCAO or 5 minutes after MCAO. Under each of these conditions, the administration of ceramide led to a 14 to 17% reduction in infarct volume compared to controls [4]. Ceramide did offer a measure of neuroprotection, but because the degree of protection represented approximately 50% of the maximal infarct reduction observed in this model after preconditioning with LPS, the results suggested that additional signaling pathways subserve tolerance.

Previous experiments performed in primary cultures of astrocytes [6] had shown that expression of intercellular adhesion molecule-1 (ICAM-1) in response to a hypoxic or a TNF-α stimulus was inhibited at the transcriptional level after preconditioning with either TNF-α or ceramide. On this basis we wondered whether preconditioning in our models involved an alteration in mechanisms controlling activation of the transcription factor, NF-κB, a critical transactivator for ICAM-1 upregulation in astrocytes [8]. The results indicate that preconditioning does not affect nuclear translocation or binding to the NFκB DNA consensus site. Preconditioning does interfere with the association of p65 and the coactivator protein, p300/CBP.

Materials and Methods

Cortical astrocyte cultures were established from 2 day-old Sprague-Dawley rats as previously described [6]. Passages 1–3 were used.

TNF-α and C-2 Ceramide Preconditioning

Astrocyte cultures were incubated for 4 h with 50 ng/ml rat-recombinant TNF-α (Chemicon International, Temecula, CA), then TNF-α-containing medium was replaced by fresh culture medium and cells were allowed to rest for 20 h, and then activated again with the same dose of TNF-a for indicated times. For ceramide preconditioning, N-acetylceramide (C-2 ceramide) (Biomol, Plymouth Meeting, PA) was added to the cultures at 10 μM 1 h prior to TNF-a addition.

Western Blots for NF-κB Studies

Preparation of cytosolic and nuclear extracts has been described elsewhere [7]. All electrophoresis buffers, 4 to 12% tris-glycine and tris-acetate mini-gels, nitrocellulose membranes, and electrophoresis equipment were from Novex (San Diego, CA). Cytosolic or nuclear extracts were boiled in equal volumes of loading buffer/ 1 mM di-thiothreithol (DTT) for 3 min and loaded on a gel at 10 μg protein/lane for determination of p65 concentrations in nuclear extracts, and at 15 μg protein/ lane for IκB determination in cytosolic fractions. Separated proteins were transferred to nitrocellulose membrane. Immunoblotting was performed as previously described [7]. Anti-IκBa and anti-p65 rabbit polyclonal antibodies were from Santa Cruz Biotechnology (SC-203 and SC-372 respectively).

Electrophoretic Mobility Shift Assay (EMSA)

EMSA was performed by using the Gel Shift Assay System (Promega) according to the manufacturer's instructions as previously described [7].

Interaction of p65 with p300 Adaptor Protein

To determine whether p65 associates with p300 upon TNF-a activation, naïve and preconditioned astrocytes were activated with TNF-a for 30 min. At the end of incubation nuclear extracts were prepared as has been described elsewhere [7]. Before immunoprecipitation, aliquots of nuclear extracts with equal amounts of protein (usually 0.5–1 mg/sample) were diluted 1:4 with the buffer used for extraction (buffer B), but containing no NaCl (20 mM HEPES, 50 mM KCl, 1 mM EDTA, 1 mM EGTA, 10% (w/v) glycerol, pH = 7.9) to adjust salt concentration to 100 mM, and then incubated with a mixture of two goat polyclonal anti-p65 antibodies (Santa Cruz Biotechnology; cat. #SC109 and SC-372) at 4 μg antibody per 1 mg protein overnight and then with 50 μl Protein G PLUS-Agarose (Santa Cruz Biotechnology) for 2 h. The beads were washed 3 times with the diluted buffer B, boiled in 30 μl of sample-loading buffer/1 mM DTT for 3 min and electrophoresed on 4–12% Tris-glycine or on tris-acetate mini gels. p300 in precipitates was identified by Western blotting with p300-specific antibody (Upstate Biotechnology, cat. #05-267). All electrophoresis buffers, mini-gels, nitrocellulose membranes, and electrophoresis equipment were from Novex (SanDiego, CA).

Results

TNF-α Preconditioning or C-2 Ceramide Treatment has no Effect on NF-κB Activation

Most of the p65/p50 heterodimers in quiescent cells are retained in cytoplasm complexed with the inhibitor of NF-κB (IκB). IκB becomes phosphorylated by a specific kinase, ubiquitinated and subsequently degraded by proteasome peptidases upon activation with TNF-α and other agonists. Removal of IκB-alpha uncovers the nuclear localization signals of subunits of NF-κB, allowing the complex to enter the nucleus and bind to DNA [1]. Initially we investigated whether inhibition of ICAM-1 transcription in tolerant cells is due to alterations in NF-κB activation. We either preconditioned astrocytes with TNF-α 24 h prior to the second addition of TNF-α or pretreated them with C-2 ceramide 1 h before TNF-α addition. We then studied the kinetics of IkB in cytosolic extracts and p65 in nuclear extracts by Western blotting. Twenty minutes after addition of TNF-α we observed in naïve cells almost complete disappearance of IκB from the cytosol and concurrent accumulation of p65 in the nucleus. In TNF-α preconditioned cells and in the cells pretreated with C-2 ceramide, IκB degradation in the cytoplasm and p65 accumulation in the nucleus were observed to follow exactly the same patterns.

TNF-α Preconditioning or C-2 Ceramide has no Effect on p65 DNA-Binding Activity

If p65 were to lose its ability to bind DNA, NF-κB-dependent transcription in tolerant cells might be inhibited. To address this possibility nave, TNF-α-preconditioned, and C-2 ceramide-pretreated astrocytes were incubated with TNF-α for 30 min and we assessed DNA-binding activity of NF-κB proteins by EMSA. We found no differences in DNA binding between nuclear extracts of naïve and tolerant cells. Supershift experiments with anti-p65 antibody demonstrated that the p65 subunit constituted most of the DNA-binding complexes.

TNF-α Preconditioning and C-2 Ceramide Prevent Interaction of p65 with p300 Adapter Protein

Antibody directed against p65 co-precipitated p300 from nuclear extracts of naïve cells activated with TNF-α. The coprecipitated p300 was identified by Western blotting with p300-specific antibody (Fig. 1). In TNF-α-preconditioned cells or in cells pretreated with C-2 ceramide, however, relatively little association between p65 and p300 was observed. It is possible that ICAM-1 gene expression is dependent on association of p65 and p300 and that in preconditioned cells the formation p65/p300 complexes is interrupted resulting in inhibition of transcription of the ICAM-1 gene.

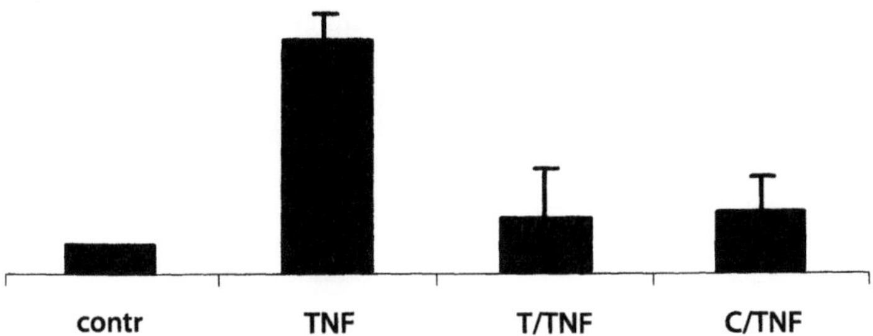

Fig. 1. Effect of preconditioning with TNF-*α* or ceramide on the interaction of p65 with p300 as shown by densitometry of Western blots. Naïve and preconditioned astrocytes were activated with TNF-*α* for 30 min. After incubation, nuclear extracts were prepared and immunoprecipitated with p65-specific antibody; by blotting with p300-specific antibody, p300 in the precipitates was identified. Contr = control untreated astrocytes; TNF = Astrocytes activated with TNF-*α* without preconditioning; T/TNF = astrocytes activated with TNF-*α* after TNF-*α* preconditioning; C/TNF = astrocytes activated with TNF-*α* after preconditioning with C-2 ceramide. Relative to controls, TNF is increased about 7-fold; T/TNF and C/TNF are increased about 1.6-fold and 1.3-fold, respectively. Each bar represents mean ± SD of three experiments.

Discussion

The results show that preconditioning primary astrocyte cultures from 2 to 3 day old Sprague Dawley rats with TNF-*α* (50 ng/ml) or ceramide (10 µM) does not affect the molecular mechanisms that lead to nuclear translocation of NF*κ*B. Also, the p65 subunit of NF*κ*B binds normally to its consensus site on DNA after the preconditioning. Preconditioning does, however, inhibit the association of p65 with the coactivator protein, p300/CBP. This may inhibit transcription of the ICAM-1 gene and may change or "fine tune" the gene profile that NF*κ*B transactivates.

NF-*κ*B is a heterodimer consisting of a 65 kDa protein (p65/RelA) and a 50 kDa protein (p50) (both belong to the Rel family of proteins) that is sequestered in cytoplasm by an anchor protein, inhibitor of NF-*κ*B (I*κ*B). When I*κ*B is phosphorylated on serines 32 and 36 by I*κ*B kinase (IKK), it becomes ubiquitinated and is subsequently degraded by proteosomal enzymes, which allows NF-*κ*B heterodimer to translocate to the nucleus (for review see [1]). The NF-*κ*B dimer binds to its DNA consensus sequence GGGRNNYYCC (R = A or G, Y = C or T) and initiates transcription in the nucleus. Only p65 is capable of trans-activation [12, 14, 16], although both subunits contain DNA-binding domains. Recent reports demonstrate that NF-*κ*B transcriptional activity can be controlled by blocking p65 phosphorylation [12] and that TNF-*α* induces phosphorylation of p65 on serine 529 [17]. The catalytic subunit of PKA (cPKA) has been shown to be part of the NF-*κ*B/I*κ*B complex in the cytoplasm of B cell line 70Z/3. In this complex, I*κ*B inhibits the catalytic activity of cPKA. I*κ*B degradation induced by LPS results in an upregulation of cPKA activity and phosphorylation p65 [18]. Phosphorylation of p65/RelA by PKA allows its interaction with the coactivator protein CBP/p300 [19].

CREB-binding protein (CBP) and p300 are highly homologous and functionally-related coactivator proteins that bind to transactivation domains of transcription factors and mediate their contact with RNA polymerase II, a holoenzyme which requires about 30 different protein cofactors for proper functioning. CBP/p300 also was shown to be a potent acetyltransferase, which can acetylate histones and transcription factors to facilitate their interaction with DNA [15]. The p65/RelA subunit of NF-κB is among many other transcription factors that have been reported to interact with p300/CBP, and p300-p65 interaction has been recently shown to regulate cell cycle progression [11] or activation of the IL-6 gene by TNF-α [15] in tumor cell lines. Similarly, activation of E-selectin and VCAM-1-CAT reporter constructs in SL-2 Schneider cells and TNF-α-induced expression of E-selectin in human umbilical vein endothelial cells is dependent on p300/CBP association with p65 [5].

References

1. Baeuerle PA (1998) IkappaB-NF-kappaB structures: at the interface of inflammation control. Cell 95:729–731
2. Chen J, Simon R (1997) Ischemic tolerance in the brain. Neurology 48:306–311
3. Chen Y, Ginis I, Hallenbeck JM (2001) The protective effect of ceramide in immature rat brain hypoxia-ischemia involves up-regulation of bcl-2 and reduction of TUNEL-positive cells. J Cereb Blood Flow Metab 21:34–40
4. Furuya K, Ginis I, Takeda H, Chen Y, Hallenbeck JM (2001) Cell permeable exogenous ceramide reduces infarct size in spontaneously hypertensive rats supporting in vitro studies that have implicated ceramide in induction of tolerance to ischemia. J Cereb Blood Flow Metab 21:226–232
5. Gerritsen ME, Williams AJ, Neish AS, Moore S, Shi Y, Collins T (1997) CREB-binding protein/p300 are transcriptional coactivators of p65. Proc Natl Acad Sci USA 94:2927–2932
6. Ginis I, Schweizer U, Brenner M, Liu J, Azzam N, Spatz M, Hallenbeck JM (1999) TNF-alpha pretreatment prevents subsequent activation of cultured brain cells with TNF-alpha and hypoxia via ceramide. Am J Physiol 276:C1171–1183
7. Ginis I, Hallenbeck JM, Liu J, Spatz M, Jaiswal R, Shohami E (2000) Tumor necrosis factor and reactive oxygen species cooperative cytotoxicity is mediated via inhibition of NF-kappaB. Mol Med 6:1028–1041
8. Lee SJ, Hou J, Benveniste EN (1998) Transcriptional regulation of intercellular adhesion molecule-1 in astrocytes involves NF-kappaB and C/EBP isoforms. J Neuroimmunol 92:196–207
9. Liu J, Ginis I, Spatz M, Hallenbeck JM (2000) Hypoxic preconditioning protects cultured neurons against hypoxic stress via TNF-alpha and ceramide. Am J Physiol Cell Physiol 278:C144–C153
10. Nawashiro H, Tasaki K, Ruetzler CA, Hallenbeck JM (1997) TNF-alpha pretreatment induces protective effects against focal cerebral ischemia in mice. J Cereb Blood Flow Metab 17:483–490
11. Perkins ND, Felzien LK, Betts JC, Leung KY, Beach DH, Nabel GJ (1997) Regulation of Nf-Kappa-B by Cyclin-Dependent Kinases Associated with the P300 Coactivator. Science 275:523–527
12. Schmitz ML, dos Santos Silva MA, Altmann H, Czisch M, Holak TA, Baeuerle PA (1994) Structural and functional analysis of the NF-kappa B p65 C terminus. An acidic and modular transactivation domain with the potential to adopt an alpha-helical conformation. J Biol Chem 269:25613–25620
13. Tasaki K, Ruetzler CA, Ohtsuki T, Martin D, Nawashiro H, Hallenbeck JM (1997) Lipopolysaccharide pre-treatment induces resistance against subsequent focal cerebral ischemic damage in spontaneously hypertensive rats. Brain Res 748:267–270
14. True AL, Rahman A, Malik AB (2000) Activation of NF-kappaB induced by H(2)O(2) and TNF-alpha and its effects on ICAM-1 expression in endothelial cells. Am J Physiol Lung Cell Mol Physiol 279:L302–L311
15. Van den Berghe W, De Bosscher K, Boone E, Plaisance S, Haegeman G (1999) The nuclear factor-kappaB engages CBP/p300 and histone acetyltransferase activity for transcriptional activation of the interleukin-6 gene promoter. J Biol Chem 274:32091–32098
16. Vasiliou V, Lee J, Pappa A, Petersen DR (2000) Involvement of p65 in the regulation of NF-kappaB in rat hepatic stellate cells during cirrhosis. Biochem Biophys Res Commun 273:546–550

17. Wang D, Baldwin AS Jr (1998) Activation of nuclear factor-kappaB-dependent transcription by tumor necrosis factor-alpha is mediated through phosphorylation of RelA/p65 on serine 529. J Biol Chem 273:29411–29416
18. Zhong H, SuYang H, Erdjument-Bromage H, Tempst P, Ghosh S (1997) The transcriptional activity of NF-kappaB is regulated by the IkappaB-associated PKAc subunit through a cyclic AMP-independent mechanism. Cell 89:413–424
19. Zhong H, Voll RE, Ghosh S (1998) Phosphorylation of NF-kappa B p65 by PKA stimulates transcriptional activity by promoting a novel bivalent interaction with the coactivator CBP/p300. Mol Cell 1:661–671
20. Zimmermann C, Ginis I, Furuya K, Klimanis D, Ruetzler C, Spatz M, Hallenbeck JM (2001) Lipopolysaccharide-induced ischemic tolerance is associated with increased levels of ceramide in brain and in plasma. Brain Res 895:59–65

Microglial Proliferation and Cell Cycle Protein Upregulation in the Rat Hippocampus Following Forebrain Ischemia

H. Kato, A. Takahashi, and Y. Itoyama

Summary. We investigated the expression of cell cycle proteins in proliferating microglial cells in the hippocampus in comparison with CA1 neurons that are destined to die after 10 min of forebrain ischemia in the rat. The animals were sacrificed 1 d, 2 d, and 7 d after ischemia. Immunohistochemistry was performed using antibodies raised against microglial response factor-1 (MRF-1, a microglia/macrophage marker), glial fibrillary acidic protein (GFAP, an astrocyte marker), and cell cycle proteins, i.e., proliferating cell nuclear antigen (PCNA), cyclin D1, and cyclin-dependent kinase-4 (cdk4). Microglial cells with immunoreactivities to PCNA, cyclin D1, and cdk4 in their nuclei started to increase in number 1 d after ischemia, especially in CA1 and dentate hilus, reached a peak after 2 d, and declined after 7 d, when microglial proliferation was striking. Limited expression of these proteins in astrocytes was observed after 7 d, when limited astroglial proliferation was seen. No remarkable expressions of these proteins were observed in CA1 neurons during the observation period. Thus, the upregulation of cell cycle proteins preceded the microglial proliferation, and may be involved in the onset of microglial activation and proliferation. However, the role of cell cycle protein expression in CA1 neuronal death was not suggested in this study.

Key words. Cerebral ischemia – cell cycle protein – microglia – astrocytes – rat

Introduction

Transient forebrain ischemia in the rat induces delayed neuronal death of CA1 neurons in the hippocampus [11, 17], as well as astrocytic and microglial activation [5, 16]. Microglial activation occurs rapidly within minutes and includes prominent cell proliferation in contrast to astroglial activation. Microglia transform into brain macrophages in response to neuronal death, and strongly activated microglia may have a detrimental effect on the survival of injured neurons [2]. On the other hand, recent studies have shown that cerebral ischemia induces the ex-

Hiroyuki Kato[1,2], Akira Takahashi[2], Yasuto Itoyama[1], Departments of Neurology[1] and Neuroendovascular Therapy[2], Tohoku University Graduate School of Medicine, Sendai, Japan

Correspondence to: Hiroyuki Kato, MD, PhD, Department of Neurology, Tohoku University School of Medicine, 1-1 Seiryo-machi, Aoba-ku, Sendai 980-8574, Japan, Tel.: +81-22-717-7189, Fax: +81-22-717-7192, E-Mail: katoh@mail.cc.tohoku.ac.jp

Maturation Phenomenon in Cerebral Ischemia V
A.M. Buchan et al. (Eds.)
© Springer-Verlag Berlin Heidelberg 2004

pression of cell cycle proteins [13, 15]. However, aberrant expression of cell cycle proteins in terminally differentiated, post-mitotic, adult neurons has been suggested to lead to neuronal death instead of DNA replication and cell division [3, 8]. To clarify the role of cell cycle proteins in neuronal death and glial activation following ischemia, we investigated the expression of cell cycle proteins in proliferating microglial cells in comparison with CA1 neurons that are destined to die after ischemia.

Materials and Methods

Induction of Cerebral Ischemia

Male adult Sprague-Dawley rats were anesthetized with 1.5% halothane in a mixture of 70% nitrous oxide and 30% oxygen. Forebrain ischemia was induced for 10 min by 2-vessel occlusion combined with hypotension, essentially as described by Smith et al. [20]. The right femoral artery was cannulated with PE50 polyethylene tubing for blood pressure monitoring and blood sampling. Bilateral common carotid arteries were occluded and the mean arterial blood pressure was maintained at 40 mmHg by withdrawal of blood. After 10 min of ischemia, the carotid arteries were recirculated and the shed blood was infused. Body temperature was maintained at approximately 37 °C. Sham-operated animals were also prepared (n = 4). The animals were sacrificed at 1 d, 2 d, and 7 d after reperfusion (n = 4 each). The brains were perfusion-fixed with 4% paraformaldehyde in 0.1 M phosphate buffer. The brains were then embedded in paraffin. Coronal sections were cut at the level of the dorsal hippocampus and used for histopathology (hematoxylin and eosin staining) and immunohistochemistry.

Immunohistochemistry

Immunohistochemistry was performed on paraffin sections using following antibodies: polyclonal anti-microglial response factor-1 (MRF-1) antibody (0.3 µg/ml) as a microglia/macrophage marker [21], monoclonal anti-glial fibrillary acidic protein (GFAP) antibody (Chemicon, 1:800) as an astrocyte marker, monoclonal anti-proliferating cell nuclear antigen (PCNA) antibody (Boehringer Mannheim, 0.5 µg/ml), monoclonal anti-cyclin D1 antibody (Santa Cruz, 2 µg/ml), and polyclonal anti-cyclin-dependent kinase-4 (cdk4) antibody (Santa Cruz, 2 µg/ml).

Immunostaining was performed using the Vectastain elite ABC kit (Vector Laboratories) according to the supplier's recommendations. After deparaffinization, the sections were treated with 10% methanol/0.3% H_2O_2 for 20 min to quench endogenous peroxidase activity. The sections were preincubated with normal serum, followed by incubation with the primary antibody overnight at 4 °C. The sections were incubated with the secondary antibody for 1 h at room temperature, then with the ABC complex for 30 min, and finally with 0.05% 3,3′-diaminobenzidine (DAB) and 0.02% H_2O_2. For double-label immunostaining, this cycle was repeated. The first cycle was visualized with DAB/Ni producing a blue/black reaction product and the second cycle was visualized with DAB producing a brown reaction product. Negative control study was performed using non-immuned IgG or by omission of the primary antibody, which produced no notable staining.

Results

CA1 Neuronal Death and Glial Activation

The CA1 neurons appeared intact on HE-stained sections 1 d after ischemia, started to die after 2 d, and almost all the neurons were damaged after 7 d, when glial proliferation was evident in the CA1 region. Scattered dentate hilar neurons were damaged by 1 d.

Microglia, as visualized by MRF-1, were uniformly distributed in control hippocampus and the morphology was that of resting microglia with ramified, thin processes and small cell bodies. After 1 d, they became activated with more stout morphology. The morphological changes became conspicuous after 2 d. After 7 d, they became amoeboid in shape and the number increased strikingly [10].

Astrocytes, as visualized with GFAP, in control hippocampus distributed evenly with resting morphology of slender processes and small cell bodies. After 1 d and more remarkably after 2 d, they became activated with reactive, hypertrophied morphology. After 7 d, reactive changes became further conspicuous, together with a slight increase in number [10].

Cell Cycle Protein Expression

In control hippocampus, PCNA immunoreactivity was detected only in the nuclei of neural progenitor cells in the subgranular zone (SGZ) and oligodendrocytes. Cyclin D1 immunoreactivity was seen in the nuclei of the neural progenitor cells in the SGZ and oligodendrocytes, and weak immunoreactivity was observed in the cytoplasm of CA1–3 pyramidal neurons and dentate granule cells. Cdk4 immuno-

Table 1. Glial proliferation and the expression of cell cycle proteins in CA1 hippocampus following forebrain ischemia

	pre	1 d	2 d	7 d
Microglial proliferation	–	–	+	++
PCNA	–	+	++	+
cyclin D1	–	+	++	+
cdk4	–	+	++	±
Astroglial proliferation	–	–	–	+
PCNA	–	–	–	±
cyclin D1	–	–	–	±
cdk4	–	–	±	+

Glial proliferation and the expression of cell cycle proteins were graded – (none), ± (slight), + (moderate), and ++ (strong). PCNA, proliferating cell nuclear antigen; cdk4, cyclin-dependent kinase-4

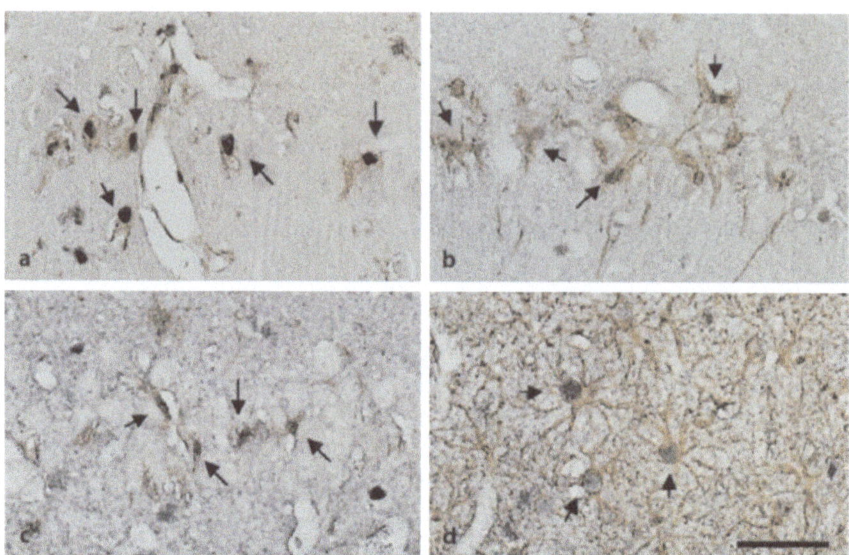

Fig. 1 a–d. Double label immunohistochemistry for cell cycle proteins and glial cells in CA1 hippocampus following 10 min of forebrain ischemia. **a–c**: Two days after ischemia, activated microglia, whose cell bodies were immunostained brown with anti-microglial response factor-1 (MRF-1) antibody, exhibited immunoreactivities (blue/black in color) in their nuclei to proliferating cell nuclear antigen (PCNA) (**a**), cyclin D1 (**b**), and cyclin-dependent kinase-4 (cdk4) (**c**). **d** After 7 d, reactive astrocytes, whose cell bodies and processes were immunostained brown with anti-glial fibrillary acidic protein (GFAP) antibody, showed immunoreactivity (blue/black) to cdk4 in their nuclei. Bar = 50 μm and applies to all.

reactivity was seen only in the nuclei of the progenitor cells in the SGZ and oligodendrocytes, and in the cytoplasm of CA2 neurons.

After ischemia, expressions of immunoreactivity to the cell cycle proteins were observed predominantly in microglia and less remarkably in astrocytes (Table 1). Activated microglia immunoreactive to PCNA, cyclin D1, and cdk4 in their nuclei started to increase in number after 1 d, especially in CA1 and dentate hilus, reached a peak after 2 d (Fig. 1), and declined after 7 d. Limited expression of the cell cycle proteins was seen in the nuclei of a small number of reactive astrocytes after 7 d (Fig. 1). No remarkable expression of cell cycle proteins were observed in CA1 neurons during the observation period, except that cyclin D1 immunoreactivity was increased in a small number of damaged CA1 neurons at 2 d. No remarkable changes were observed in oligodendrocytes. The number of neural progenitor cells in the SGZ increased in number after 7 d.

Discussion

Microglia and the Cell Cycle Proteins

The present study demonstrated an up-regulation of cell cycle proteins (PCNA, cyclin D1, and cdk4) predominantly in activated microglia in the hippocampus after ischemia. This up-regulation, which started after 1 d and reached a peak after 2 d, preceded the proliferation of microglia that started after 2 d and was striking after 7 d. Microglia exhibited early widespread morphological activation within a day, which was not restricted to the areas of neuronal damage (CA1 and dentate hilus). When neuronal death took place, microglia were further activated into macrophages and proliferated, as reported previously [9, 10]. Microglial cells have been reported to exhibit mitosis after ischemia [6].

Wiesser et al. [24] investigated the expression of cyclin D1 mRNA with in situ hybridization following transient forebrain ischemia in the rat, and found a transient increase in cyclin D1 mRNA in microglia in CA1 after 2–3 d. Cyclin D1 belongs to a class of proteins that activate the cyclin-dependent kinases including cdk4 [1]. Phosphorylation of other proteins such as the retinoblastoma protein by the cyclin-activated cdk triggers the progression of the cell cycle, switching from the G1 phase to the S phase of the cell cycle [18]. PCNA is involved in both DNA replication and DNA repair, and has been used as a general marker of dividing cells [14]. Thus, transient activation of cell cycle proteins is thought to be involved in the preparation and execution of the microglial proliferation and possibly in the transformation into phagocytic macrophages.

Astrocytes and the Cell Cycle Proteins

Astrocytes were also activated after ischemia exhibiting hypertrophied morphology (reactive astrocytes). The astroglial activation occurred later than microglial activation, and their proliferation was limited and seen only after 7 d. Earlier papers have shown that astrocytes do not proliferate in early postischemic periods [6,16]. In accordance with this, we did not observe cell cycle protein expression in astrocytes in the early reperfusion period (1–2 d), but a limited expression of cell cycle proteins was observed in reactive astrocytes after 7 d. It is likely that astrocytes proliferate later than this time point, at a chronic, restorative stage between 2 and 5 weeks after ischemia [16].

Neuronal Death and the Cell Cycle Proteins

There is a hypothesis that aberrant expression of the cell cycle proteins in terminally differentiated, adult neurons leads to apoptotic neuronal death [3, 7, 8]. However, our findings showed no cell cycle protein expression in CA1 neurons before their death, except for minor expression of cyclin D1 in a small number of damaged CA1 neurons after 2 d, which does not account for the death of the majority of CA1 neurons. Therefore, our result does not support this hypothesis in CA1 neurons following ischemia.

Earlier papers have shown in a variety of experimental models of neuronal death especially in vitro that neuronal death is accompanied by the induction of cyclin D1 [4, 8, 12]. Osuga et al. [15] reported cyclin D1 and cdk4 expression in neurons following brief focal cerebral ischemia in mice. However, previous papers, in accordance with ours, reported microglial, but not neuronal, expression of cyclin D1 in CA1 following global cerebral ischemia [19, 24], although another paper reported a conflicting result showing an expression of cyclin D1 in CA1 neurons and reactive astrocytes following ischemia [22]. To our knowledge, there have been no reports on cdk4 expression in CA1 neurons following ischemia. Tomasevic et al. [23] reported the immunohistochemical changes in PCNA in CA1 neurons after forebrain ischemia. After an antigen retrieval procedure with microwave heating, CA1 neurons became immunoreactive to PCNA, and the PCNA immunoreactivity in CA1 neurons decreased after ischemia. We did not perform the antigen retrieval procedure because microglial cells exhibited striking immunoreactivity to PCNA without it. In any event, PCNA did not increase in CA1 neurons after ischemia in both studies.

Conclusions

This study showed that cell cycle proteins were upregulated after ischemia, predominantly in microglia, which was likely to play a role in microglial activation and proliferation. Astrocytes that displayed only limited, late proliferation exhibited limited, late expression of cell cycle proteins in their nuclei. Thus, the cell cycle proteins may play a critical role in the activation and proliferation of glial cells after ischemia. However, this study did not detect cell cycle protein expression in CA1 neurons, and did not support the hypothesis that aberrant expression of cell cycle proteins leads to neuronal death of CA1 neurons after ischemia.

Acknowledgements. This study was supported in part by Grant-in-Aid for Scientific Research (13670627) from the Japan Society for the Promotion of Science (HK). The authors wish to acknowledge Drs. Shuuitsu Tanaka and Tatsuro Koike at Molecular Neurobiology Laboratory, Graduate Program in Biological Sciences, Hokkaido University, Sapporo, Japan, for providing us with the anti-MRF-1 antibody.

References

1. Baldin V, Lukas J, Marcote MJ, Pagano M, Draetta G (1993) Cyclin D1 is a nuclear protein required for cell cycle progression in G1. Genes Dev 7:812–821
2. Banati RB, Gehrmann J, Schubert P, Kreutzberg GW (1993) Cytotoxicity of microglia. Glia 7:111–118
3. Copani A, Uberti D, Sortino MA, Bruno V, Nicoletti F, Memo M (2001) Activation of cell-cycle-associated proteins in neuronal death: a mandatory or dispensable path? Trends Neurosci 24:25–31

4. Freeman RS, Estus S, Johnson EM Jr (1994) Analysis of cell cycle-related gene expression in postmitotic neurons: selective induction of cyclin D1 during programmed cell death. Neuron 12:343–355
5. Gehrmann J, Bonnekoh P, Miyazawa T, Hossmann K-A (1992) Immunohistochemical study of an early microglial activation in ischemia. J Cereb Blood Flow Metab 12:257–269
6. Gehrmann J, Bonnekoh P, Miyazawa T, Oschlies U, Dux E, Hossmann K-A, Kreutzberg GW (1992) The microglial reaction in the rat hippocampus following global ischemia: immuno-electron microscopy. Acta Neuropathol 84:588–595
7. Heintz N (1993) Cell death and the cell cycle: a relationship between transformation and neu-rodegeneration? Trends Biochem Sci 18:157–159
8. Herrup K, Busser JC (1995) The induction of multiple cell cycle events precedes target-related neuronal death. Development 121:2385–2395
9. Kato H, Kogure K, Araki T, Itoyama Y (1995) Graded expression of immunomolecules on acti-vated microglia in the hippocampus following ischemia in a rat model of ischemic tolerance. Brain Res 694:85–93
10. Kato H, Tanaka S, Oikawa T, Koike T, Takahashi A, Itoyama Y (2000) Expression of microglial response factor-1 in microglia and macrophages following cerebral ischemia in the rat. Brain Res 882:206–211
11. Kirino T, Tamura A, Sano K (1984) Delayed neuronal death in the rat hippocampus following transient forebrain ischemia. Acta Neuropathol 64:139–147
12. Kranenburg O, van der Eb AJ, Zantema A (1996) Cyclin D1 is an essential mediator of apoptot-ic neuronal cell death. EMBO J 15:46–54
13. Li Y, Chopp M, Powers C, Jiang N (1997) Immunoreactivity of cyclin D1/cdk4 in neurons and oligodendrocytes after cerebral ischemia in rat. J Cereb Blood Flow Metab 17:846–856
14. Morris GF, Mathews MB (1989) Regulation of proliferating cell nuclear antigen during the cell cycle. J Biol Chem 264:13856–13864
15. Osuga H, Osuga S, Wang F, Fetni R, Hogan MJ, Slack RS, Hakim AM, Ikeda JE, Park DS (2000) Cyclin-dependent kinases as a therapeutic target for stroke. Proc Natl Acad Sci USA 97:10254–10259
16. Petito CK, Morgello S, Felix JC, Lesser ML (1990) The two patterns of reactive astrocytosis in postischemic rat brain. J Cereb Blood Flow Metab 10:850–859
17. Pulsinelli WA, Brierley JB, Plum F (1982) Temporal profile of neuronal damage in a model of transient forebrain ischemia. Ann Neurol 11:491–498
18. Sherr CJ (1994) G1 phase progression: cycling on cue. Cell 79:551–555
19. Small DL, Monette R, Fournier M-C, Zurakowski B, Fiander H, Morley P (2001) Characteriza-tion of cyclin D1 expression in a rat model of cerebral ischemia. Brain Res 900:26–37
20. Smith ML, Bendek G, Dahlgren N, Rosen I, Wieloch T, Siesjö BK (1984) Models for studying long-term recovery following forebrain ischemia in the rat. 2. A 2-vessel occlusion model. Acta Neurol Scand 69:385–401
21. Tanaka S, Suzuki K, Watanabe M, Matsuda A, Tone S, Koike T (1998) Upregulation of a new microglial gene, mrf-1, in response to programmed neuronal cell death and degeneration. J Neurosci 18:6358–6369
22. Timsit S, Rivera S, Ouaghi P, Guischard F, Tremblay E, Ben-Ari Y, Khrestchatisky M (1999) In-creased cyclin D1 in vulnerable neurons in the hippocampus after ischaemia and epilepsy: a modulator of in vivo programmed cell death? Eur J Neurosci 11:263–278
23. Tomasevic G, Kamme F, Wieloch T (1998) Changes in proliferating cell nuclear antigen, a pro-tein involved in DNA repair, in vulnerable hippocampal neurons following global cerebral isch-emia. Mol Brain Res 60:168–176
24. Wiessner C, Brink I, Lorenz P, Neumann-Haefelin T, Vogel P, Yamashita K (1996) Cyclin D1 messenger RNA is induced in microglia rather than neurons following transient forebrain isch-aemia. Neuroscience 72:947–958

A Role for Cerebrovascular Endothelium
in Ischemia and Reperfusion

M. Spatz, Y. Chen, S. Golech, A. Strasser, J. Bembry, F. A. Lenz,
R. Mechoulam, and R. M. McCarron

Key words. 2-Arachidonoyl glycerol – brain endothelium – cerebral blood flow – endothelin-1 – hydrogen peroxide (H_2O_2) – ischemia – reperfusion

Introduction

The unique features of brain capillary and microvascular endothelial cells embrace specialized transport and carrier systems as well as enzymes associated with various metabolic pathways that are essential to sustain the dynamic homeostasis of the brain [2, 39]. Many of these processes are thought to be neuronally regulated [47]. It is now widely accepted that cerebral vascular endothelial cells not only constitute a permeability barrier to ions and organic molecules (e.g., water, electrolytes, proteins, neurotransmitters), but they are also an important secretory organ. BCEC produce agents and factors that may be involved in autocrine and paracrine regulation of the microvascular function of the brain [30, 41, 44, 50]. In general, the substances (e.g., prostacyclin, nitric oxide (NO), and adenosine) produced by these cells are considered to be cytoprotective. However, several other agents which are formed in endothelial cells (i.e., endothelin-1 (ET-1), angiotensin II, thromboxane, leukotrienes, platelet-activating factor and superoxide radicals) impair perfusion, alter BBB permeability and/or mediate cellular injury when released in excess. Many of these substances which modulate the secretory functions and reactivities of the vascular endothelium can also be released from adjacent cellular elements including other vascular cells, circulating blood cells and brain cells.

In the past few years, we have focused on in-vivo and in-vitro studies of ET-1, NO, and free radical species and their roles in cerebral ischemia [6, 44–47]. Endothelin-1 and NO, the most potent vasoconstrictor and vasodilator respectively, are the main contributors responsible for controlling the vascular tone and blood flow [8, 14, 30, 41]. They have been identified as major players in vascular diseases involving

Maria Spatz[1], Ye Chen[1], Suzanne Golech[2], Alois Strasser[1,2], Joilet Bembry[1], Frederick A. Lenz[3], Raphael Mechoulam[4], Richard M. McCarron[2]
[1] Stroke Branch, National Institutes of Neurological Disorders and Stroke (NINDS), National Institutes of Health (NIH), Bethesda, MD, USA
[2] Resuscitative Medicine Department, Naval Medical Research Center, Bethesda, MD, USA
[3] Johns Hopkins University School of Medicine, Baltimore, MD, USA
[4] Hebrew University, Jerusalem, Israel
Correspondence: Maria Spatz, M.D., National Institutes of Health, NINDS, Stroke Branch, 36 Convent Drive, MSC 4128, Bethesda, Maryland 20892-4128, USA, Tel.: (301) 496-8112, Fax: (301) 402-2769, E-Mail: spatzm@ninds.nih.gov

Maturation Phenomenon in Cerebral Ischemia V
A. M. Buchan et al. (Eds.)
© Springer-Verlag Berlin Heidelberg 2004

not only the brain, but also the heart, lung and kidney. In the brain, ischemic release of ET-1 and the relative deficiency of NO are associated with reduced cerebral blood flow (CBF) and ischemic injury that can be ameliorated by intravenous treatment with ET-1 receptor (ETA) antagonists. In human brain endothelial cells (HBEC), the interplay between ET-1 and NO involves binding and activation of ETA receptors, Ca^{2+} mobilization, cytoskeletal rearrangements and vasodilator-stimulated phosphoprotein (VASP) changes which are all linked to the c-GMP/c-GMP kinase system. This communication will briefly summarize recent experimental data identifying a new NO-independent pathway involving a novel endocannabinoid that abrogates many of these ET-1-induced endothelial changes. This substance has been demonstrated to have both vasoactive and cytoprotective properties. Lately we also investigated the possibility of additional factors such as reactive oxygen species (ROS) that are formed and released during ischemia/reperfusion and could influence the postischemic CBF [3, 17, 18, 24, 26]. This report describes a possible endothelial mechanism responsible for the observed amelioration of postischemic hypoperfusion by nitroxide (TEMPO), a free radical scavenger.

Materials and Methods

In-Vivo Studies

The model for the induction of global ischemia was the same as previously reported [4]. Separate groups of 3 month old female Mongolian gerbil (4 to 8 animals in each) were used for these studies. Cerebral ischemia with reperfusion was induced by bilateral carotid artery occlusion (15 min) with release and reperfusion (60 min) in spontaneously ventilated gerbils under halothane (1.5%) and N_2O/O_2 (2:1) or pentobarbital (20 mg/kg) anesthesia; respective sham-operated animals served as controls. Systemic blood pressure (SBP) was measured with blood pressure transducer (Universal Harvard Oscillograph), head (temporal muscle) and rectal temperatures were continuously monitored and rectal temperature was maintained at 37–38 °C with a thermostatically regulated heating lamp throughout the entire experimental procedure. Cerebral blood flow was also continuously monitored by transcranial laser Doppler flowmetry (Laserflo Model BPM 403A, TSl; St. Paul. MN). The treatment consisted of intraperitoneal (i.p.) administration of 50 mg/kg 4-hydroxy-2,2,6,6-tetramethylpiperidine-N-oxyl (TEMPO) immediately before inducing ischemia, or intravenous (i.v.) administration of 25 mg/kg TEMPO 5 min after release of occlusion; solvent alone was used in sham animals. All animal procedures were conducted in strict accordance with the NIH Guide for the Care and Use of Laboratory Animals.

In Vitro Studies

These experiments were conducted to examine: 1) a possible interaction between 2-AG (a potential vasodilator) and the potent vasoconstrictor ET-1, and 2) the effect of TEMPO on the endothelial Ca^{2+} content and cytoskeleton changes (F-actin, vimentin and/or VASP). In experiments presented here, HBEC were incubated with indicated concentrations of various agents at specific times as described in the Results section.

Cell Culture

Isolated brain microvessels and capillaries were used for the dissociation and cultivation of endothelial cell as previously described [13]. The endothelial cell lines (derived from at least six different brains; passages 7–15) were >95% Factor VIII-positive. Cells were grown on gelatin (1%)-coated 24-well Linbro plates at 37 °C in humidified 5% CO_2/air for 48 h [45].

$[Ca^{2+}]i$ Measurements and Inositol Phosphates Analyses

After washing (3×) in solution containing 137 mM NaCl, 5 mM KCl, 1 mM $MgCl_2$, 25 mM sorbitol, 10 mM HEPES, and 3 mM $CaCl_2$ (pH 7.0), HBEC were incubated with 2.5 µM fluorescent probe Fluo-3/AM for 90 min at 37 °C. Fluorescence was measured using a fluorescein filter pair (excitation, 485 ± 20 nm; emission, 530 ± 20 nm) as previously described [14]. Changes in $[Ca^{2+}]i$ are expressed as fluorescence intensity ratio determined as: [Experimental fluorescence value (F_e) – basal fluorescence value (F_0)]/F_0 × 100%; the fluorescence change in presence of ET-1 alone was considered as 100% [6].

Inositol-4,5-biphosphate (IP_2) and inositol-1,4,5-triphosphate (IP_3) formation were determined by a modified technique as previously described [48].

Immunocytochemistry

The expressions of actin, vimentin, and VASP were determined by immunocyto-chemical analyses of HBEC (grown on coverslips) that were fixed (4% paraformaldehyde), permeabilized (0.1% Triton X-100) and stained with phalloidin, or monoclonal antibodies to vimentin or VASP, respectively, as previously described [6, 7].

Western Blotting

HBEC grown on 35 mm Petri dishes were exposed to tested agents as indicated. Cells were then harvested and lysed; proteins were electrophoresed and transferred onto PVDF membranes, which were probed overnight with VASP antibody as previously described [7].

Statistics

Results were analyzed by Student's t test or one-way ANOVA followed by Fisher's protective least squares difference (PLSD) test.

Results

Characterization of 2-AG Effect on Unstimulated and ET-1-Stimulated Intracellular Ca^{2+}

The intracellular Ca^{2+} content of HBEC was elevated by the addition of 2-AG ($EC_{50} = 50\ \mu M$). The level of Ca^{2+} gradually rose starting at 30 s (4%) and continuing for up to 30 min (19%). This effect was not observed in Ca^{2+}-free medium indicating that the 2-AG-induced increases in the uptake of Ca^{2+} likely utilized extracellular sources. The selective CB1-receptor antagonist, SR141716A (1 μM) inhibited (35%) the 2-AG-stimulated Ca^{2+} uptake; no effect was observed by treatment with the CB1-receptor antagonist alone. The HBEC Ca^{2+} levels induced by 20 nM ET-1 were the same in the presence or absence of BAPTA-AM or in Ca^{2+}-free medium, which is indicative of Ca^{2+} mobilization as opposed to Ca^{2+} uptake. As

Table 1. Summary of effects of ET-1 and 2-AG on Ca^{2+} mobilization in HBEC

Treatment		Intracellular Ca^{2+} mobilization (% Inhibition)
20 nM ET-1	None	0
	+ 0.1 μM 2-AG	19.0±0.7
	+ 1 μM 2-AG	37.0±2.6
	+ 25 μM 2-AG	39.0±1.9
	+ 50 μM 2-AG	60.8±6.9
20 nM ET-1 + 50 μM 2-AG	None	60.8±6.9
	+ 10 μM L-NAME	60.6±10.0
	+ 100 μM L-NAME	62.0±9.6
	+ 10 μM Indomethacin	61.4±12.2
	+ 0.1 μM SR141716A	31.9±1.5
	+ 1 M SR141716A	15.0±1.4
	+ 200 nM BIS	18.1±1.5
	+ 200 nM Quinine	39.5±2.8
	+ 100 nM Apamine	47.2±3.5
	+ 100 nM Charybdotoxin	56.3±6.1
	+ 15 nM KCl	34.3±3.0
	+ 50 nM KCl	34.8±2.5

The data indicate responses of human brain microvascular endothelial cell (HBMEC) and capillary endothelial cells (HBEC) cultures. Confluent cultures were treated with 20 nM ET-1 alone or in the presence of indicated concentrations of the following compounds: the putative vasorelaxant 2-arachidonoylglycerol (2-AG); nitric oxide synthase inhibitor, L-nitro-arginine methylester (L-NAME); cyclooxygenase inhibitor, indomethacin; a CB1-receptor antagonist (SR141746A); a PKC inhibitor, bisindolylmaleimide (BIS); inhibitor of Ca^{2+}-dependent K^+ channels (Quinine), a selective blocker of small conductance Ca^{2+}-activate K^+ channels (apamine), or a blocker of several K^+-channel subtypes (charybdotoxin); or potassium chloride (KCl) as previously described[Circ Res]. % Inhibition was calculated by the formula: $([Ca^{2+}]_i$ changes (ET-1) $- [Ca^{2+}]_i$ changes (ET-1 plus additions))/ $([Ca^{2+}]_i$ changes (ET-1)$\times 100$.

shown in Table 1, 2-AG reduced the ET-1-stimulated Ca^{2+} mobilization in a dose-dependent manner. This inhibiton was not affected by the inhibition of nitric oxide synthase (NOS), cyclooxygenase, or lipoxygenase by N^G-nitro-L-arginine methylester (L-NAME), indomethacin, or nordihydroguaiaretic acid. However, SR141716A dose-dependently prevented the 2-AG-induced decrease of ET-1-induced Ca^{2+} mobilization. In addition to affecting the ET-1-induced response, 2-AG also had an effect (30% reduction) on Ca^{2+} mobilization induced by Mas7 (an analog of masterpan, a peptide from wasp venom and known to activate G protein). Also, 2-AG inhibited the formation of IP_2 (70%) and IP_3 (51%). All of these effects implicate the involvement of G-protein and certain second messenger in 2-AG modulation of ET-1-stimulated Ca^{2+} mobilization.

Substances implicated in hyperpolarization of cell membranes (selective inhibitors of K^+ channels or high $[K^+]$) were used to further elucidate possible pathways responsible for the 2-AG effect on ET-1-induced Ca^{2+} mobilization. Quinine, apamin or charybdotoxin (selective inhibitors of Ca^{2+}-dependent K^+ channels, small-conductance Ca^{2+}-activated channels, and blockers of several K^+ channel subtypes, respectively) partially prevented the 2-AG-induced decrease of ET-1-stimulated Ca^{2+} mobilization. Quabain (inhibitor of Na^+K^+-ATPase) and BaCl (inhibitor of inwardly rectifying K^+ channels) were ineffective. High doses of K^+ (15 or 50 mM) partly prevented the 2-AG effect on ET-1-induced Ca^{2+} mobilization; a similar effect was observed with bisindolylmaleimide (BIS) but not with H_8 or ODQ, indicating the involvement of protein C kinase but not cAMP kinase or the activation of guanyl cyclase.

Cytoskeleton and VASP Response to 2-AG and ET-1

The endothelial cytoskeletal F-actin and vimentin filaments were affected by treatment with 2-AG and/or ET-1. 2-AG reduced while ET-1 increased the thickness of F-actin and vimentin fibers. The 2-AG-induced rarification of these fibers is visible in untreated and ET-1-treated HBEC (Fig. 1). VASP was easily visible at the terminal segments of the F-actin filaments. Similar results were obtained with vimentin (not shown).

Due to the lack of commercially available antibodies specific for the phosphorylated form of VASP, it was not possible to morphologically differentiate between the nonphosphorylated and phosphorylated VASP. Nonetheless, 2-AG-induced phosphorylation was clearly demonstrated by Western blot analysis (Fig. 2). This effect was blocked by cAMP inhibition (i.e., treatment with H_8) but not by inhibitors of cGMP or guanylyl cyclase (i.e., treatment with H_7 or ODQ, respectively) (results now shown).

The Effect of Nitroxide: Studies *In Vivo* and *In Vitro*

To test ROS involvement in postischemic hypoperfusion, the gerbils subjected to global brain ischemia and reperfusion were treated with TEMPO. The data (Fig. 3) clearly indicates that administration of either 50 mg/kg TEMPO (i.p.) prior to the

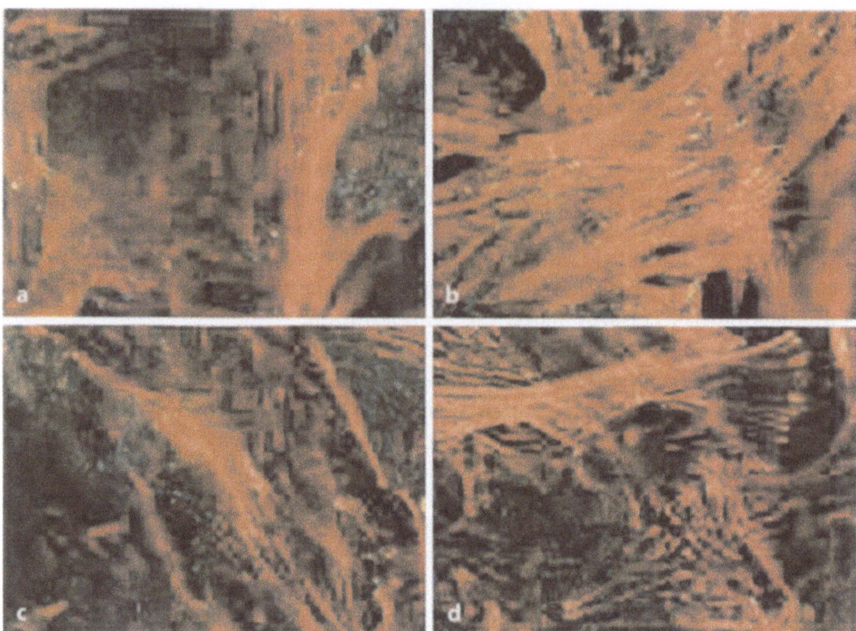

Fig. 1. Cytoskeleton F-actin and VASP in HBEC. Cultured human brain endothelial cells grown on glass cover slips (and processed as described in Materials and Methods section) were treated with medium alone (**a**), 20 µM ET-1 (**b**), 50 µM 2-AG (**c**), or 50 µM 2-AG and 20 µM ET-1 (**d**). Note the 2-AG-induced reduction in the thickness of F-actin filaments in **c** and **d**, as compared to **a** and **b**, respectively.

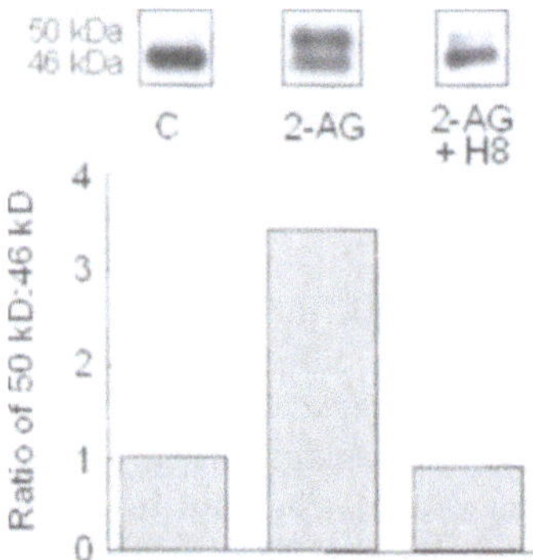

Fig. 2. Western blot of 2-AG effects on phosphorylation of VASP. HBEC samples were treated with either 2-AG (50 µM) alone or pretreated with H8 (10 µM) and processed as described in Materials and Methods section. The phosphorylated VASP is indicated by the shift from 46 to 50 kDa. The data shown are from a representative experiment; bar graph indicates the relative levels of VASP.

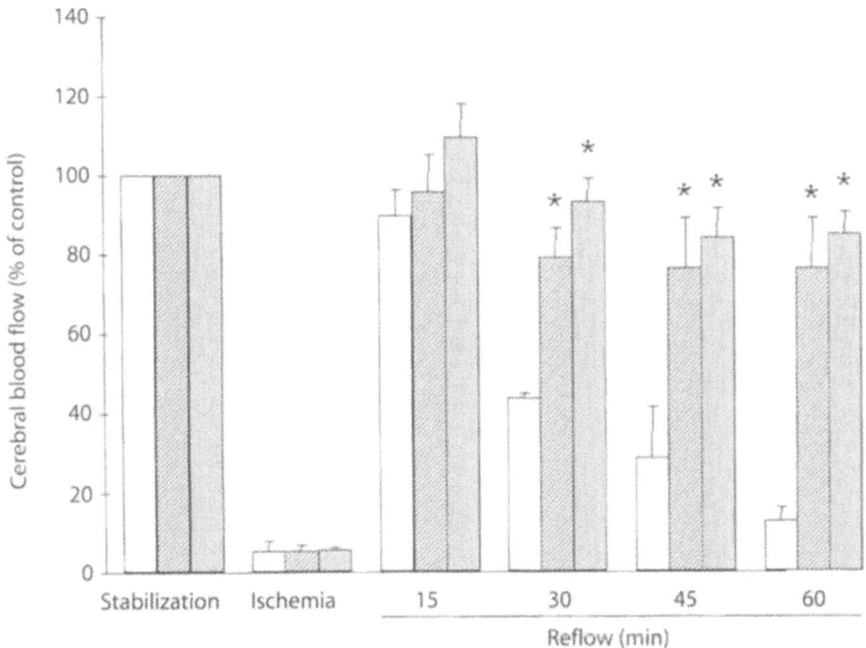

Fig. 3. Effect of TEMPO on changes in cerebral blood flow (CBF) induced by global brain ischemia. Gerbils subjected to 15 min bilateral carotid artery occlusion (15 min) and reflow (60 min) received either saline (□); TEMPO (50 mg/kg) injected intraperitoneal immediately prior to occlusion (▨); or administrated by intravenous injection (5 min) after the release of occlusion (■). The data are presented as mean ± SEM of 4–8 animals in each group (*, p < 0.05); although CBF values were recorded every min, only selective representative data are shown.

release or 25 mg/kg TEMPO (i.v.) after the release of bilateral carotid artery occlusion prevented the postischemic hypoperfusion. To elucidate the possible participation of endothelium in this event, experiments were designed to examine the effect of H_2O_2 (ROS) on Ca^{2+} content and cytoskeleton in HBEC. H_2O_2 stimulated Ca^{2+} mobilization in dose-dependent manner; the level of Ca^{2+} mobilization induced by 10 mM H_2O_2 equaled that of 20 nM ET-1 (results not shown). ET-1 treatment had no additive effect on H_2O_2-stimulated Ca^{2+} mobilization (results not shown). TEMPO dose-dependently reduced the H_2O_2-induced Ca^{2+} mobilization in HBEC (Fig. 4). H_2O_2 also induced cytoskeletal rearrangements in HBEC; this was manifested by increased thickness of vimentin filaments and no apparent changes in F-actin filaments (results not shown). Preliminary studies indicate that TEMPO pretreatment of HBEC modified the H_2O_2-induced alterations of vimentin.

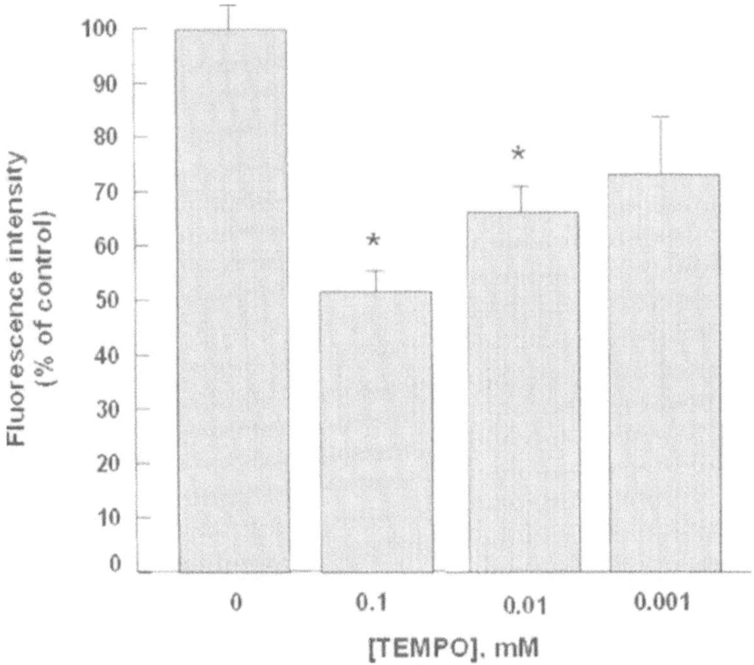

Fig. 4. Effect of TEMPO on H_2O_2-induced Ca^{2+} mobilization. Fluorescence intensity of HBEC exposed to 10 mM H_2O_2 alone are expressed as 100%. TEMPO dose-dependently reduced the H_2O_2-induced Ca^{2+} mobilization (*, $p < 0.05$).

Discussion

The findings presented here clearly indicate that 2-AG interacts with ET-1 and is able to abolish the capacity of ET-1 to induce Ca^{2+} mobilization and cytoskeleton (F-actin and vimentin) rearrangements in HBEC. The interplay between 2-AG and ET-1 is mediated by CB1-receptors, G-proteins, phosphoinositol signal transduction pathway, Ca^{2+}-activated K^+ channels and KCl. These results indicate that the interaction between 2-AG and ET-1 is similar to that reported for ET-1 and NO, but involve different mechanisms.

2-AG one of two endogenously occurring cannabinoids (anandamide is the other) which has neuromodulary and cardiovascular effects has been implicated in vasodilation [10, 23, 27, 28, 38]. It is a derivate of arachidonic acid and can be enzymatically hydrolyzed back to arachidonic acid. 2-AG structurally resembles eicosanoids but they differ in their biosynthetic pathways [35]. 2-AG is produced in gut, brain, blood and vascular cells [10, 23, 27, 28]; it (as well as anandamide) activates CB1- and CB2-receptors. CB1 receptors have been demonstrated in the brain and peripheral organs whereas CB2 receptors were identified in the periphery, particularly in immune cells [10, 23, 28, 49]. It has been suggested that activation of cholinergic receptors mediate 2-AG in parallel to NO and in this way contribute to endothelium-dependent relaxation [27].

Although 2-AG (or anadamide) induce vasodilation, their effectiveness as endothelium-derived relaxing (EDRF) and hyperpolarizing factor (EDHF) agents as well as their mechanism of action remain controversial [5, 10, 38]. Reports suggest that the vasodilatory effect of these substances is not a direct one but it is mediated by other factors (i.e., NO, cyclooxygenase or lipoxygenase) depending on the particular species as well as the relative vessel size. There is currently no information regarding their putative interactions with vasoconstrictors and their ultimate contribution to controlling vascular homeostasis. The results presented here are to best of our knowledge the first demonstration of interaction between 2-AG and a vasoconstrictor (i.e., ET-1). This interaction suggests an alternative pathway in regulating vasoactivity which may especially be operative when NO-mediated cGMP-dependent pathway of relaxation is impaired (i.e., trauma, ischemia, hypertension, diabetes mellitus, hypercholesterolemia) [8, 30, 34].

The cannabinoids were also shown to be cytoprotective in experimental models of brain trauma and ischemia [19, 27, 28, 31, 33]. Previous in-vitro studies indicate that 2-AG impedes the lipopolysaccharide (LPS)-induced formation of ROS in murine macrophages [13]; it also reduces the serum levels of $TNF\alpha$ in LPS-treated mice [13]. 2-AG has been implicated in inhibiting glutamate release [16, 42]. All of these factors affected by 2-AG may play a role in the mechanism of cytoprotection induced by 2-AG. The recent findings discussed here suggest that 2-AG counteraction of ET-1-induced effects on endothelium may be another possible mechanism involved in the observed cytoprotection of 2-AG.

The findings presented in this study also indicate that TEMPO is beneficial in preventing the postischemic hypoperfusion induced by global ischemia. A similar effect on the CBF which was also associated with a reduction in infarct volume after focal ischemia was seen with nitroxide [36, 37]. Previous studies demonstrated that the nitroxide treatment ameliorated the increased BBB permeability and edema formation seen in brain ischemia and trauma [1, 51]. The nitroxide has properties of a stable radical that can react with various ROS. It has a low toxicity and high membrane permeability (including BBB permeability); it also mimics superoxide activity, thereby protecting cells from oxidative stress. As shown here, nitroxide also modifies the H_2O_2-induced Ca^{2+} mobilization and cytoskeleton (vimentin) arrangement in HBEC. This effect is similar to that observed with ET-1 alone [6, 7]. The possible involvement of ET-1 in these effects is supported by finding that exogenous and endogenous H_2O_2 induce the production of ET-1 in human mesanglial cells [20]. H_2O_2 is one of several ROS oxidants (others include hypochlorite (HOCL), hydroxyl, and superoxide (O_2^-), which is generated during ischemia reperfusion injury [17, 18]. These substances can induce cellular damage in a variety of cell types and tissues. In aortic EC, for example, H_2O_2 impairs NO production through partial inactivation of eNOS cofactors [21]. If such effects of H_2O_2 (i.e., reduced eNOS activity and increased ET-1 production) occur in HBEC, this may contribute to disequilibrium between vasodilators and vasoconstrictors resulting in postischemic hypoperfusion.

Even though the precise mechanisms responsible for ET-1-, 2-AG- or H_2O_2-mediated effects on the endothelium remain unknown, it is clear that cytoskeletal filaments are altered by these substances. These results strongly suggest that cytoskeletal structures play an important physiological role in the function of endothelium. Previous reports demonstrating the prevention of ischemic injury by the modulation of actin filaments in the endothelial cytoskeleton supports this contention [25].

At this time, it should be pointed out that the present study examined the ET-1-, 2-AG- or H_2O_2-mediated effects on both endothelial Ca^{2+} levels and cytoskeleton. These two parameters represent key factors involved in the activation or inhibition of cellular contractile elements. The involvement of Ca^{2+} in vascular responses to vasoactive agents was principally investigated in muscle cells or endothelium derived from large vessels. The NO-induced modulation of ET-1-stimulated Ca^{2+} mobilization (including cGMP regulation of intracellular Ca^{2+} content) was previously reported in smooth muscle, platelets, leukocytes and CHO cells transfected with ETA receptor cDNA [9, 12, 15, 32]. Reports also indicate that some components of the signal transduction pathway (i.e., cGMP or cAMP) are linked to cytoskeletal changes and involve vasodilator-stimulated phosphoprotein (VASP) and cytoskeletal F-actin [40]. The assembly and disassembly of actin are regulated by Ca^{2+} and phosphoinositides, which are themselves modified by upstream signals (i.e., small guanosine nucleotide triphosphatases (GTPases), protein kinases, and ion channels) [22]. In addition, the observed alignment of VASP with actin filaments was suggested to play a role in hindering or advancing the assembly of actin [40].

The results of this study demonstrate that small vessel endothelium (i.e., HBEC) also possess the intrinsic capacity to respond to vasoactive mediators. Such endothelial reactivity is important not only for the interactions with smooth muscle, but also for maintaining the endothelial cellular structure and function (e.g., membrane polarity, tight junctions, cellular adhesions and signal transduction). Thus, the described responses of endothelium mediated by various signal transduction pathways emphasize their dominant role in controlling and regulating functional integrity. It is this role that is compromised in vascular diseases of the brain (i.e., ischemia and trauma) [25, 29, 47]. An awareness of existing multiple pathways for interaction between the vasoconstrictors at the site of the vascular endothelium and vasodilators may provide insight to mechanisms for regulating vascular tone, blood flow and barrier permeability in addition to having various therapeutic implications.

References

1. Beit-Yannai E, Zhang R, Trembovler V, Samuni A, Shohami E (1996) Cerebroprotective effect of stable nitroxide radicals in closed head injury in the rat. Brain Res 717:22–28
2. Betz AL, Goldstein GW (1986) Specialized properties and solute transport in brain capillaries. Annu Rev Physiol 48:241–250
3. Chan PH (2001) Reactive oxygen radicals in signaling and damage in the ischemic brain. J Cereb Blood Flow Metab 21:2–14
4. Chang CJ, Ishii H, Yamamoto H, Yamamoto T, Spatz M (1993) Effects of cerebral ischemia on regional dopamine release and D1 and D2 receptors. J Neurochem 60:1483–1490
5. Chataigneau T, Feletou M, Thollon C, Villeneuve N, Vilaine JP, Duhault J, Vanhoutte PM (1998) Cannabinoid CB1 receptor and endothelium-dependent hyperpolarization in guinea-pig carotid, rat mesenteric and porcine coronary arteries. Br J Pharmacol 123:968–974
6. Chen Y, McCarron RM, Bembry J, Ruetzler C, Azzam N, Lenz FA, Spatz M (1999) Nitric oxide modulates endothelin 1-induced Ca^{2+} mobilization and cytoskeletal F-actin filaments in human cerebromicrovascular endothelial cells. J Cereb Blood Flow Metab 19:133–138

7. Chen Y, McCarron RM, Ohara Y, Bembry J, Azzam N, Lenz FA, Shohami E, Mechoulam R, Spatz M (2000) Human brain capillary endothelium: 2-arachidonoglycerol (endocannabinoid) interacts with endothelin-1. Circ Res 87:323–327

8. Cohen RA, Vanhoutte PM (1995) Endothelium-dependent hyperpolarization. Beyond nitric oxide and cyclic GMP. Circulation 92:3337–3349

9. Cornwell TL, Lincoln TM (1989) Regulation of intracellular Ca2+ levels in cultured vascular smooth muscle cells: reduction of Ca2+ by atriopeptin and 8-bromo-cyclic GMP is mediated by cyclic GMP-dependent protein kinase. J Biol Chem 264:1145–1155

10. Di Marzo V (1998) 'Endocannabinoids' and other fatty acid derivatives with cannabimimetic properties: biochemistry and possible physiopathological relevance. Biochim Biophys Acta 1392:153–175

11. Faraci FM (1992) Regulation of the cerebral circulation by endothelium. Pharmacol Ther 56:1–22

12. Riesco A, Caramelo C, Blum G, Monton M, Gallego MJ, Casado S, Lopez Farre A (1993) Nitric oxide-generating system as an autocrine mechanism in human polymorphonuclear leucocytes. Biochem J 292:791–796

13. Gallily R, Breuer A, Mechoulam R (2000) 2-Arachidonylglycerol, an endogenous cannabinoid, inhibits tumor necrosis factor-alpha production in murine macrophages, and in mice. Eur J Pharmacol 6:R5–7

14. Gellai M, De Wolf R, Fletcher T, Nambi P (1997) Contribution of endogenous endothelin-1 to the maintenance of vascular tone: role of nitric oxide. Pharmacology 55:299–308

15. Goligorsky MS, Tsukahara H, Magazine H, Andersen TT, Malik AB, Bahou WF (1994) Termination of endothelin signaling: role of nitric oxide. J Cell Physiol 158:485–494

16. Hajos N, Ledent C, Freund TF (2001) Novel cannabinoid-sensitive receptor mediates inhibition of glutamatergic synaptic transmission in the hippocampus. Neuroscience 106:1–4

17. Hall ED, Braughler JM (1989) Central nervous system trauma and stroke. II. Physiological and pharmacological evidence for involvement of oxygen radicals and lipid peroxidation. Free Radic Biol Med 6:303–313

18. Halliwell B (1991) Reactive oxygen species in living systems: source, biochemistry, and role in human disease. Am J Med 91:14S–22S

19. Hampson AJ, Grimaldi M, Axelrod J, Wink D (1998) Cannabidiol and (–)Delta9-tetrahydrocannabinol are neuroprotective antioxidants. Proc Natl Acad Sci USA 95:8268–8273

20. Hughes AK, Stricklett PK, Padilla E, Kohan DE (1996) Effect of reactive oxygen species on endothelin-1 production by human mesangial cells. Kidney Int 49:181–189

21. Jaimes EA, Sweeney C, Raij L (2001) Effects of the reactive oxygen species hydrogen peroxide and hypochlorite on endothelial nitric oxide production. Hypertension 38:877–883

22. Janmey PA (1994) Phosphoinositides and calcium as regulators of cellular actin assembly and disassembly. Annu Rev Physiol 56:169–191

23. Jarai Z, Wagner JA, Goparaju SK, Wang L, Razdan RK, Sugiura T, Zimmer AM, Bonner TI, Zimmer A, Kunos G (2000) Cardiovascular effects of 2-arachidonoyl glycerol in anesthetized mice. Hypertension 35:679–684

24. Kumura E, Yoshimine T, Iwatsuki KI, Yamanaka K, Tanaka S, Hayakawa T, Shiga T, Kosaka H (1996) Generation of nitric oxide and superoxide during reperfusion after focal cerebral ischemia in rats. Am J Physiol 270:C748–775

25. Laufs U, Endres M, Stagliano N, Amin-Hanjani S, Chui DS, Yang SX, Simoncini T, Yamada M, Rabkin E, Allen PG, Huang PL, Bohm M, Schoen FJ, Moskowitz MA, Liao JK (2000) Neuroprotection mediated by changes in the endothelial actin cytoskeleton. J Clin Invest 106:15–24

26. McCord JM (1987) Oxygen-derived radicals: a link between reperfusion injury and inflammation. Fed Proc 46:2402–2406

27. Mechoulam R, Fride E, Ben-Shabat S, Meiri U, Horowitz M (1998) Carbachol, an acetylcholine receptor agonist, enhances production in rat aorta of 2-arachidonoyl glycerol, a hypotensive endocannabinoid. Eur J Pharmacol 362:R1–3

28. Mechoulam R, Spatz M, Shohami E (2002) Endocannabinoids and neuroprotection. Sci STKE 2002:RE5

29. Molitoris BA, Leiser J, Wagner MC (1997) Role of the actin cytoskeleton in ischemia-induced cell injury and repair. Pediatr Nephrol 11:761–767

30. Moncada S, Palmer RM, Higgs EA (1991) Nitric oxide: physiology, pathophysiology, and pharmacology. Pharmacol Rev 43:109–142

31. Nagayama T, Sinor AD, Simon RP, Chen J, Graham SH, Jin K, Greenberg DA (1999) Cannabinoids and neuroprotection in global and focal cerebral ischemia and in neuronal cultures. J Neurosci 19:2987–2995

32. Okishio M, Ohkawa S, Ichimori Y, Kondo K (1992) Interaction between endothelium-derived relaxing factors, S-nitrosothiols, and endothelin-1 on Ca2+ mobilization in rat vascular smooth muscle cells. Biochem Biophys Res Commun 183:849–855

33. Panikashvili D, Simeonidou C, Ben-Shabat S, Hanus L, Breuer A, Mechoulam R, Shohami E (2001) An endogenous cannabinoid (2-AG) is neuroprotective after brain injury. Nature 413:527–531

34. Panza JA, Quyyumi AA, Brush JE Jr, Epstein SE (1990) Abnormal endothelium-dependent vascular relaxation in patients with essential hypertension. N Engl J Med 323:22–27

35. Piomelli D, Beltramo M, Giuffrida A, Stella N (1998) Endogenous cannabinoid signaling. Neurobiol Dis 5:462–473

36. Pluta RM, Rak R, Wink DA, Woodward JJ, Khaldi A, Oldfield EH, Watson JC (2001) Effects of nitric oxide on reactive oxygen species production and infarction size after brain reperfusion injury. Neurosurgery 48:884–892; discussion 892–893

37. Rak R, Chao DL, Pluta RM, Mitchell JB, Oldfield EH, Watson JC (2000) Neuroprotection by the stable nitroxide Tempol during reperfusion in a rat model of transient focal ischemia. J Neurosurg 92:646–651

38. Randall MD, Kendall DA (1998) Endocannabinoids: a new class of vasoactive substances. Trends Pharmacol Sci 19:55–58

39. Rapport SI (1976) Blood-Brain Barrier in Physiology and Medicine. Raven Press, New York

40. Reinhard M, Halbrugge M, Scheer U, Wiegand C, Jockusch BM, Walter U (1992) The 46/50 kDa phosphoprotein VASP purified from human platelets is a novel protein associated with actin filaments and focal contacts. EMBO J 11:2063–2070

41. Rubanyi GM, Polokoff MA (1994) Endothelins: molecular biology, biochemistry, pharmacology, physiology, and pathophysiology. Pharmacol Rev 46:325–415

42. Schlicker E, Kathmann M (2001) Modulation of transmitter release via presynaptic cannabinoid receptors. Trends Pharmacol Sci 22:565–572

43. Schmidt A, Hall NM (1998) Signaling to the actin cytoskeletal. Annu Rev Cell Dev Biol 14:305–338

44. Spatz M, Bacic F, Merkel N, Bembry J, McCarron RM (1993) The regulatory functions of cerebral microvascular endothelium. In: CNS Barriers and Modern CSF Diagnostics of Klaus Felgenhauer, Manfred Holzgraefe, and Hilmar W. Prange. Weinheim, New York, Basel, Cambridge, Tokyo, pp 41–59

45. Spatz M, Kawai N, Merkel N, Bembry J, McCarron RM (1997) Functional properties of cultured endothelial cells derived from large microvessels of human brain. Am J Physiol 272:C231–C239

46. Spatz M, Stanimirovic D, Strasser A, McCarron RM (1995) Nitro-L-arginine augments the endothelin-1 content of cerebrospinal fluid induced by cerebral ischemia. Brain Res 684:99–102

47. Spatz M, Yasuma Y, Strasser A, McCarron RM (1996) Cerebral postischemic hypoperfusion is mediated by ETA receptors. Brain Res 726:242–246

48. Stanimirovic DB, Yamamoto T, Uematsu S, Spatz M (1994) Endothelin-1 receptor binding and cellular signal transduction in cultured human brain endothelial cells. J Neurochem 62:592–601

49. Sugiura T, Kodaka T, Nakane S, Kishimoto S, Kondo S, Waku K (1998) Detection of an endogenous cannabimimetic molecule, 2-arachidonoylglycerol, and cannabinoid CB1 receptor mRNA in human vascular cells: is 2-arachidonoylglycerol a possible vasomodulator? Biochem Biophys Res Commun 243:838–843

50. Vane JR, Botting RM (1994) Mediators from the endothelial cell and their participation in inflammation. Int J Tissue React 16:19–49

51. Yasuma Y, Strasser A, Ruetzel C, McCarron RM, Spatz M (1997) The effect of nitric oxide inhibition on ischemic brain edema. Acta Neurochir Suppl 70:202–205

Neural Stem Cells:
Developmental Lessons May Yield Therapeutic Insights

Evan Y. Snyder

An intriguing phenomenon with possible therapeutic dividends has begun to emerge from our observations of the behavior of neural stem cell (NSC) clones in various mouse models of CNS injury and degeneration.

During phases of active neurodegeneration, factors seem to be transiently elaborated to which NSCs may respond by migrating (even long distances) to degenerating regions and differentiating specifically towards replacement of dying neural cells. NSCs may "attempt" to emulate in the brain what hematopoietic stem cells do in the periphery: repopulate and reconstitute ablated regions. These "repair mechanism" may reflect the re-expression of basic developmental principles (particularly during particular temporal "windows" following injury) that may be harnessed for therapeutic ends. In addition, NSCs in their native state (or genetically-engineered) may serve as vehicles for protein delivery and appear capable of simultaneous cell replacement and gene therapy (e.g., with factors that might enhance differentiation, neurite outgrowth, connectivity, neuroprotection).

When combined with certain synthetic biomaterials, NSCs may be even more effective in "engineering" the damaged CNS towards reconstitution.

Harvard Medical School, Please note new lab contact information and e-mail, Harvard Institutes of Medicine, Neurology, Beth Israel-Deaconess Medical Center, 77 Avenue Louis Pasteur, Room 855, Boston, MA 02115, Phone: 617-667-0562, Fax: 617-667-0811, Lab Phone: 617-667-0868, Page: (still via Children's Hospital, Boston), 617-355-6369, beeper #1244, E-Mail: esnyder1@caregroup.harvard.edu
Part-Time Assistant: Paula Cohen, Phone: 617-667-2940, Fax: 617-667-7175, E-Mail: pcohen@caregroup.harvard.edu

Maturation Phenomenon in Cerebral Ischemia V
A.M. Buchan et al. (Eds.)
© Springer-Verlag Berlin Heidelberg 2004

III Clinical Trials/Neuroprotection

Experimental Neuroprotection:
Translation to Human Stroke Trials

P. A. Barber, B. Bruederlin, and A. M. Buchan

Summary. For a drug to be effective *in vivo*, *in-vitro* protection must be translated to studies of both global (transient forebrain ischemia) to discern "cellular" neuroprotection (cytoprotection) and focal ischemia to evaluate "parenchymal" neuroprotection (infarct reduction). Drug concentrations that suggest that the mechanism is active *in vitro* must translate to drug levels that achieve cytoprotection and infarct volume reductions which are not only significant but sustained (indefatigable). Initiation of experimental therapy should be delayed in order to have clinical relevance. Most importantly, the "effective dose" plasma level reached in rodents to achieve a neuroprotective effect must be matched in humans, in the absence of overt toxicity.

To date there are a number of shortcomings with "cytoprotective drugs" – in rats, in particular, their inability to protect white matter and striatum. The drugs have to be given early and the cytoprotective dose in rats has yet to be achieved in humans.

This review covers mechanisms of cell death following both focal and global ischemia such as excitotoxicity, acid and Ca^{++} overload, free radical-mediated injury, inflammation and apoptosis.

Drugs which block NMDA and AMPA receptors, as well as downstream events, have now made it to the clinic and they will be critically reviewed. Hypothermia following both global and focal ischemia is protective and at least for global ischemia has now been translated two successful clinical trials in patients resuscitated from cardiac arrest. For the first time neuroprotection has been demonstrated in patients who achieved better neurological outcomes. The review will finish with some rules for the translation of experimental work to clinical trials.

Key words. Neuroprotection – cytoprotection – t-PA – hypothermia – excitotoxicity – apoptosis – clinical stroke trials

Philip A. Barber, Barbara Bruederlin, Alastair M. Buchan

Corresponding author: Dr. Alastair Buchan, Professor of Neurology, Department of Clinical Neurosciences, University of Calgary, Rm 1162, 1403–29 Street N.W., Calgary, Alberta, Canada T2N 2T9, Phone: [403] 944-1581, Fax: [403] 944-1602, E-Mail: abuchan@ucalgary.ca

Maturation Phenomenon in Cerebral Ischemia V
A. M. Buchan et al. (Eds.)
© Springer-Verlag Berlin Heidelberg 2004

Introduction

The best neuroprotectant will always be the most expedient return of oxygen and glucose. Attempts to restore circulation within the first three hours (with a relatively intact CT brain scan) are highly effective, confirming the concept that reperfusion will salvage brain tissue. Thrombolysis induced by the administration of exogenous t-PA will reduce disability for stroke patients presenting within 3 h of a significant hemispheric ischemic stroke [87]. Out of every thousand stroke patients treated with IV t-PA, an additional 160 will recover to an independent status within three months. For patients treated between three and six hours there may be a signal of efficacy but neither the clinical trials [51] nor the subsequent meta-analysis has yet to give us a robust answer [125]. The fact that patients benefit within three hours, but not between three and six, is not because of an excess of hemorrhages in the three to six hour group [125], but because of a lack of cell recovery following brain resuscitation with delayed reperfusion. At six hours, reperfusion is too late to restore susceptible brain cells. A successful adjuvant neuroprotectant must either increase the time window to allow more patients to be effectively reperfused, perhaps up to six hours, or reduce the injury sustained during ischemia. In the face of "reperfusion injury" neuroprotection must sustain those cells which are put at risk in the hours and days that follow successful reperfusion.

Traditionally evidence for neuroprotection is sought first in *in-vitro* models of neuronal cell culture exposed to either anoxia/aglycemia or glutamate insults. Promising candidates are then developed *in vivo* in rodent and murine models of both global and focal ischemia. Currently the evidence that neuroprotection works, lending support for clinical trials, is based on the *in-vivo* models of focal stroke. This paper will review "cytoprotection" from models of global ischemia and "infarct reduction", will review mechanisms of injury, and will illustrate the successful translation of hypothermia in global ischemia to positive clinical efficacy trials. In 2002, for the first time neuroprotection in animal models has been translated to man [7, 57]. Hypothermia used late in the postischemic phase achieved not only cytoprotection but has also reduced focal infarct volume. Successful clinical trials of hypothermia in focal stroke are now keenly anticipated.

Problems with Drug Testing for Neuroprotection

Drug testing in animal models has suffered from a lack of physiological control resulting in spurious outcomes [11]. In the ensuing clinical trials there have also been a series of unexpected physiological perturbations which have resulted in deleterious side effects either leading to excess toxicity or causing futility in terms of improving outcome.

A second major problem has been the difficulty of achieving drug levels in humans comparable to those which achieved neuroprotection in animal models. A third difficulty is that, in the animal models, there is a slow maturation of injury during the first few days. As a result, observed "neuroprotection" in reality is only a mirage, i.e. all there is, is a postponement of injury [25]. In humans, where three month clinical outcomes are the gold standard, early neuroprotection will likely have evaporated in tandem with the slow maturation of injury.

Global Models: Reduction in Selective Neuronal Death (Cytoprotection)

Two rodent models of global, or more accurately, severe forebrain ischemia are routinely used: the 4-VO transient severe forebrain ischemia model [96, 97] and the 2-VO plus hypotension model [112]. Both have been used to examine selective CA_1 death. Murine models of global ischemia have been difficult, if not impossible, to achieve.

These models, properly used, create a transient oligaemia (no blood flow) to the hippocampus, cortex and striatum during ischemia followed, after 5, 10, 15 or even 30 minutes, by a prompt recovery of both blood flow and energy levels (ATP) [98]. The forebrain ischemic insult although brief is very severe, importantly, drugs given following reperfusion cannot influence the degree of blood flow reduction during the ischemic insult. In this model, hypothermia even 6 to 24 h after reperfusion allows for long-term recovery of cells [26]. There is a prolonged interval of depressed energy in the hippocampal neurons prior to cell death [98]. Given the extraordinary sensitivity of CA_1 cells compared with CA_3 cells, any agent that reverses this selective vulnerability could be defined as neuroprotectant. Drugs which prevent CA_1 cell death are likely to attenuate cortical and striatal injury resulting in reduced infarction following either permanent or transient focal ischemia.

Focal Models: Infarct Volume Reduction (Organotypic Protection)

Mechanical models of focal ischemia use either an intraluminal suture, which induces severe ischemia in the striatum and relatively milder ischemia in the cortex [55], or a more distal extravascular clip, which may spare the striatum but results in more severe cortical ischemia [15]. These vascular occlusions can be either permanent or transient and unlike autologous clot-embolic models followed by lysis, the timing of reperfusion can be accurately controlled [63]. In suture models, the striatum undergoes rapid necrosis, following which there is a gradual recruitment of the cortex to the infarct, the speed of which depends on the duration and severity of the ischemic insult [55]. Much of the cortex will receive at least partial increases in cerebral blood flow from collaterals. Clip models which employ either a clip or a ligation of the artery in the rhinal fissure result in a different pattern of injury. Distal models can be engineered to produce precortical injury [15]. With permanent ischemia, an infarction evolves within 24 h. In reperfusion models transient ischemia results in less injury; if a critical ischemic threshold is exceeded then the infarct volume in the cortex will mature, but over a longer time interval. Ultimately this injury will achieve the same volume of infarction as that seen with a permanent occlusion [15]. The infarction in the cortex is associated with a breakdown in the blood brain barrier and with edema, causing swelling [56]. Components of excitotoxicity, apoptosis and a subsequent neuroinflammatory reac-

Table 1. Neuroprotection in the rat

- should involve neuroprotection of CA_1 cells following transient forebrain ischemia (cytoprotection)
- should result in lasting cortical protection (organotypic protection)
- should protect the striatum and white matter
- should be achieved with comparable drug levels to those achieved in the clinic
- should be obtained with late administration, post ischemic following transient ischemia

tion all work in parallel, resulting in necrosis and the full evolution or maturation of the infarct.

In lisencephalic species such as mice and rats there is very little white matter, the striatum is hard to protect, and while 50% infarct volume reductions are often seen, this reduction in injury may not be static. In the same way that brief ischemia results in less infarction, if a longer interval is allowed to elapse this small infarct will evolve, recruiting the penumbral zones. When infarct volume reductions are seen at 24 or 48 h, whether a drug-induced protection can be maintained out to 7–28 days or even to three months remains unknown, although it has been demonstrated for post-ischemic hypothermic protection [24].

Mechanisms of Cell Death Overview

Critical reductions in cerebral blood flow result in the inhibition of protein synthesis, the loss of membrane potential, the loss of high energy ATP and terminal anoxic depolarization [59]. Energy failure is clearly a prerequisite to cell loss. In humans, using PET metabolic imaging, it is possible to examine regions where there may be increased oxygen utilization ($CMRO_2$), suggesting potentially critical hypoperfusion (misery perfusion), and, with hyperacute MRI imaging, regions that fall within a PWI-DWI mismatch. Much of the data from animal model work with neuroprotectants is derived from severely ischemic tissue where there is a loss of all electric potential, a loss of energy and a loss of protein synthesis. Penumbra has been defined as an area in which there is still residual energy but protein synthesis is temporarily inhibited [59]. The recruitment of the penumbra to the infarct relates to a secondary failure of energy with a secondary loss of blood brain barrier integrity and of ion homeostasis followed by activation of either apoptotic or inflammatory processes [73].

Global models make it clear that both the loss of energy and the loss of ion homeostasis are fully recoverable. The reasons for the secondary decline in CA_1 energy and ion homeostasis are not understood, but may relate to the failure of the recovery of protein synthesis.

Excitotoxicity

The acute insult is associated with a massive release of extracellular glutamate. Calcium entering the N-type calcium channels promotes the synaptic release of glutamate, which in turn acts on the post-synaptic glutamate receptors, namely the NMDA, AMPA and the metabotropic receptors [21]. The NMDA receptor has been studied in great detail and drugs have been developed which act not only on the glutamate site and its linked ion channel, but also the polyamine and glycine allosteric sites. During depolarization there is a release of the magnesium block of the ion channel. Glutamate, in concert with glycine and polyamine, will stimulate calcium entry into post-synaptic neurons through the NMDA channel. The AMPA channel was initially thought to be less significant, as it gates sodium rather than calcium, but evidence from the sub-unit examination of the heteromeric channel receptor complex suggests that following ischemia, there is a loss of GluR2, making the post-ischemic AMPA channel highly conductive of calcium [93]. The metabotropic receptor, by stimulating second messengers, may activate events in a similar way to calcium entry, resulting in the endogenous release of high levels of intracellular calcium.

Most of the clinical trials have tested the hypothesis that NMDA antagonism, either non-competitively or by blocking the glycine or polyamine sites, would reduce clinical injury and result in better neurological outcomes at three months. All of these trials have failed [16].

The corollary is that agonizing inhibitory receptors such as serotonin and the GABA A receptors will abrogate the levels of calcium seen in the intracellular compartment, but these trials have also faltered.

High levels of calcium, which can injure mitochondria releasing cytochrome C [113], which initiates apoptotic pathways, can also stimulate enzymes resulting in lipolysis, proteolysis and endonuclease activation, which will also result in apoptosis. Calcium activates death processes [21], and so, by preventing calcium influx, the multitude of downstream events could be arrested prior to the initiation of injurious parallel pathways. If one pathway is inhibited increased activity of another will result, so caution must be exercised in dealing with processes that involve a plethora of parallel pathways.

Rise in (Ca^{2+}) Homeostasis

Extracellular Ca^{2+} concentrations are several thousand times greater than intracellular concentrations and the mechanisms that control this excessive gradient are energy dependent. Calcium enters the cell predominantly through two types of channel: voltage controlled and receptor operated [83]. There are several agonists that increase permeability to Ca^{2+} when bound to specific receptors, namely excitatory amino acids (glutamate, nucleotides, and cyclic nucleotides).

Ca^{2+} influx through voltage-sensitive Ca^{2+} channels further depolarises the membrane, resulting in excitatory amino acids being released into the extracellular space. Energy-dependent processes such as glutamate re-uptake are impaired, further increasing extracellular glutamate, resulting in prolonged activation of membrane glutamate receptors, and Ca^{2+} influx [110].

The export of Ca^{2+} from neurons into the extracellular environment is linked directly or indirectly to the utilization of energy [108, 109]. In the former category is an adenosine triphosphate (ATP)-dependent calcium pump and in the second is a Na^+/Ca^{2+} exchanger which utilizes energy stored in the sodium electrochemical gradient. The exchanger can operate in either direction so that when Na^+ gradient falls its function can reverse so it imports, rather than exports, Ca^{2+}. Release of Ca^{2+} from mitochondria is either Na^+-dependent or -independent. Substantial amounts of intracellular Ca^{2+} are stored in the neuron, bound mainly in the mitochondria and endoplasmic reticulum. The accumulation of Ca^{2+} in the endoplasmic reticulum requires energy to fuel an ATP-consuming pump.

The consequences of the unregulated rise in intracellular cytoplasmic Ca^{2+} during ischemia are linked with glutamate excitotoxicity to a number of biochemical processes that result in further detriment to ischemic brain tissue [110].

During ischemia other Ca^{2+}-mediated reactions are accelerated, such as proteolysis and the assembly/reassembly of microtubules [6]. Cytoskeletal proteins are degraded when Ca^{2+} rises intracellularly. Lipolysis produces an increase in free fatty acids (FFA) which are potentially neurotoxic, and the oxidation of arachidonic acid (catalysed by cyclo-oxygenase and lipo-oxygenase) yields prostaglandins, leukotrienes, and thromboxane A_2. These agents, together with platelet-activating factor (PAF), can damage cellular membranes and act additionally as chemoattractants [110]. This activation also produces oxygen-free radicals. These reactions can produce sustained damage to the cell membranes and receptors, further accelerating the toxic effects of glutamate by impaired uptake. Phosphorylation of proteins are Ca^{2+} mediated, resulting in alteration in protein kinase and phosphatase activity. These provoke changes in ion membrane and receptor activity, affecting gene transcription and translation, unmasking processes leading to programmed cell death (apoptosis).

Acidosis

Acidosis contributes not only to tissue damage during ischemia but also inhibits recovery during re-oxygenation, through mechanisms including edema formation, inhibition of H^+ extrusion, inhibition of lactate oxidation and impairment of mitochondrial respiration [108, 109]. Cellular acidosis may promote intracellular edema formation by inducing Na^+ and Cl^- accumulation in the cell via coupled Na^+/H^+ and Cl^-/hydrogen carbonate (HCO_3^-) exchange; acidosis activates Na^+/H^+ exchange and H^+ leaks back into the cell via the Cl^-/HCO_3^- antiporter causing accumulation of Na^+ and Cl^- in the cell, with water entering the cell by osmosis. Furthermore, extracellular acidosis outcompetes Na^+ at the external site of the Na^+/H^+ antiporter, thereby retarding or preventing H^+ movement from the cells. Acidosis also blocks a lactate oxidase form of the lactate dehydrogenase complex, retarding both oxidation of lactate accumulated during ischemia and oxidative phosphorylation in isolated mitochondria, impairing ATP production. Acidosis enhances the production of free radicals, and will trigger DNA fragmentation.

Free Radicals

Several Ca^{2+}-dependent enzymes trigger reactions which produce reactive oxygen species (ROS). These events may be initiated during ischemia, if there is some residual blood flow, but are greatly accelerated during reperfusion. Reactions result in the oxidation of arachidonic acid (by cyclo-oxygenase and lipo-oxygenase), oxidation of xanthine and hypoxanthine to O_2^- and H_2O_2 (via xanthine oxidase), and the formation of NO (by nitric oxide synthase). ROS are produced by the mitochondrial respiratory chain and since mitochondria contain superoxide dismutase, the result is the formation of O_2^- and H_2O_2. Superoxide (O_2^-) is formed during the operation of Complex I and Complex II of the mitochondrial electron transport chain.

The most common free radicals produced during cerebral ischemia are O_2^- and OH, and these, like other free radicals, react with and exacerbate damage of proteins, DNA, and lipids, particularly the fatty acid component of the cell membrane, causing changes in the permeability and integrity of the cell membranes (lipid peroxidation).

Free radical mediated disruption of the inner mitochondrial membrane and the oxidation of proteins that mediate electron transport impair their function. The mitochondrial membrane then becomes leaky and eventually mitochondria become overloaded with Ca^{2+}, resulting in impaired ATP production and ultimately energy and membrane failure [17]. Cytochrome C is released from the mitochondria and has been suggested as a trigger for apoptosis [34].

Free radical scavengers, such as tocopherol and vitamin C, and enzymes which metabolize free radicals (superoxide dismutase), maintain a balance.

Free radicals are also generated during ischemia by the release of iron from ferritin stores within ischemic neurons [29]. As the cerebrospinal fluid has a low concentration of ferritin-binding proteins, much of the iron released from damaged brain cells remains unbound and is therefore available to catalyze the generation of radical hydroxyl (-OH), leading to iron-induced lipid peroxidation. This process is further compounded by the fact that the central nervous system is relatively depleted of superoxide dismutase, which scavenges -OH and controls the release of iron from intracellular stores.

Activated leucocytes produce oxygen-free radicals. During ischemia these inflammatory cells become localized to the brain microvessels through the action of P-selectin. This exposes the microvascular endothelium to high levels of oxygen-free radicals and causes oxidative damage to specific local sites, contributing to the breakdown of the blood brain barrier and the resulting brain edema.

Inflammation

Early accumulation of neutrophils in ischemic brain has been confirmed histologically [19, 23, 32, 50, 52, 100, 104], by biochemical studies [4, 5], and in radiolabeled leukocyte studies [39, 95]. Brain ischemia is associated with expression of a number of inflammatory mediators (IL-1 and TNF-a) and chemokines (IL-8 for neutrophils, MCP-1 for monocytes, RANTES, LTB4, IP-10), and with the up-regulation of adhesion molecules (ICAM-1, selectins) which maintain leukocyte recruitment to the vascular endothelium [43]. Cytokines, such as tumour necrosis factor *a* (TNF*a) and interleukin 1β*, are responsible for the accumulation of inflammatory cells in the injured brain and may affect the survival of damaged neurons [30]. During ischemia, cytokines attract leukocytes and stimulate the production of adhesion receptors on leukocytes and endothelial cells. Leukocytes promote infarction through their toxic by-products, phagocytic action and by the immune reaction. This inflammatory reaction not only contributes to lipid-membrane peroxidation, but also exacerbates the degree of tissue injury caused by the rheologic effects of "sticky" leukocytes in the blood, which interferes with microvascular perfusion [30]. Injury is also mediated through the production of leukocyte neurotoxic products [30, 64, 65]. The signalling mechanisms involved have not yet been defined, but are quite clearly strongly mediated by both TNF-*a* and IL-1β.

Increased expression of TNF-*a* RNA occurs rapidly at 1–3 h after middle cerebral artery occlusion (MCAo) in rats [76, 119]. Levels of cortical TNF-*a* mRNA peak at 12 h, and persist for 5 days; TNF-*a* mRNA expression precedes neutrophil migration into the brain parenchyma.

Like TNF-*a*, IL-1β also has many pro-inflammatory properties, and receptors for this cytokine have been demonstrated in the CNS [35, 105, 114]. It is produced in microglia, astrocytes, neurons, and endothelium. It has been demonstrated in focal transient cerebral ischemia in the rat [84, 129], and the exacerbation of stroke injury has been demonstrated following the administration of exogenous IL-1β [77]. IL-1 receptor antagonist is a naturally occurring inhibitor of IL-1 activity, while exogenous administration of IL-1 antagonists attenuates infarct size [8, 99].

Cytokines induce the expression of the adhesion receptors such as intracellular adhesion molecule-1 (ICAM-1), vascular cell adhesion molecule-1 (VCAM-1), and endothelial cell adhesion molecule-1 (ECAM-1). Adhesion receptors interact with complementary surface receptors on neutrophils. The neutrophils in turn adhere to the endothelium, cross the vascular wall and enter the brain parenchyma. Macrophages and monocytes follow neutrophils and migrate into ischemic brain. Some of the benefit of anti-inflammatory drugs in ischemia may be due to mitigation of hyperthermia, since hypothermia itself also reduces inflammation. Anti-neutrophil benefits might also be accrued by an improvement in post-ischemic blood flow [43, 77].

Production of toxic mediators by activated inflammatory cells and injured neurons has important consequences. Infiltrating neutrophils produce inducible nitric oxide synthase (NOS), an enzyme that can produce toxic amounts of nitric oxide (NO). The pathogenic importance of NO is reflected by the observation that pharmacological inhibitors of inducible NOS reduce ischemia. The inflammatory process might also be linked to apoptosis because antibodies against adhesion molecules attenuate post-ischemic inflammation and reduce apoptotic cell death in ischemic brain [33, 62].

Early Gene Expression

Focal ischemia is a powerful stimulus to elicit genomic responses from the brain in the form of multiple early gene expression. In addition to the genes expressed for TNF-α and IL-1β, many other inflammatory genes are expressed during focal ischemia. Depending on the severity of the ischemia and the intrinsic nature of the neuronal populations, it is thought that a stress response and changes in gene expression are elicited by ischemia, which may be vital to cell survival and repair. Following focal ischemia, immediate early genes are the first to be regulated, as shown by the identification of c-fos, c-jun and zinc finger gene, but the expression is only transient [61, 122].

A second phase consists of the expression of heat shock proteins.

The third phase is comprised by increased cytokine gene expression as previously described for TNF-α and IL-1β, but also includes IL-6 [120] and IL-1 ra [123]. Genes for chemokines such as IL-8 [75], IP-10 [124] and MCO-1 [121] are also elevated, and probably play an important role in neutrophil and mononuclear cell infiltration. The rolling and adhesion of leukocytes on surface endothelium in the region of ischemic tissue are dependent on the coordinated expression of adhesion molecules (ligands and receptors) on both the inflammatory cells and the endothelium. Some of the responsible molecules include ICAM-1, ELAM-1, and P-selectin on the endothelial side, and CD11/CD18, MAC-1, and LFA-1 on the leukocyte side [49]. After MCAo in rats, both ICAM-1 [118] and ELAM-1 [120] mRNA were expressed during the first 1–3 h. After stroke in primates, both P-selectin and ICAM-1 have been shown to be upregulated on the postcapillary microvessels in the ischemic penumbra [89]. In addition, growth factors [61] such as nerve growth factor, brain-derived nerve growth factor, and tumour suppresser gene p53 [74, 126] are also increased. This third phase commences approximately 1 h post ischemia and peaks at 12 h, before subsiding at 2–5 days.

The fourth phase of new gene expression may be an inflammatory reaction to ischemic brain tissue. Proteolytic enzymes (metalloproteinases) are increased and are implicated in the remodeling of the extracellular matrix [101, 102] and their endogenous proteinase inhibitor, TMPI. The expression of the enzymes is associated with an infiltration of inflammatory cells and may cause secondary injury and repair of injured brain.

The fifth phase involves the transcription of mediators such as transforming growth factor-β and osteopontin [41], which play a role in tissue remodeling.

Although TNF-α precipitates leukocyte endothelium recruitment and infiltration into injured brain it is not directly toxic to neurons [46, 94]. Some investigators have even suggested that it provides a protective effect on the neurons. Studies using TNF-receptor knock-out mice (p55 and p750) in cerebral ischemia suggest increased sensitivity to brain ischemia, highlighting the potential beneficial effects of TNF-α [9, 105]. This has also been supported by the observations that repeated exogenous administration of both TNF-α and IL-1β induce ischemic tolerance [88, 90]. While these studies contradict much of the data, which shows that TNF-α and IL-1β augmentation is injurious, blocking these cytokines can be neuroprotective. These cytokines are subsequently involved in the later repair and remodeling of damaged brain. Anti-inflammatory interventions that limit the degree of damage have been shown to interfere with nervous regeneration and recovery [58]. Certain strategies may be required to intervene early in brain inflammation to reduce injury and neurodegeneration; other interventions may be needed to facilitate repair and recovery of regenerating processes after CNS injury [36, 69]. The timing of specific interventions is critical.

Apoptosis

The arbitrary and misleading classification of cell death following cerebral ischemia as either necrotic or apoptotic by histological criteria is best defined by the continuum of biochemical processes, with necrosis and apoptosis at either end [80]. Ischemic cell death within the brain is an active process that requires energy, involves new gene expression, and is caspase mediated [34, 70]. Apoptosis, or programmed cell death, occurs in the nervous system during development. It is mediated by DNA cleavage and by other autolytic processes causing nuclear shrinkage, chromatin clumping, cytoplasmic blebbing, and ultimate cell death.

During ischemia brain cells that are exposed to excessive glutamate-receptor activation, Ca^{2+} overload, oxygen radicals, or those with mitochondria and DNA damage can die by a form of apoptosis (ischemic brain cell death). The weighting of apoptosis depends on the nature of the stimulus, the type of cell and the stage it has reached in its development. Necrosis is the predominant mechanism that follows acute permanent focal vascular occlusion, or global ischemia, whereas in milder injury death may more closely resemble apoptosis. The pathway of apoptosis is initiated by early involvement of mitochondria, which in turn leads to triggering the genes for proteolytic processes such as the caspases, as well as genes that suppress or augment cell death. Three major classes of genes control apoptosis. These include: (1) those that prevent cell death, like bcl-2 and bcl-xL; (2) those that promote cell death, like bax and bcl-xs, and (3) those that promote the engulfment and destruction of cells. Drugs that block caspases [73] or enhance the action of bcl-2 [1] may protect against ischemic injury in certain situations. Of more than twelve that have been identified, caspases 1 and 3 seem to have a pivotal role in ischemia-mediated apoptosis.

A sharp increase in the permeability of the mitochondrial inner membrane occurs during ischemia, termed the *mitochondrial permeability transition* (MPT). This has been known to occur in necrosis, but has also been observed in apoptosis [71, 85]. The MPT leads to membrane depolarization, uncoupling of oxidative phosphorylation, release of intra-mitochondrial ions, and swelling which, when severe, ruptures the outer membrane releasing macromolecules into the cytoplasm. If membrane depolarization is incomplete, cellular recovery is still possible. This release of apoptogenic substances activates a cascade of proteolytic caspase activity [40], leading to destruction of key structural and functional proteins, causing ultimate cell death. The bcl-2 family of proteins are intimately involved in this mitochondrial release of killer molecules [48]. If the programmed path is blocked, ATP depletion due to energy failure will force the cell to necrosis instead.

During ischemic acidosis, ATP depletion, depolarization, and Ca^{2+} overload are processes that ultimately will damage the mitochondria. MPT has been observed within one hour following ischemia [91]. The inducibility of MPT depends on the cell type and brain region, for instance, CA_1 hippocampus is more sensitive to ischemia than cortex, which correlates with the known susceptibility of these areas to ischemic damage [44].

The release of mitochondrial components, such as cytochrome C, has been observed in focal ischemia [45] and from hippocampal CA cells following global ischemia [2, 86]. Another mitochondrial component, caspase-9, has also been found to be released from the mitochondria of ischemic brain in dogs [67]. In addition, apoptotic cell death requires the expression of new genes including those related to the Bcl-2 family and to P53. Bcl-2 is an integral membrane protein, which is found predominately in the mitochondrial outer membrane, which is targeted by translocation of Bax following a cell death signal. Bax leads to the formation of a pore through which killer molecules such as cytochrome C may escape. Transcription factors such as P53 lie within a network of inter-related genes associated with cell death and can lead to upregulation of target genes such as Bax and those involved in generating ROS [111].

In global ischemia, hippocampal Bax levels were increased by 60% following 6 h and remained so for 72 h. In this study, Bax appeared to favour apoptosis by binding the anti-apoptotic species Bcl-x [3]. Other studies have shown an imbalance of pro-apoptotic Bax, concomitant with a decrease in the anti-apoptotic Bcl-2 and Bcl-x proteins [66].

Knock-out mice have been used to define the role of the Bcl-2 family members in ischemic cell death. Over-expression of anti-apoptotic Bcl-2 reduced focal brain infarct by 42% at 7 days after middle cerebral artery occlusion [81].

In focal ischemia, a disturbance in the ratio of Bax/Bcl-2 levels has been seen in rat cortex at 6 h at the border of the lesion [47] or overall at 24 h [38] and at 3–12 h in mice [82], which is consistent with the findings of increased Bcl-2 expression in surviving neurons at 24 h [20].

P53 has been reported to be elevated following focal cerebral ischemia [22], and both P53 mRNA and protein have been shown to be elevated at the periphery of the focal infarct [74, 92, 126]. In the absence of the P53 gene, there was reduced infarct volume [27]. It appears that increased p53 is transcriptionally active since p53 target genes are also upregulated following focal ischemia, for example Bax, Gadd45 [18, 60], Mdm2 [116], and $p21^{WAF/Cip1}$ [117], although elevated p21 has not been found by others [92].

Caspases are a family of cysteine proteases, involved in regulating cytokine maturation and apoptosis [28], of which 14 have been identified. Caspases can be classified, according to their amino acid sequence homology, into 3 groups, i.e. caspase-1, -3 and -9 subfamilies. The caspase-1 subfamily includes -1, -4, -5 and –11. Since the predominant defect in caspase-1 knock-out mice is the inability to process prointerleukin-1β, the major function of caspase-1 is believed to be regulating cytokine maturation [68]. On the other hand, caspase-1 is responsible for the activation of executioner caspases, directly involved in apoptosis progression [31]. Further evidence that caspase-1 activation could play a pivotal role in ischemic neurodegeneration comes from caspase-1 –/– mice and mice expressing a dominant-negative mutant caspase-1 gene. Both of these genetically modified mouse strains are more resistant to ischemic insult than wild-type mice [54, 106]. The effects of caspase inhibition on cerebral ischemia induced neurodegeneration was shown by using the non-selective inhibitors z-VAD.fmk and z-DVED.fmk. Both inhibitors induced a significant neuroprotection in mouse models of transient cerebral ischemia [42, 53, 79], and z-VAD.fmk was neuroprotective also in transient and permanent models of ischemia in the rat [53, 73, 78].

Peptide-based caspase inhibitors prevent neuronal loss. During transient focal ischemia, levels of neuronal apoptosis inhibitory protein (NAIP) become elevated.

Clinical Translation

Problems with NMDA Antagonists

Experimentally, NMDA antagonists have been the prototypic neuroprotectants. Despite this, they do not protect the white matter (there are no NMDA receptors in white matter) and they have never protected the striatum in experimental focal ischemia. In the laboratory, they have not been shown to be effective if given late, i.e. during reperfusion. In normothermically controlled animals, the cytoprotective effect of CA_1 cells is lost [11, 13]. The focal infarct reductions are in part related to changes in vasoactivity (increased rCBF) and increases in tachycardia and blood pressure (also resulting in a relative change of cerebral blood flow) [127].

The side effects have included cardiotoxicity, psychotomimetic reactions and, of concern to neuropathologists, the appearance of vacuolations in the limbic cortex in, not only post-ischemic, but also naïve, brain [10].

AMPA Antagonists

AMPA antagonists are potent neuroprotectants, which have been demonstrated to both protect CA_1 [12, 107] and to reduce infarct volumes, even when given quite late during reperfusion following transient ischemia or late in the process of permanent focal ischemia [14]. These compounds block either the AMPA channel or the AMPA receptor; those post-ischemic receptors are permissive to calcium [128]. Of importance, these agents have been given very late, up to 12–24 h after global ischemia [72] and in the first hour or two after reperfusion following focal ischemia and have still resulted in protection [128]. This protection has not been demonstrated to be indefatigable [25] and although recurrent administration of the drug might result in this, because of the toxicity (mainly nephrotoxicity) [128], these compounds which are in the main insoluble have been stuck in development. The Yamanouchi compound YM872 is now in active clinical development [130].[1]

---

[1] Note to Alastair from Barbara "This cited paper in References is no longer relevant. Suggested replacements would either be #'d 114 (i.e. Takahashi 2002) or 64 (i.e. Kawasaki-Yatsugi 2000)".

Downstream Events

Following the loss of energy in the form of ATP, the depolarization and the transient acidosis, recovery is associated with a variety of putative mechanisms of maturation of the injury. As discussed, these include parallel pathways: a stimulation of apoptotic pathways, free radical mediated injury, a loss of membrane stability, and a loss of cytokine activity. During or following cell death there is evidence (causal or secondary) for a deleterious neuro-inflammatory response. Agents have therefore been developed which test most of these putative mechanisms but many of the preclinical prerequisites have not been met before these trials have gone into development. In particular, trials have studied lipid peroxidation inhibitors and antibodies to adhesion molecules, but as yet no trials have been done in the clinic that studied drugs that block apoptosis [16].

Hypothermia

Following Global Ischemia

Induction of mild hypothermia, deployed even 12 to 24 h after transient forebrain ischemia can produce striking cytoprotection, not only in the early reperfusion, but also sustained for at least 28 days [26]. So long as the duration of hypothermia is prolonged long term survival can be observed even with delayed hypothermia. This implies that events seen in the first 12–24 h are not necessarily causally linked to the execution of these cells since the effects are reversible. This long-term survival is seen not just with histological outcome measures but is also seen with electrophysiological observation and with behavioural outcomes [37]. Neuroprotection depends on not only the depth of the hypothermia but its duration, although the longer the delay following reperfusion to the intervention with hypothermia, the lesser the possibility of long-term neuronal survival. Recently, a group in Australia and a group in Germany have shown that with prolonged post ischemic moderate hypothermia, mimicking our experimental work [26], not only survival but neurological outcomes are improved. This therefore represents the first translation of experimental neuroprotection to a successful clinical trial. Managing successfully resuscitated cardiac arrest patients now requires the use of mild hypothermia in the first 48-72 h following resuscitation [7, 57].

Following Focal Ischemia

While post ischemic hypothermia of sufficient duration will protect CA1 cells and experimentally improve neurological outcome, it has been more difficult to translate hypothermic intervention into the clinic following embolic stroke. Cooling is easier in an intensive care setting, particularly if the patient has been intubated and is unconscious. In focal ischemia, when patients are critically ill or are deliberately anaethesized and intubated for procedures such as intraarterial thrombolysis, it is possible to cool, but for most patients coming into the stroke units cooling presents a challenge because of the shivering and homeostatic responses induced by artificial cooling. In laboratory studies, hypothermia, with a similar paradigm to that which prevented CA1 cell death, has been shown to produce long term benefit in terms of infarct volume reduction and improved behaviour. Hypothermia was induced following reperfusion after 90 minutes of normothermic ischemia and functional outcome was assessed using a staircase test 30 days after ischemia. The staircase test, which is sensitive to motor cortex injury, can reveal consistent deficits unlike some other measures which are less sensitive over the long term and which can show recovery despite extensive brain injury. Despite being delayed, prolonged hypothermia has for the first time demonstrated long term functional and histological protection following transient focal ischemia [24].

Rules for Translation – Biomarkers/Surrogates

These rules are proposed in order to allow the translation in neuroprotection from *in-vitro* experimental work, through *in-vivo* animal work, to human clinical trials. The cellular basis of protection should extrapolate to the organ and, importantly with the concentration of agent working *in-vitro*, should translate to similar cellular level with drug levels both in serum and in brain to induce a pharmacological effect. The cellular protection (following global ischemia) should be seen and extended to infarct reduction in models of focal ischemia, again with similar drug levels as seen to induce the biochemical effect *in-vitro*. Long term cell survival and long term infarct volume reduction should be seen and associated with a behavioural surrogate outcome. In terms of translating to the human, biomarkers such as those seen with MR imaging would prove useful, particularly at the Phase II level. The drugs, if they are used in combination, should be given in a logical sequence and should not be toxic in combination. The trials should be stratified, not only by time, but also by tissue window, using imaging as a means of selection and stratification in the randomized trials.

Conclusion

Although many drugs influencing excitotoxicity, free radical expression, neuroinflammation and apoptosis have been studied in the laboratory, none of them have yet proven successful in human clinical trials. Hypothermia used post ischemically has shown protection in both global and focal models and the outcome is improved at both the histological and behavioural level and has proven to be enduring. Long term behavioural outcome following global ischemia has now been translated to successful neurological outcome in patients resuscitated from cardiac arrest. The translation of neuroprotection has therefore been established in man (using post ischemic hypothermia) but is yet to be demonstrated for any drug. Even hypothermia has of yet not been shown to be beneficial following focal ischemia in randomized clinical trials. Surrogates at the Phase II level, which allow one to determine that the mechanism of action is inhibited by antagonists, are needed before proceeding with large randomized Phase III trials.

Acknowledgements. We gratefully acknowledge the assistance of Annley Wilson for the preparation of this manuscript.

Canadian Institutes of Health Research [CIHR], Heart and Stroke Foundation of Alberta, NWT and Nunavut [HSF], Alberta Heritage Foundation for Medical Research [AHFMR].

References

1. Adams JM, Cory S (1998) The Bcl-2 protein family: arbiters of cell survival. Science 281:1322–1326
2. Antonawich FJ (1999) Translocation of cytochrome C following transient global ischemia in the gerbil. Neurosci Lett 274:123–126
3. Antonawich FJ, Krajewski S, Reed JC, Davis JN (1998) Bcl-x(1) Bax interaction after transient global ischemia. J Cereb Blood Flow Metab 18:882–886
4. Barone FC, Hillegass ZM, Rzimas MN, Schmidt DB, Folley JJ, White RF, Price WJ, Feuerstein GZ, et al (1991) Polymorphonuclear leukocyte infiltration into cerebral focal ischemic tissue: myeloperoxidase activity and histologic verification. J Neurosci Res 29:336–345
5. Barone FC, Hillegass ZM, Rzimas MN, Schmidt DB, Folley JJ, White RF, Price WJ, Feuerstein GZ, Clark RK, Griswald DE, Sarau HM (1995) Time related changes in myeloperoxidase activity and leukotriene B4 receptor binding reflect leukocyte influx in cerebral focal stroke. Mol Chem Neuropathol 24:13–30
6. Baudry M, Bundman MC, Smith EK, Lynch GS (1981) Micromolar calcium stimulates proteolysis and glutamate binding in rat brain synaptic membranes. Science 212:937–938
7. Bernard SA, Gray TW, Buist MD, Jones BM, Silvester W, Gutteridge G, Smith K (2002) Treatment of comatose survivors of out-of-hospital cardiac arrest with induced hypothermia. N Engl J Med 346:557–563
8. Betz AL, Yang G-Y, Davidson BL (1995) Attenuation of stroke size in rats using an adenoviral vector to induce overexpression of interleukin-1 receptor antagonist in brain. J Cereb Blood Flow Metab 15:547–551
9. Bruce AJ, Boling W, Kindy MS, Pesheni J, Kraemer PJ, Carpenter MK, Holtsberg FW, Mattson MP (1996) Altered neuronal and microglial responses to excitotoxic and ischemic brain in mice lacking TNF receptors. Nature Med 2:788–794
10. Buchan AM (1990) Do NMDA antagonists protect against cerebral ischemia: are clinical trials warranted? Cerebrovascular & Brain Metabolism Reviews 2:1–26
11. Buchan AM, Pulsinelli WA (1990) Hypothermia but not the N-methyl-D-aspartate antagonist, MK-801, attenuates neuronal damage in gerbils subjected to transient global ischemia. J Neurosci 10:311–316
12. Buchan AM, Li H, Cho SH, Pulsinelli WA (1991) Blockade of the AMPA receptor prevents CA$_1$ hippocampal injury following severe but transient forebrain ischemia in adult rats. Neurosci Let 132:255–258
13. Buchan AM, Li H, Pulsinelli WA (1991) N-Methyl-D-Aspartate antagonists, MK801, fails to protect against neuronal damage caused by transient severe forebrain ischemia in adult rats. J Neuroscience 11:1049–1056
14. Buchan AM, Xue D, Huang ZG, Smith KE, Lesiuk H (1991) Delayed AMPA receptor blockade reduces cerebral infarction induced by focal ischemia. NeuroReport 2:473–476
15. Buchan AM, Xue D, Slivka A (1992) A new model of temporary focal neocortical ischemia in the rat. Stroke 23:273–279
16. Buchan AM (2001) "Neuroprotective stroke trials: a ten year dry season". Update in Intensive Care and Emergency Medicine 37:236–251
17. Budd SL (1998) Mechanisms of neuronal damage in brain hypoxia/ischemia: focus on the role of mitochondrial calcium accumulation. Pharmacol Ther 80:203–229
18. Charriaut-Marlangue C, Richard E, Ben-Ari Y (1999) DNA damage and DNA damage-inducible protein Gadd45 following ischemia in the P7 neonatal rat. Brain Res Dev Brain Res 116:133–140
19. Chen H, Chopp M, Bodzin G (1992) Neutropenia reduces the volume of cerebral infarct after transient middle cerebral artery occlusion in the rat. Neurosci Res Comm 11:93–99
20. Chen J, Graham SH, Chan PH, Lan J, Zhou RL, Simon RP (1995) Bcl-2 is expressed in neurons that survive focal ischemia in the rat. NeuroReport 6:394–398

21. Choi DW (1988) Calcium-mediated neurotoxicity: relationship to specific channel types and role in ischemic damage. Trends Neurosci 11:465–469
22. Chopp M, Li Y, Zhang ZG, Freytag SO (1992) P53 expression in brain after middle cerebral artery occlusion in the rat. Biochem Biophys Res Comm 182:1201–1207
23. Clark RK, Lee EV, Fish CJ, White RF, Price WJ, Jonak GZ, Feuerstein GZ, Barone FC (1993) Development of tissue damage, inflammation and resolution following stroke: an immunohistochemical and quantitative planimetric study. Brain Res Bull 35:623–639
24. Colbourne F, Corbett D, Zhao Z, Yang J, Buchan AM (2000) Prolonged but delayed postischemic hypothermia: a long-term outcome study in the rat middle cerebral artery occlusion model. Cereb Blood Flow Metab 20:1702–1708
25. Colbourne F, Li H, Buchan AM (1999) Continuing post-ischemic neuronal death in CA1: Influence of ischemia duration, and cytoprotective doses of NBQX and SNX-111. Stroke 30:662–668
26. Colbourne F, Li H, Buchan AM (1999) Indefatigable CA1 sector neuroprotection with mild hypothermia induced 6 h after severe forebrain ischemia in rats. J Cereb Blood Flow Metab 19:742–749
27. Crumrine RC, Thomas AL, Morgan PF (1994) Attenuation of p53 expression protects against focal ischemic damage in transgenic mice. J Cereb Blood Flow Metab 14:887–891
28. Cryns V, Yuan J (1998) Proteases to die for. Genes Dev 12:1551–1570
29. Davalos A, Fernandez-Real JM, Rilart M (1994) Iron related damage in acute ischemic stroke. Stroke 24:1543–1546
30. Del Zoppo GJ, Schmid-Schonbein GW, Mori E, Copeland BR, Chang CM (1991) Polymorphonuclear leukocytes occlude capillaries following middle cerebral artery occlusion and reperfusion in baboons. Stroke 22:1276–1283
31. Denner L (1999) Caspases in apoptic death. Exp Opin Invest Drugs 8:37–50
32. Dereski MO, Chopp M, Knight RA, Chen H, Garcia JH (1992) Focal cerebral ischemia in the rat: temporal profile of neutrophil responses. Neuroscience Res Comm 11:179–186
33. Dereski MO, Chopp M, Knight RA, Rodolosi LC, Garcia JH (1993) The heterogeneous temporal evolution of focal ischemic neuronal damage in the rat. Acta Neuropathol [Berl] 85:327–333
34. Dirnagl U, Iadecola C, Moskowitz MA (1999) Pathobiology of ischaemic stroke: an integrated view. Trends Neurosci 22:391–397
35. Dinarello CA (1996) Biology of interleukin-1. FASEB J 2:108–115
36. Dopp J, de Vellis J (1998) Strategies or therapeutic manipulation of cytokines and their receptors in inflammatory neurodegenerative diseases. Ment Retard Devel Disabil Res Rev 4:200–211
37. Dong H, Moody-Corbett F, Colbourne F, Pittman Q, Corbett D (2001) Electrophysiological properties of CA1 neurons protected by postischemic hypothermia in gerbils. Stroke 32:788–795
38. Dubal DB, Shughrue PJ, Wilson ME, Merchenthaler I, Wise PM (1999) Estradiol modulates bcl-2 in cerebral ischemia: a potential role for estrogen receptors. J Neurosci 19:494–498
39. Dukta AJ, Kochanek PM, Hallenbeck JM (1989) Influence of granulocytopenia on canine cerebral ischemia induced by embolism. Stroke 20:390–395
40. Eldadah BA, Faden AI (2000) Caspase pathways, neuronal apoptosis and CNS injury. J Neurotrauma 17:811–829
41. Ellison JA, Vellier JJ, Spera PA, Jonak ZL, Wang XK, Barone FC, Feuerstein GZ (1998) Osteopontin and its integrin receptor avb3 are upregulated during formation of the glial scar following focal stroke. Stroke 29:1698–1707
42. Endes M, Namura S, Shimizu-Sasamota M, Waeber C, Zhang L, Gomez-Isla T, Hyman BT, Moskowitz MA (1998) Attenuation of delayed neuronal death after mild focal ischemia in mice by inhibition of the caspase family. J Cereb Blood Flow Metab 18:238–247
43. Feuerstein GZ, Wang XK, Barone FC (1998) Inflammatory mediators of ischemic injury: cytokine gene regulation in stroke. In: Ginsberg MD, Bogousslowsky J (eds) Cerebrovascular disease: pathophysiology, diagnosis and management. Blackwell Science, pp 507–531
44. Friberg H, Connern C, Halestrap AP, Wieloch T (1999) Differences in the activation of the mitochondrial permeability transition among brain regions in the rat correlate with selective vulnerability. J Neurochem 72:2488–2497
45. Fujimura M, Morita-Fujimura Y, Kawase M, Copin JC, Calagui B, Epstein CJ, Chan PH (1999) Manganese superoxide dismutase mediates the early release of mitochondrial cytochrome C and subsequent DNA fragmentation after permanent focal cerebral ischemia in mice. J Neurosci 19:3414–3422
46. Garcia JE, Nonner D, Ross, Barrett JN (1992) Neurotoxic components in normal serum. Exp Neurol 118:309–316
47. Gillardon F, Lenz C, Waschke KF, Krajewski S, Reed JC, Zimmermann M, Kuschinsky W (1996) Altered expression of Bcl-2, Bcl-X, and c-Fos colocalizes with DNA fragmentation and

ischemic cell damage following middle cerebral artery occlusion in rats. Mol Brain Res 40:254–260

48. Graham SH, Chen J, Clark RSB (2000) Bcl-2 family gene products in cerebral ischemia and traumatic brain injury. J Neurotrauma 17:831–841
49. Granger DN, Kvietys PR (1993) Leukocyte-endothelial adhesion induced by ischemia and reperfusion. Can J Physiol Pharmacol 71:67–75
50. Guillian D, Chen J, Ingman JE, George JK, Noponen M (1989) The role of mononuclear phagocytes in wound healing after traumatic injury to adult mammalian brain. J Neurosci 9:4410–4429
51. Hacke W, Kaste M, Fieschi C, von Kummer R, Davalos A, Meier D, Larrue V, Bluhmki E, Davis S, Dounan G, Schneider D, Diez-Tejedor E, Trouillas P (1998) Randomised double blind placebo-controlled trial of thrombolytic therapy with intravenous alteplase in acute ischaemic stroke [ECASS II]. Lancet 352:1245–1251
52. Hallenbeck JM, Dutka AJ, Tanishima T, Kochanek PM, Kumaroo KK, Thompson CB, Obrenovitch TP, Contreras TJ (1986) Polymorphonuclear leukocyte accumulation in brain regions with low blood flow during the early postischemic period. Stroke 17:246–253
53. Hara H, Fink K, Endres M, Friedlander RM, Gaglardini V, Juan J, Moskowitz MA (1997a) Attenuation of transient focal cerebral ischemia injury in transgenic mice expressing a mutant ICE inhibitory protein. J Cereb Blood Flow Metab 17:370–375
54. Hara H, Friedlander RM, Gagliardini V, Ayata C, Fink K, Huang Z, Shimizu-Sasamata, Ayata C, Fink K, Huang Z, Shimizu-Sasamata M, Yuan J, Moskowitz MA (1997b) Inhibition of interleukin 1: Converting enzyme family proteases reduces ischemic and excitotoxic neuronal damage. Proc Natl Acad Sci USA 94:2007–2012
55. Hata R, Maeda K, Hermann D, Mies G, Hossmann K-A (2000) Evolution of brain infarction after transient focal cerebral ischemia in mice. Cereb Blood Flow Metab 20:937–946
56. Huang ZG, Xue D, Preston E, Karbalai H, Buchan AM (1999) Biphasic opening of the blood-brain barrier following transient focal ischemia: effects of hypothermia. The Cana Neurol Sci 26:4, 298–304
57. The Hypothermia After Cardiac Arrest (HACA) Study Group (2002) Mild therapeutic hypothermia to improve the neurologic outcome after cardiac arrest. N Engl J Med 346:549–556
58. Hirschberg DL, Yoles E, Belkin M, Shwartz M (1994) Inflammation after axonal injury has conflicting consequences for recovery of function: rescue of spared axons is impaired but regeneration is supported. J Neuroimmunol 50:9–16
59. Hossman KA, Hata R, Maeda K, Trapp T, Mies G (2003) Transgenic Mutants for the Investigation of Molecular Stroke Mechanisms. Maturation Phenomenon in Cerebral Ischemia, Vol V
60. Hou ST, Tu Y, Buchan AM, Huang Z, Preston E, Rasquinha I, Robertson GS, MacManus JP (1997) Increases in DNA lesions and the DNA damage indicator Gadd45 following transient cerebral ischemia. Biochem Cell Biol 75:383–392
61. Hsu CY, An G, Liu JS, Xue JJ, He YY, Lin TN (1993) Expression of immediate early gene and growth factor mRNAs in a focal cerebral ischemia model in the rat. Stroke 24:178
62. Iadecola C, Zhang F, Casey R, Nagayama M, Ross ME (1997) Delayed reduction of ischemic brain injury and neurological deficits in mice lacking the inducible nitric oxide synthase gene. J Neurosci 17:9157–9164
63. Kaplan B, Brint S, Tanabe J, Jacewicz M, Wang XJ, Pulsinelli W (1991) Temporal thresholds for neocortical infarction in rats subjected to reversible focal cerebral ischemia. Stroke 22:1032–1039
64. Kochanek PM, Hallenback JM (1992) Polymorphonuclear leukocytes and monocytes/macrophages in the pathogenesis of cerebral ischemia and stroke. Stroke 23:1367–1379
65. Kogure K, Yamasaki Y, Matsuo Y, Kato H, Onodera H (1996) Inflammation of the brain after ischemia. Acta Neurochir 66 (suppl):40–45
66. Krajewski S, Mai JK, Krajewska M, Sikorska M, Mossakowski MJ, Reed JC (1995) Upregulation of bax protein levels in neurons following cerebral ischemia. J Neurosci 15:6364–6376
67. Krajewski S, Krajewska M, Ellerby LM, Welsh K, Xie Z, Deveraux QL, Salvesen GS, Bredesen DE, Rosenthal RE, Fiskum G, Reed JC (1999) Release of caspase-9 from mitochondria during neuronal apoptosis and cerebral ischemia. Proc Natl Acad Sci USA 11:5752–5757
68. Kuida K, Loppke JA, Ku G, Harding MW, Livingston DJ, Su MSS, Flavell RA (1995) Altered cytokine export and apoptosis in mice deficient in interleukin 1β converting enzyme. Science 217:2000–2002
69. Lazarov-Spiegler O, Rapalino O, Agranov G, Schwartz M (1998) Restricted inflammatory reaction in CNS: A key impediment to axonal regeneration? Molec Med Today 4:337–342
70. Lee JL, Zipfel GJ, Choi DW (1999) The changing landscape of ischaemic brain injury mechanisms. Nature 399 (suppl):A7–A14

71. Lemasters JJ, Qian T, Elmore SP, Trost LC, Nishimura Y, Herman B, Bradham CA, Brenner DA, Nieminen AL (1998) Confocal microscopy of the mitochondrial permeability transition in necrotic cell killing, apoptosis and autophagy. Biofactors 8:283–285
72. Li H, Buchan AM (1993) Treatment with an AMPA antagonist 12 h following severe normothermic forebrain ischemia prevents CA1 neuronal injury. J Cereb Blood Flow Metab 13:933–939
73. Li H, Colbourne F, Sun P, Zhao Z, Buchan AM (2000) Caspase inhibitors reduce neuronal injury after focal but not global cerebral ischemia in rats. Stroke 31:176–182
74. Li Y, Chopp M, Zhang ZG, Zaloga C, Niewenhuis L, Gantam S (1994) P53-immunoreactive protein and P53 mRNA expression after transient middle cerebral artery occlusion in rats. Stroke 25:849–856
75. Liu T, Young PR, McDonnell PC, White RF, Barone FC, Feuerstein GZ (1993) Cytokine-induced neutrophil chemoattractant mRNA expressed in cerebral ischemia. Neurosci Lett 164:125–128
76. Liu T, Clark RK, McDonnell PC, Young PR, White RF, Barone FC, Feuerstein GZ (1994) Tumor necrosis factor-alpha expression in ischemic neurons. Stroke 25:1481–1488
77. Loddick SA, Rothwell NJ (1996) Neuroprotective effects of human recombinant interleukin-1 receptor. J Cereb Blood Flow Metab 16:932–940
78. Loddick SA, MacKenzie A, Rothwell NJ (1996) An ICE inhibitor z-VAD-dbc attenuates ischaemic brain damage in the rat. NeuroReport 7:1465–1468
79. Ma J, Endes M, Moskowitz MA (1998) Synergistic effects of caspase inhibitors and MK-801 in brain injury after transient focal cerebral ischemia in mice. Br J Pharmacol 124:756–762
80. MacManus JP, Buchan AM (2000) Apoptosis after experimental stroke: fact or fashion? J Neurotrauma 17:899–913
81. Martinou JC, Dubois-Dauphin M, Staple JK, Rodriguez I, Frankowski H, Missotten M, Albertini P, Talabot D, Catsicas S, Pietra C, et al (1994) Overexpression of BCL-2 in transgenic mice protects neurons from naturally occurring cell death and experimental ischemia. Neuron 13:1017–1030
82. Matsushita K, Matsuyama T, Kitagawa K, Matsumoto M, Yanagihara T, Sugita M (1998) Alterations of Bcl-2 family proteins precede cytoskeletal proteolysis in the penumbra, but not in infarct centres following focal cerebral ischemia in mice. Neuroscience 83:439–448
83. Miller RJ (1987) Multiple calcium channels and neuronal function. Science 235:46–53
84. Minami M, Kyraishi Y, Yabunchi K, Yamazaki A, Satoh M (1992) Induction of interleukin-1β mRNA in rat brain after transient forebrain ischemia. J Neurochem 58:390–392
85. Murphy AN, Fiskum G, Beal MF (1999) Mitochondria in neurodegeneration: bioenergetic function in cell life and death. J Cereb Blood Flow Metab 19:231–245
86. Nakatsuka H, Ohta S, Tanaka J, Toku K, Kumon Y, Maeda N, Sakanaka M, Sakaki S (1999) Release of cytochrome C from mitochondria to cytosol in gerbil hippocampal CA1 neurons after transient forebrain ischemia. Brain Res 849:216–219
87. National Institute of Neurological Disorders and Stroke rt-PA Stroke Study Group (1995) Tissue plasminogen activator for acute ischemic stroke. New Engl J Med 333:1581–1587
88. Nawashiro H, Tasaki K, Ruetzlav CA, Hallenbeck JM (1997) TNF-alpha pretreatment induces protective effects against focal cerebral ischemia in mice. J Cereb Blood Flow Metab 17:483–490
89. Okada Y, Copeland BR, Mori E, Tung MM, Thomas WS, del Zoppo GJ (1994) P-selectin and intercellular adhesion molecule-1 expression after focal brain ischemia and reperfusion. Stroke 25:202–211
90. Okutsuki T, Reutzler CA, Tasaki K, Hallenbeck JM (1996) Induction of tolerance to ischemia by preconditioned interleukin-1α in gerbil hippocampal neurons. J Cereb Blood Flow Metab 16:1137–1142
91. Ouyang YB, Kuroda S, Kristian T, Siesjo BK (1997) Release of mitochondrial aspartate aminotransferase (mast) following transient focal cerebral ischemia suggests the opening of a mitochondrial permeability transition pore. Neurosci Res Commun 20:167–173
92. Park SW, Kim YB, Hwang SN, Choi DY, Kwon JT, Min BK, Suk JS (2000) The effects of N-methyl-N-nitrosourea and azoxymethane on focal cerebral infarction and the expression of p53, p21 proteins. Brain Res 855:298–306
93. Pellegrini-Giampietro DE, Zukin RS, Bennett MVL, Cho S, Pulsinelli WA (1992) Switch in glutamate receptor subunit gene expression in CA_1 subfield of hippocampus following global ischemia in rats. Proc Natl Acad Sci USA 89:10499–10503
94. Piani D, Spranger M, Frei K, Schaffner A, Fontana A (1992) Macrophage-induced cytotoxicity of N-methyl-D-aspartate receptor positive neurons involves excitatory amino acids rather than reactive oxygen intermediates and cytokines. Eur J Immunol 22:2429–2436

95. Pozzilli C, Lenzi GL, Argentino C, Carolei A, Raura M, Signor A, Bozzao L, Pozzilli P (1985) Imaging of leukocyte infiltration in lumen cerebral infarcts. Stroke 16:251–255
96. Pulsinelli WA, Brierly JB (1979) A new model of bilateral hemispheric ischemia in the unanesthetized rat. Stroke 10:267–272
97. Pulsinelli WA, Buchan AM (1988) The four-vessel occlusion rat model: method for complete occlusion of vertebral arteries and control of collateral circulation. Stroke 19:913–914
98. Pulsinelli WA, Duffy TE (1983) Regional energy balance in rat brain after transient forebrain ischemia. J Neurochem 40:1500–1503
99. Relton JK, Martin D, Thompson RC, Russell DA (1996) Peripheral administration of interleukin-1 receptor antagonist inhibits brain damage after focal cerebral ischemia in the rat. Exp Neurol 138:206–213
100. Ritter L, Coull B, Davisgomen G, McDonough P (1998) Leukocytes accumulate in the cerebral microcirculation during the first hour of reperfusion following stroke. FASEB J 12:188
101. Romanic AM, White RF, Arleth AJ, Ohlstein EO, Barone FC (1998) Matrix metalloproteinase expression increases following cerebral ischemia: inhibition of MMP-9 reduces infarct size. Stroke 29:1020–1030
102. Rosenberg GA, Navratil M, Barone F, Feuerstein G (1996) Proteolytic cascade enzymes increase in focal cerebral ischemia in rat. J Cereb Blood Flow Metab 16:360–366
103. Rothwell NJ (1991) Functions and mechanisms of interleukin-1 in the brain. Trends Pharmacol Sci 12:430–435
104. Rothwell NJ (1997) Cytokines and acute neurodegeneration. Mol Psych 2:120–121
105. Rothwell NJ, Luheshi GN (1996) Brain TNF: damage limitation or damaged reputation? Nature Med 2:746–747
106. Schielke GP, Yang GY, Shivers BD, Betz AL (1998) Reduced ischemic brain injury in interleukin-1 beta converting enzyme-deficient mice. J Cereb Blood Flow Metab 18:180–185
107. Sheardown MJ, Nielson EØ, Hansen AJ, Jacobsen P, Honre T (1990) 2,3-Dihydroxy-6-nitro-7-sulfamoyl-benzo[F] quinoxaline: a neuroprotectant for cerebral ischemia. Science 247:571–574
108. Siesjö BK (1992a) Pathophysiology and treatment of focal cerebral ischemia. Part I: Pathophysiology. J Neurosurg 77:169–184
109. Siesjö BK (1992b) Pathophysiology and treatment of focal cerebral ischemia. Part II: Pathophysiology. J Neurosurg 77:351–354
110. Siesjö BK, Kristian T, Katsura K (1998) Overview of bioenergetic failure and metabolic cascades in brain ischemia. In: Ginsberg MD, Bogousslavsky J (eds) Cerebrovascular disease. Blackwell Science, pp 3–13
111. Sionov RV, Haupt Y (1999) The cellular response to p53: the decision between life and death. Oncogene 18:6145–6157
112. Smith ML, Bendek G, Dahlgren N, Rosen I, Wieloch T, Siesjö BK (1984) Models for studying long-term recovery following forebrain ischemia in the rat: a 2-vessel occlusion model. Acta Neurol Scand 69:385–401
113. Sugawara T, Fujimura M, Morita-Fujimura Y, Kawase M, Chan PH (1999) Mitochondrial release of cytochrome C corresponds to the selective vulnerability of hippocampal CA1 neurons in rats after transient global cerebral ischemia. J Neurosci 19:RC39:1–6
114. Takao T, Tracey DE, Mitchel WM, De Souza EB (1990) Interleukin-1 receptors in mouse brain: characterisation and neuronal localisation. Endocrinology 127:3070–3078
115. Tomasevic G, Kamme F, Stubberod P, Wieloch M, Wieloch T (1999) The tumor suppressor p53 and its response gene p21 WAFI/Cip1 are not markers of neuronal death following transient global cerebral ischemia. Neuroscience 90:781–792
116. Tu Y, Hou ST, Huang Z, Robertson GS, MacManus JP (1998) Increased Mdm2 expression in rat brain after transient middle cerebral artery occlusion. J Cereb Blood Flow Metab 18:658–669
117. van Lookeren Campagne M, Gill R (1998) Increased expression of cyclin G1 and p21WAF1/CIP1 in neurons following transient forebrain ischemia: comparison with early DNA damage. J Neurosci Res 53:279–296
118. Wang X, Siren AL, Liu Y, Yue TL, Barone FC, Feuerstein GZ (1994b) Upregulation of intracellular adhesion molecule-1 (ICAM-1) on brain microvascular endothelial cells in rat ischemic cortex. Mol Brain Res 26:61–68
119. Wang X, Yue T-L, Barone FC, White RF, Gagnon RC, Feuerstein GZ (1994) Concomitant cortical expression of TNFα and IL-1β mRNA follows early response gene expression in transient focal ischemia. Mol Chem Neuropathol 23:103–114
120. Wang X, Yue TL, Barone FC, Feuerstein GZ (1995a) Demonstration of increased endothelial-leukocyte adhesion molecule-1 mRNA expression in rat ischemic cortex. Stroke 26:1665–1669

121. Wang X, Yue TL, Barone FC, Feuerstein GZ (1995b) Monocyte chemoattractant protein-1 (MCP-1) mRNA expression in rat ischemic cortex. Stroke 26:661–666
122. Wang X, Yue TL, Young PR, Barone FC, Feuerstein GZ (1995) Expression of interleukin-6, c-fos, and zif268 mRNA in rat ischemic cortex. J Cereb Blood Flow Metab 15:166–171
123. Wang X, Barone FC, Aiyar NV, Feuerstein GZ (1997) Interleukin-1 receptor and receptor antagonist gene expression after focal stroke in rats. Stroke 28:155–162
124. Wang X, Ellison JA, Siren AL, Lysko PG, Yue TL, Barone FC, Shatzman A, Feuerstein GZ (1998) Prolonged expression of interferon inducible protein 10 in ischemic cortex after permanent occlusion of the middle cerebral artery in rat. J Neurochem 71:1194–1204
125. Wardlaw JM, Yamaguchi T, del Zoppo G (1999) Thrombolytic therapy versus control in acute ischaemic stroke. Stroke module of the Cochrane Database of Systematic Reviews. Available in the Cochrane Library (database on disk and CDROM). Oxford: Update software
126. Watanabe H, Ohta S, Kumon Y, Sakaki S, Sakanaka M (1999) Increase in p53 protein expression following cortical infarction in the spontaneously hypertensive rat. Brain Res 837:38–45
127. Xue D, Huang ZG, Smith KE, Buchan AM (1992) Immediate or delayed mild hypothermia prevents focal cerebral infarction. Brain Research 587:66–72
128. Xue D, Huang ZG, Barnes K, Lesiuk HJ, Smith KE, Buchan AM (1994) Delayed treatment with AMPA, but not NMDA, antagonists reduces neocortical infarction. J Cereb Blood Flow Metab 14:251–261
129. Yabucchi K, Minami M, Katsumata S, Yamazaki A, Satoh M (1994) An in situ hybridization study on interleukin-1β mRNA induced by transient forebrain ischemia in the rat brain. Mol Brain Res 26:135–142
130. Yao H, Ibayashi S, Nakane H, Cai H, Uchimura H, Fujishima M (1997) AMPA receptor antagonist, YM90K, reduces infarct volume in thrombotic distal middle cerebral artery occlusion in spontaneously hypertensive rats. Brain Res 753:80–85

IV Factors and Mechanisms Enhancing Susceptibility or Tolerance (Glia)

Acid-Sensitive Ion Channels in Brain:
New Modulation of Ischemic Injury

R. P. Simon

Key words. Apoptosis – neurodegeneration – brain – rat – neuron – stroke

Acid-Sensing Ion Channels

Acid-sensing ion channels (ASICs) are members of the Deg/ENaC superfamily of ion channels [21]. Members of this family include the constitutively active epithelial sodium channels (ENaCs) in mammals, the stretch activated degenerins (Mec-4, -10; Unc-105) in *C. elegans*, and the ligand-gated FMRF-amide channel in *H. aspersa*. All are inhibited by amiloride, are permeable to Na$^+$, and share a common structure consisting of short intracellular amino- and carboxy-terminal domains connected to transmembrane segments separated by a large extracellular loop.

ASICs are expressed throughout the mammalian nervous system. To date, six subunits have been identified and cloned. Using heterologous expression systems, channel activation, kinetics, and conductance properties have been determined, and several subunits have been shown to associate into heteromultimeric channel complexes with properties distinct from those of homomultimeric channel complexes [2, 3, 15]. The subunit composition of channels *in vivo* has not been determined.

ASICs as Proton-Gated Channels

Four of the ASIC subunits form functional homomultimeric channels that are activated by acidic pH to conduct a sodium-selective cation current. The pH of half-maximal activation (pH$_{0.5}$) of these channels differs: ASIC 1, pH$_{0.5}$ = 6.2 [20]; ASIC 1B, a splice variant with a unique 171 amino acid N-terminal, pH$_{0.5}$ = 5.9 [6]; ASIC2a, pH$_{0.5}$ = 4.4 [Price, 1996 #8; 22]; and ASIC 3, pH$_{0.5}$ = 6.5 [19]. A fifth subunit, ASIC2b is not activated by pH, but has been shown to associate with other subunits and modulate their activity [15].

ASICs expressed by sensory neurons in the periphery have been implicated in the perception of pain during tissue acidosis [4, 13], particularly in the ischemic myocardium where ASICs induce anginal pain [17]. The presence of ASICs in the

Correspondence to: Roger P. Simon, MD, Director and Chair, R. S. Dow Neurobiology Laboratories, Legacy Research, 1225 NE 2nd Avenue, Portland, OR 97232, Tel.: (503) 413-5454, Fax: (503) 413-5465, E-Mail: rsimon@downeurobiology.org

Maturation Phenomenon in Cerebral Ischemia V
A. M. Buchan et al. (Eds.)
© Springer-Verlag Berlin Heidelberg 2004

brain, which lacks nociceptors, suggests that these channels may have functions in the setting of acidosis beyond nociception. In the normal brain, ASICs may act as neurotransmitters, responding to the marked, local fluctuations in extracellular cerebral pH known to accompany synaptic activity [7, 12]. ASICs may also act as mediators of pathological stimuli, responding to the acidosis that accompanies epilepsy, trauma, and, the focus of this proposal, stroke [23].

ASICs as Mechano-Gated Channels

ASICs display significant homology to mechanosensing channels in *C. elegans*, mutations of which cause neurodegeneration. Recently, ASIC2a/b expression was detected immunohistochemically in the lanceolate nerve endings that lie adjacent to hair follicles. Disruption of the mouse ASIC2a/b gene reduced sensitivity to hair movement [16]. ASIC2 may be linked to the cytoskeleton through interaction with the protein stomatin [14], a putative member of the mechanotransducing complex in mammals [9]. ASIC2 expression has also been detected in Meisner corpuscles of glabrous skin and other mechanosensory terminals [11]. A mechanosensing role for ASICs in the brain has not been addressed. Both ASIC2a and ASIC2b are expressed throughout the brain, however, and could function as mechanically gated channels that regulate cell volume [10]. The potential role of ASICs as mechanosensors responding to organelle swelling during ischemia will be addressed by studies proposed herein.

Global Ischemia Induces Expression of ASIC2a Protein in Neurons that do not Sustain DNA Damage

We chose the ASIC2a protein as the focus of our studies because transcripts for ASIC2a have been detected almost solely in the brain [21]. When expressed in heterologous systems, the homomultimeric channel is activated half-maximally at $pH_{0.5} = 4.4$ and conducts a transient, sodium-selective current. Mutation of residue Gly 430 to a bulky amino acid raises $pH_{0.5}$ to 6.7, abolishes inactivation, and causes cell death [1, 5, 22]. Mutation of the same residue in nematode homologs causes neurodegeneration.

Western blotting was used to assess the effect of global ischemia on ASIC2a expression (Fig. 1). Blots of lysates prepared from the hippocampi of sham and ischemic rats show upregulation of a band of 80 kD when probed with an antibody to ASIC2a. Little protein was detected in the sham lane. After reperfusion from global ischemia, expression of ASIC2a was maximal at 8 h and persisted at 72 h.

To examine further the expression pattern of ASIC2a, immunohistochemistry of rat brain tissue was performed with the same antibody (Fig. 2). In sham tissue, a low level of bright green staining was observed in the CA1 (panel A) and CA3 (panel B) sectors of the hippocampus and in the cortex (panel C). Four hours after ischemia, bold staining was observed in many cells of the cortex. The pattern persisted at 8 h; at 24 h antibody also labeled nerve cell processes. In the CA1 and CA3 region, the cytoplasm of most cells was stained at 8 h.

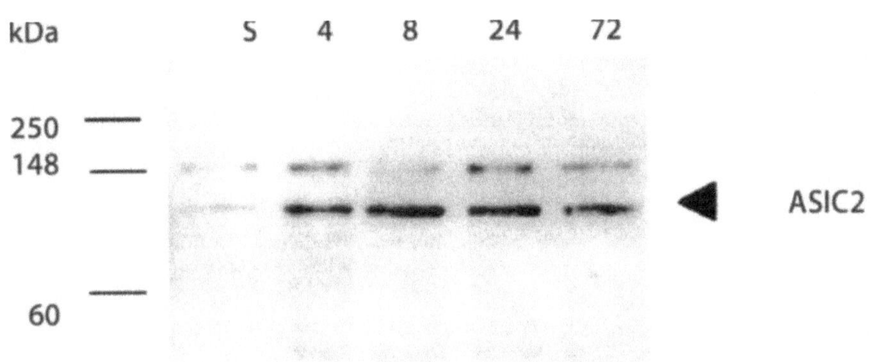

Fig. 1. Global cerebral ischemia induces upregulation of ASIC2a in brain. Representative Western blot (n = 3 per lane) showing increased protein for ASIC2a. Time is shown in hours post 15 minutes global ischemia.

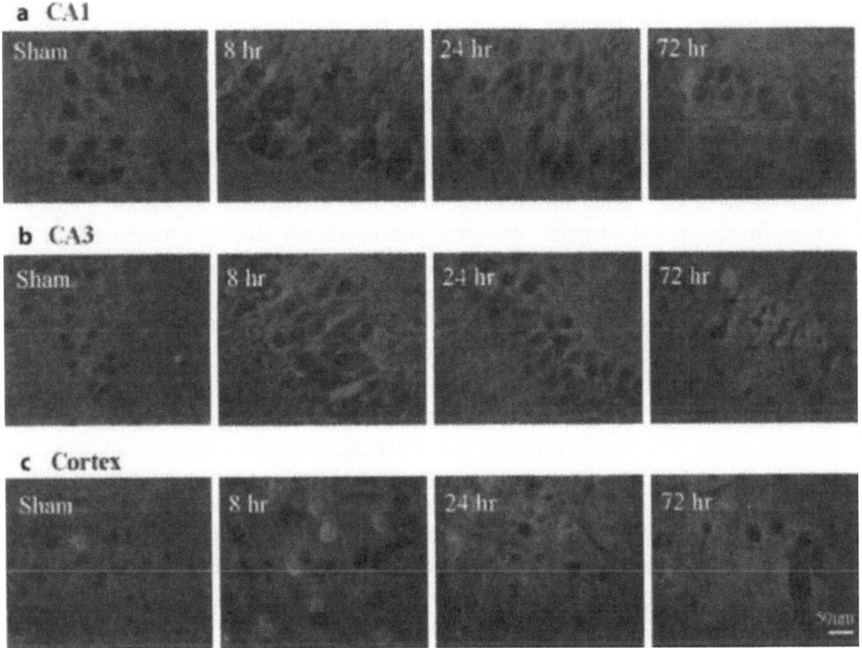

Fig. 2a–c. ASIC2a immunohistochemistry following ischemia. ASIC2a antibody staining of brain sections 8–72 h post global ischemia. Note upregulation of ASIC2a in both hippocampus and cortex.

To determine the cellular distribution of ASIC2a in ischemic tissue, triple fluorescence-labeling experiments were performed with nuclear dye Hoechst 33258 and antibodies to ASIC2a and neuronal marker NeuN or ASIC2a and astrocytic marker GFAP. ASIC2a and NeuN immunoreactivity co-localized, indicating that global ischemia induces ASIC2a expression in neurons. ASIC2a and GFAP immunoreactivity did not co-localize (data not shown).

To determine whether upregulation of ASIC2a in selectively vulnerable regions of ischemic brain occurs in cells destined to survive or apoptose, we double labeled tissue for ASIC2a and DNA strand breaks. Klenow fragment-mediated labeling of DNA single and double strand breaks with protruding 5′ termini revealed widespread DNA damage in cells of the CA1 sector 72 h after reperfusion from ischemia. The antibody to ASIC2a labeled scattered cells, but did non co-localize to cells with DNA damage. We have seen this pattern previously in studies of ischemic cell death suppressor proteins, such as Bcl-w, where the death suppressor gene product was not found in cells with DNA damage.

Some ASIC Protein Expression Localizes to the Mitochondria

In our earliest immunohistochemical studies of ASIC protein expression in ischemic rat brain we used a polyclonal antibody (ab973, raised by us) designed to recognize a sequence of nineteen amino acids – LLGDIGGQMGLFIGASSILT – preceding the second transmembrane domain of each of the ASIC subunits. We noticed that some of the ASIC immunoreactivity had a punctate appearance suggesting a membrane-bound protein. We observed similar staining patterns in studies of Bcl-2 and Bcl-w, anti-apoptotic proteins present in the mitochondrial membrane after ischemia [24]. We subsequently determined the sub-cellular localization of ASIC immunoreactivity using double-label immunohistochemistry with antibodies to either the mitochondrial-membrane bound enzyme CoxIV or the endoplasmic reticulum (ER) protein Grp94 (data not shown). In normal and ischemic tissue, we observed some co-localization of CoxIV and ASIC immunoreactivity, but no co-localization of ASIC and Grp94 immunoreactivity. Our results suggest that a portion of the ASIC protein present in neurons is located at the mitochondrial membrane.

To begin to assess which of the ASIC subunits might be expressed at the mitochondria, we did double-labeling immunohistochemistry with antibodies to ASIC2a and CoxIV. Little co-localization was observed in normal brain or in ischemic brain 4 and 72 h after reperfusion, suggesting that a subunit other than ASIC2a is present in mitochondria.

Oxygen Glucose Deprivation Alters the Properties of Acid-Induced Currents in Cultured Cortical Neurons

Acid-evoked currents in cultured cortical neurons have been previously described [18]. Shifts in extracellular pH from 7.4 to < 7.0 evoke sodium currents that are inhibited by amiloride. Intercellular variation in $pH_{0.5}$ suggests ASIC subunits are differentially expressed in the cells. To assess the effect of ischemia on ASIC cur-

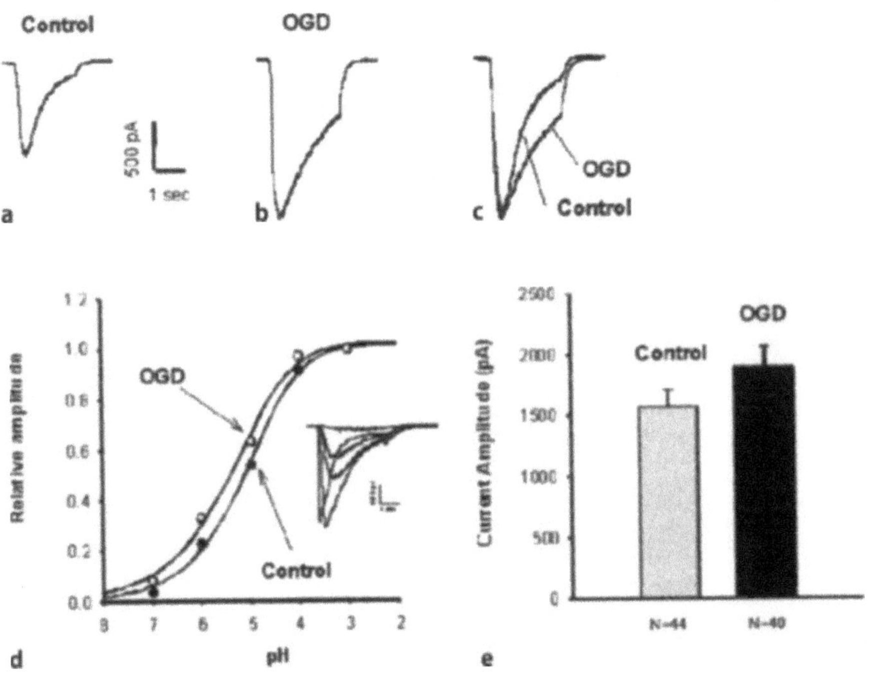

Fig. 3a–e. Oxygen-glucose deprivation (OGD) potentiates acid-sensing currents in cultured cortical neurons. **a** Representative trace showing the acid-sensing current in control neuron. Current was activated by pH drop from 7.4 to 6.0. **b** Acid-sensing current in neuron treated with 1 h OGD. **c** Overlay of two currents showing the slow down of desensitization of acid-sensing current after OGD treatment. **d** pH-dependency of acid-sensing current in control and OGD-treated neurons. **e** Summary data showing the increase of amplitude of acid-sensing current following OGD.

rent we used oxygen and glucose deprivation (OGD) as an in-vitro model of ischemia/hypoxia. Dissociated cultures of mouse neocortical neurons were subjected to one hour of OGD and allowed to recover for ten hours. We then assessed ASIC activity by patch-clamp recording, using a multi-barreled perfusion system to change pH rapidly from 7.4 to 6.0 and 5.0. The acid-evoked current observed in ischemic cells was larger than the current observed in normal cells and contained a sustained component (Fig. 3 A–C, E). A shift in pH sensitivity was also observed (Fig. 3 D). Our results indicate that OGD causes a change in ASIC activity. This might reflect changes in subunit expression. For instance, the ASIC1a comprising channel is activated at $pH_{0.5} = 6.2$, whereas the ASIC3-comprising channel is activated at $pH_{0.5} = 6.5$ and shows a slower, sustained component when activated at lower pH. The change in acid-evoked current might also reflect a change in subunit stoichiometry. Co-expression of ASIC subunits has been shown to generate channels with novel pH sensitivities, kinetics, and ion selectivities [3, 15]. Finally, OGD might alter the properties of existing channels by modulating protein interactions or phosphorylation states.

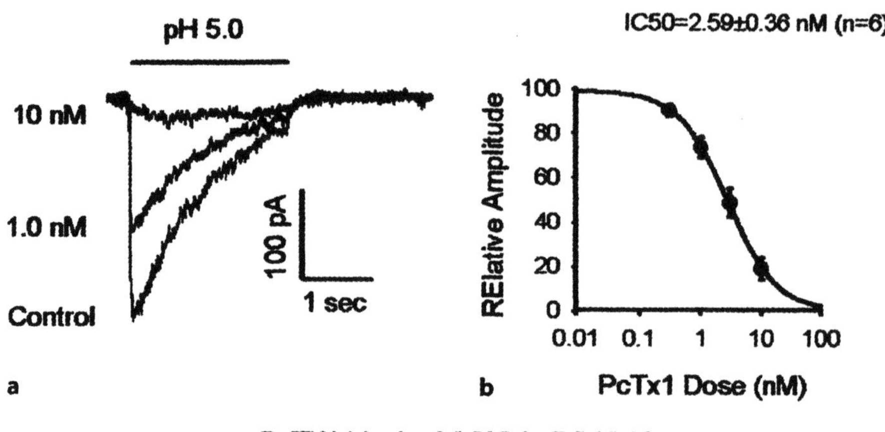

Fig. 4 a, b. Dose-dependent blockade of acid-sensing current in PC12 cells by PcTX1. **a** Representative traces showing the blockade of acid-sensing current by PcTX1 at 1.0 and 10 nM. Current was activated by a pH drop from 7.4 to 5.0 with a holding potential of −60 mV. **b** Dose-inhibition curve of PcTX1 blockade of acid-sensing current. IC_{50} for PcTX1 block was 2.59 ± 0.36 nM. n = 6.

Neurotoxin PcTX1 Blocks Acid-Evoked Current in PC12 Cells

Nerve growth factor (NGF)-differentiated PC12 cells are a neuronal cell line commonly used to study membrane receptor and ion channel functions. In electrophysiology studies of PC12 cells, we observed a transiently activated current upon changing the pH from 7.4 to 5.0. This observation suggested that PC12 cells likely expressed ASIC1a subunits. This was consistent with RT-PCR detection of ASIC1a transcripts (data not shown). Next, we assayed the ability of a tarantula venom containing Psalmotoxin 1 (PcTX1, Invertebrate Biologics, Los Gatos, CA) to block acid-evoked channel activity. PcTX1 was recently isolated from the venom of *Psalmopoeus cambridgei* and shown to inhibit specifically the channel activity of ASIC1a, but not ASIC1b, ASIC2a, or ASIC3 [8]. We found that 10 nM PcTX1 was sufficient to abolish 90% of acid-evoked current (Fig. 4).

Conclusion

Thus, acid-sensitive ion channels exist in brain where their physiologic and pathophysiologic roles are unlikely to be those of nociception or mechanosensing, but may offer a novel modulatory site for the treatment of ischemic brain injury. The evolving story suggests that the channel activity is more responsive in the setting of modeled ischemia, suggesting a modulatory component beyond pH. In-vivo studies suggest that the ASIC2a subunit may offer neuroprotection, perhaps by heteromeric modulation of ASIC1a channel function. The possibility that these channels might be localized in the mitochondrial compartment offers an additional focus of biological intervention in ischemia.

References

1. Adams CM, Snyder PM, Price MP, et al (1998) Protons activate brain Na+ channel 1 by inducing a conformational change that expresses a residue associated with neurodegeneration. J Biol Chem 273(46):30204
2. Babinski K, Catarsi S, Biagini G, et al (2000) Mammalian ASIC2a and ASIC3 subunits co-assemble into heteromeric proton- gated channels sensitive to Gd3+. J Biol Chem 275(37):28519
3. Bassilana F, Champigny G, Waldmann R, et al (1997) The acid-sensitive ionic channel subunit ASIC and the mammalian degenerin MDEG form a heteromultimeric H+-gated Na+ channel with novel properties. J Biol Chem 272(46):28819
4. Bevan S, Yeats J (1991) Protons activate a cation conductance in a sub-population of rat dorsal root ganglion neurones. J Physiol 433:145
5. Champigny G, Voilley N, Waldmann R, et al (1998) Mutations causing neurodegeneration in Caenorhabditis elegans drastically alter the pH sensitivity and inactivation of the mammalian H+-gated Na+ channel MDEG1. J Biol Chem 273(25):15418
6. Chen CC, England S, Akopian AN, et al (1998) A sensory neuron-specific, proton-gated ion channel. Proc Natl Acad Sci USA 95(17):10240
7. Chesler M, Kaila K (1992) Modulation of pH by neuronal activity. Trends Neurosci 15(10):396
8. Escoubas P, De Weille JR, Lecoq A, et al (2000) Isolation of a tarantula toxin specific for a class of proton-gated Na+ channels. J Biol Chem 275(33):25116
9. Fricke B, Lints R, Stewart G, et al (2000) Epithelial Na+ channels and stomatin are expressed in rat trigeminal mechanosensory neurons. Cell Tissue Res 299(3):327
10. Garcia-Anoveros J, Derfler B, Neville-Golden J, et al (1997) BNaC1 and BNaC2 constitute a new family of human neuronal sodium channels related to degenerins and epithelial sodium channels. Proc Natl Acad Sci USA 94(4):1459
11. Garcia-Anoveros J, Samad T, Woolf CJ, et al (2000) The Deg/ENaC channel BNaC1 is located mechanosensory terminals of DRG neurons. Society for Neuroscience Abstracts 26:938
12. Krishtal OA, Osipchuk YV, Shelest TN, et al (1987) Rapid extracellular pH transients related to synaptic transmission in rat hippocampal slices. Brain Res 436(2):352
13. Krishtal OA, Pidoplichko VI (1981) A receptor for protons in the membrane of sensory neurons may participate in nociception. Neuroscience 6(12):2599
14. Lewin GR (2000) ASIC2a/b may interact with stomatin in mechanosensory cells. New Orleans
15. Lingueglia E, de Weille JR, Bassilana F, et al (1997) A modulatory subunit of acid sensing ion channels in brain and dorsal root ganglion cells. J Biol Chem 272(47):29778
16. Price MP, Lewin GR, McIlwrath SL, et al (2000) The mammalian sodium channel BNC1 is required for normal touch sensation. Nature 407(6807):1007
17. Sutherland SP, Benson CJ, Adelman JP, et al (2000) Acid-sensing ion channel 3 matches the acid-gated current in cardiac ischemia-sensing neurons. Proc Natl Acad Sci USA epub
18. Varming T (1999) Proton-gated ion channels in cultured mouse cortical neurons. Neuropharmacology 38(12):1875
19. Waldmann R, Bassilana F, de Weille J, et al (1997) Molecular cloning of a non-inactivating proton-gated Na+ channel specific for sensory neurons. J Biol Chem 272(34):20975
20. Waldmann R, Champigny G, Bassilana F, et al (1997) A proton-gated cation channel involved in acid-sensing. Nature 386(6621):173
21. Waldmann R, Champigny G, Lingueglia E, et al (1999) H(+)-gated cation channels. Ann NY Acad Sci 868:67
22. Waldmann R, Champigny G, Voilley N, et al (1996) The mammalian degenerin MDEG, an amiloride-sensitive cation channel activated by mutations causing neurodegeneration in Caenorhabditis elegans. J Biol Chem 271(18):10433
23. Xiong ZQ, Saggau P, Stringer JL (2000) Activity-dependent intracellular acidification correlates with the duration of seizure activity. J Neurosci 20(4):1290
24. Yan C, Chen J, Chen D, et al (2000) Overexpression of the cell death suppressor Bcl-w in ischemic brain: implications for a neuroprotective role via the mitochondrial pathway. J Cereb Blood Flow Metab 20(3):620

In-Vitro Elucidation of Mechanisms Underlying Cell Swelling and Death of Nerve and Glial Cells

N. Plesnila, F. Ringel, F. Staub, E. Stohr, C. C. Chang, and A. Baethmann

Summary. Notwithstanding that the control of a normal cell volume seems to assume a high cell biological priority, as demonstrated by the spontaneous normalisation of the cell volume following anosmotic exposure, the cellular swelling response may also have important functions to support the homeostasis. Prominent examples are the clearance of excessive K^+ or glutamate concentrations from the interstitial compartment in order to defend a normal neuronal function. This requires not only physiological transmembrane Na^+- and K^+-concentration gradients, but also an absence, or at least very low levels of excitatory transmitter compounds in the perisynaptic (interstitial) compartment. Yet, cell swelling may also reflect sequelae of cell injury, may be even an early indication of impending cell death. For example, cell swelling from arachidonic acid or higher levels of acidosis may eventually merge into an irreversible insult of a viable cell.

Nevertheless, the development of the acidosis-induced cell swelling following a dose-response characteristic is associated with activation of a variety of ion-transport and -exchange mechanisms, which rather may be viewed as a homeostasis mechanism to ensure survival of the cell under these circumstances. If, however, an overwhelming insult, such as continued interruption or poor quality of the cerebral blood flow had occurred, formation of cell swelling is attributable to the failure of the basic elements of cell volume control involving enhanced leakage of Na^+-ions into the cell. Inhibition of the active Na^+-pump and break down of the energy metabolism. Simultaneously, one must admit that this concept certainly is too simple to explain the plethora of mechanisms and phenomena associated with cytotoxic cell swelling and cell damage in the brain, particularly under clinically relevant conditions.

Key words. Cytotoxic brain edema – glial and nerve swelling – irreversible cell damage – trauma – ischemia

Nikolaus Plesnila, Florian Ringel, Frank Staub, Edith Stohr, Raymond Chang and Alexander Baethmann
Institute for Surgical Research, Klinikum of the Ludwig-Maximilians University, Großhadern, 81377 Munich, Germany

Correspondence to: Dr. N. Plesnila, M.D., Institute for Surgical Research, Klinikum of the University of Munich-Großhadern, Marchioninistr. 27, 81377 München, Germany, Tel.: 0049-89-7095-4352, Fax: 0049-89-7095-4353, E-Mail: plesnila@icf.med.uni-muenchen.de

Maturation Phenomenon in Cerebral Ischemia V
A.M. Buchan et al. (Eds.)
© Springer-Verlag Berlin Heidelberg 2004

Introduction

Cell swelling in the brain parenchyma, particularly of astrocyte processes, or elements of the myelin sheath, traditionally has been considered as manifestations of *cytotoxic brain edema* occurring under a variety of pathological conditions [1]. Without opening of the blood-brain barrier and influx of plasma-like *vasogenic edema* fluid, *cytotoxic edema* is characterized as an increase of the cell volume by fluid, which is derived from the extracellular space and, of course, the blood vessels. As a result the intracellular compartment expands at the expense of a shrinking interstitial space. This has been reliably demonstrated by various procedures in the laboratory, by measurement of the brain tissue impedance, or electronmicroscopy utilizing suitable fixation precautions [1]. More recently, assessment of the *Apparent Diffusion Coefficient (ADC)* by the nuclear magnetic resonance technology is considered to provide respective information [e.g. 8]. The ADC measurement, however, must be subjected to careful interpretations, as thereby in essence the mobility of water molecules is studied without being able to identify where respective changes take place.

As to cerebral ischemia, a complete interruption of the cerebral blood flow may result within a short period (1–2 minutes) in a massive translocation of extracellular fluid into cells, whereas the brain volume in total may not be affected, as no additional fluid can enter the brain through the cerebral circulation [7]. However, once the recirculation is initiated a massive influx of water into the brain occurs – through the intact blood-brain barrier. This is attributable to the fact that, during the circulatory arrest period, the osmotic pressure, or concentration, respectively of brain parenchyma is markedly increasing, i.e. by no less than approximately 50 mosmol over that of blood providing a powerful osmotic gradient underlying the net fluid shift [7]. Since an osmotic concentration difference of only 1 mosmol is equivalent to nearly 20 mmHg, the resulting osmolarity increase of ischemic brain tissue generates an osmotic pressure differential between blood and tissue of no less than ca. 1000 mmHg – an enormous force which is driving fluid into the brain immediately upon onset of recirculation.

This notwithstanding, this type of ischemic brain edema mostly affecting the intracellular compartment during recirculation does not necessarily reflect cell damage, as it is reversible provided the post-ischemic blood flow is adequate. There are other forms of cell swelling induced by mediator agents, as by arachidonic acid, or by acidosis, where cell swelling may be attributable to a failure of function or activation of homeostasis mechanisms. As to the latter, cell swelling from e.g. an increased interstitial glutamate- or K^+-ion level is increasingly viewed as manifestation of an activation of homeostasis mechanisms, i.e. as an attempt to normalize the extracellular environment by clearance of agents exerting toxic properties, if accumulating in this compartment. As to increased K^+-ion levels in the interstitial space, the resulting cell swelling is attributable to activation of cellular uptake mechanisms for these ions, in order to normalize the extracellular homeostasis as a requirement for a normal physiological neuronal activity. Altogether, cell swelling in the brain does, therefore, not always reflect damage of nerve or glial cells, rather it may indicate activation of mechanisms providing for a normalisation of the homeostasis. With regard to an increased glutamate- or K^+-

ion level, the reestablishment of a physiologically low K^+- or amino acid concentration in the extracellular compartment obviously has a higher priority than the defense of a normal cell volume.

The understanding of cell swelling in the brain including the physiological mechanisms of cell volume control has been markedly facilitated by the availability of respective in-vivo and in-vitro procedures to quantitatively measure cell volume and its changes, even under dynamic conditions. Obviously, quantitative measures of the cell volume can be more easily obtained under in-vitro conditions, e.g. by using suspended single cells than by studying the brain in-vivo, particularly in view of the highly complex structure of the neuropil in the grey matter of the brain. This not withstanding, estimates of changes of the extra and intracellular compartment can also be made under in-vivo conditions, albeit with limitations. For example, employment of electronmicroscopy is usually limited to parts of cells or cell processes, and moreover, burdened by fixation artifacts [21].

Approaches to study global changes of the cell volume in brain parenchyma were frequently based on the distribution measurement of indicator compounds confined to the extracellular compartment, or on measurements of the electrical tissue impedance. The latter method takes advantage of the properties of brain parenchyma to conduct a given electrical current preferentially through the extracellular compartment. It is thereby possible to observe changes of the electrical conductivity either by an increase or decrease of this fluid space [19]. The global methods allow for general assumptions on the extra-/intracellular fluid distribution, whereas these procedures do not lend themselves to conclusions on the cell type, which is affected by swelling or shrinking, respectively. If, however, these methods are combined with electronmicroscopical procedures, highly valuable information can be obtained. A case in point is the metabolic (or cytotoxic) type of brain edema induced by an experimental energy failure secondary to administration of the metabolic inhibitor 6-aminonicotinamide. Thereby, the brain water content is markedly increased together with the electrical tissue impedance, which indicates the development of cytotoxic brain edema. Respective electronmicroscopical investigations provided evidence for a marked swelling of glial processes under these circumstances [4]. Corresponding experiments were carried out by using dinitrophenol, in order to uncouple the oxidative phosphorylation, again to induce energy failure. The resulting brain edema studied as an increase of the water content was also associated with a shrinking of the extracellular compartment suggestive of an increase of the intracellular space, which was concluded from the distribution measurement of the extracellular volume marker ^{35}S-thiosulphate [2]. Although these experiments have confirmed that swelling of brain cells can be induced and objectively measured under in-vivo conditions, a detailed evaluation was not possible, which would require more subtle technologies together with utilization of defined cell types.

Such a model was developed in our laboratory by Kempski and co-workers years ago [11]. It made possible to study cells from central nervous tissue in single cell suspension with selective changes of pertinent parameters of the environment. Respective studies were made to elucidate both the mechanisms of cell volume control on the one hand side, and cell swelling-inducing properties of pathophysiological relevant mediator agents, as glutamate, arachidonic acid, or acidosis on the other. For that purpose the cell volume response was followed under dynamic conditions using cell volume measurements in short intervals. With this procedure, the volume of intact (i.e. viable cells) is measured by a highly sensitive procedure – electrical cellvolumetry utilizing hydrodynamic focussing. The method was in essence derived from the classical Coulter procedure [9]. The technology makes it possible to detect changes of the cell volume as small as only 1–2%.

Mechanisms of Cell Volume Control

In order to understand the host of mechanisms, which could induce cell swelling under clinically relevant pathological conditions, as hypoxia, glutamate overflow, acidosis, release of toxic mediator agents, etc. the physiology of cell volume regulation must be understood in details. A valuable concept is the double DONNAN equilibrium (or pump-leak model) governing the electrolyte- and fluid exchange between cells and the environment, and providing for a stable cell volume. The basic principles of the DONNAN equilibrium can be summarized as follows [14]:
1. Low membrane permeability for Na^+-ions under resting conditions.
2. Maintenance (or normalisation, respectively) of the normal intra-/extracellular ion-concentration gradients and, thus, fluid distribution by membrane pumps and transporters, especially the Na^+/K^+-ATPase.
3. Supply of metabolic energy, particularly of ATP, to fuel the active Na^+-pump mechanisms which is indispensable, e.g. for the active export of Na^+-ions against their steep electro-chemical gradient out of the cell.

On the basis of these processes cell swelling can be explained in essence by the following disturbances:
1. Increase of the membrane permeability for Na^+-ions leading to a net influx of this ion species, which would overwhelm their active extrusion
2. Inhibition of the active Na^+-pump (Na^+/K^+-ATPase)
3. Failure of the energy metabolism to provide fuel for the active Na^+-pump.

These basic hypotheses have been studied and confirmed by a variety of in-vivo and in-vitro experiments. An important exception of these mechanisms seems to be provided by cell swelling resulting from an osmotic imbalance between the extra- and intracellular space. Accordingly, an abrupt dilution of the extracellular osmolarity as from water intoxication, which rapidly causes overwhelming cell swelling does not fit into the general concept given above. Neither the Na^+-permeability of the cell membrane is increased, nor the active Na^+-pump inhibited or the energy metabolism failing. Instead, the acute decrease of the extracellular osmotic concentration, e.g. by water intoxication, generates an osmotic gradient between the intra- and extracellular compartment, which is driving water into the cell through the intact cell membrane – which is facilitated by the high permeability of the cell membrane to the small water molecules also under normal conditions. For that reason, in almost all organs with few exceptions (e.g. the kidney) no osmotic concentration difference can be found between cells and their extracellular environment. This would immediately result in a net fluid shift causing either shrinking or swelling of cells. Aside from the peculiarity of osmotic cell swelling or osmotic brain edema in that case, the shift of fluid from an osmotic gradient is important nevertheless, as practically all manifestations of cell swelling in one way or the other must finally be attributed to the generation of an osmotic concentration difference – mostly to an increase of osmotic solutes in the cell providing for a net fluid shift along this gradient.

Taken together, aside from the specificities of the osmotically induced cell swelling from water intoxication – under clinical circumstances from acute hyponatremia of various origins – practically all other forms of cell swelling in the brain, indicating cell injury or activation of defence mechanisms are based on a net influx of osmotically active solutes followed by a net increase of the cellular water. Subsequently few aspects of cellular edema in brain tissue are discussed under clinically pertinent conditions.

Cell Swelling Resulting from Disturbances of the Pump-Leak Equilibrium

Under certain conditions the membrane permeability for Na^+-ions in the brain is increased, most often secondary to release and accumulation of excitatory transmitters in the extracellular compartment, as glutamate. It has been observed that the addition of glutamate to glial cells *in vitro*, brain slices or cerebral cortex *in vivo* induces cellular edema [cf. 5]. Actually, a transient swelling of nerve fibres has even been observed under normal conditions during synaptic excitation. This, however, is rapidly reversible for obvious reasons. In addition, an increase in the cell volume from an enhanced intracellular influx of Na^+-ions is also observed in the absence of an increased membrane permeability for this ion species. For example, a variety of compounds or ions are utilizing the transmembrane Na^+-ion gradient as a direct source to energize their extra- to intracellular transport. Accordingly, the uphill influx of such a compound into the cells is coupled to a concurrent downhill influx of Na^+-ions which, of course, raises the osmotic solute concentration in the cell, thereby the cell water content. A case in point is the clearance of extracellular glutamate through high-affinity uptake by the glial cells. Due to the steep intra- to extracellular concentration gradient the uptake of glutamate is energy dependent – so to say fuelled by a simultaneous downhill influx of Na^+-ions into the glial cells along their concentration gradient. Accordingly, the transport of 1 mol glutamate into the cell is accompanied by an influx of 2–3 mol Na^+-ions raising the cellular Na^+-concentration (together with glutamate) as a mechanism of the resulting cell swelling [cf. 5]. The cell volume increase from this process is, therefore, not a pathological phenomenon, but rather a compensatory response to restore the extracellular homeostasis. Glial swelling observed during exposure to high extracellular K^+-levels, e.g. from excessive neuronal excitation, or following interruption of blood flow, can also be considered as a result of their clearance function, which again raises the osmotic solute concentration in the cell underlying cell swelling. Altogether, the swelling of glial cells from accumulation of transmitter compounds or active pumping of excess K^+-ions in the interstitial compartment is not attributable to disturbances of the pump-leak equilibrium.

Cell Swelling from Lactic Acid

Lactacidosis is an important phenomenon in e.g. cerebral ischemia or trauma or other adverse conditions eventually resulting in cell swelling and cell death [13]. Measurements of the tissue's pH under respective circumstances have confirmed that the severity of acidosis, which could be enhanced by a high blood glucose level, suffices to induce cell swelling and cell death. On the basis of our own experiments a cell swelling threshold from acidosis of suspended glial cells has been identified at pH 7 – 6.8, whereas under acute conditions cell viability was affected only when the pH fell to 5.6 or below [18]. Both cell swelling and irreversible cell injury were exhibiting a dose-response behaviour, once these pH thresholds were passed.

Cell swelling from acidosis has been considered as a defence mechanism for the maintenance of a normal intracellular pH (pH_i). More recent investigations, however, indicate that activation of respective processes, particularly of the Na^+/H^+-antiporter does not suffice to maintain the intracellular pH at a normal level, which actually during acidosis is also liable to decrease [15, 16]. Nevertheless, a certain degree of protection by the activation of the Na^+/H^+-exchanger can be assumed on the basis of experiments that inhibition of this transporter, e.g. by amiloride, attenuates the acidosis-induced cell swelling, while adversely affecting pH_i. Albeit indirectly this is confirming protection by activation of the Na^+/H^+-antiporter. This is further supported by experimental data demonstrating a stronger decrease of pH_i from acidosis under conditions, when Na^+-ions are omitted from the suspension medium (without affecting osmolarity) as compared with the same level of acidosis under maintenance of a normal Na^+-concentration in the medium. It may be assumed that the activation of the antiporter requiring availability of extracellular Na^+-ions at least limits the severity of the cellular acidosis. Another interesting observation is concerned with the dependence of the cell volume response at the same degree of acidosis on the type of acid administered. Accordingly, the degree of swelling induced from lactic acid exposure at a given pH is far more pronounced as compared to that from sulfuric acid exposure [18]. In addition, experiments on the pH-threshold of irreversible cell injury from acidosis using primary cultured astrocytes were demonstrating that, at pH 5.6 or below, lactic acid induced a progressive decline of the cell viability, whereas cells were surviving at this pH when sulfuric acid was administered instead [18].

Experiments of this laboratory provide further information as to the mechanisms of intracellular acidification from acid exposure of suspended glial cells. Accordingly, evidence is available that the acidotic exposure activates also anion exchangers, resulting in a discharge of HCO_3^- ions against an intracellular accumulation of Cl^--ions. Further, H^+-ions may enter the cells through specific channels in the cell membrane which can be blocked by Zn^+-ions [16]. The resulting accumulation of protons in the cytosol in turn activates the above mentioned exchange mechanisms resulting in an accumulation of Na^+-ions. The most prominent example is the Na^+/H^+-antiporter. Finally, lactic acid and other carbonic acids may enter the cell as non-dissociated protonated molecules, which is facilitated by a decreasing pH according to the dissociation equilibrium of the acid. The movement of non-dissociated lactic acid molecules through the cell membrane is facilitated by their lipophilicity as compared to the water- soluble negatively charged lactate anion. Once, however, the protonated lactic acid molecule is inside the cell, it dissociates into H^+-ions and lactate anions, which not only enhances intracellular acidification but also the osmotic load, since polar lactate anions are trapped in the cytosol. Altogether, the cell swelling and damage from lactic acid demonstrates *par excellence* the ambivalent nature of the cellular response mechanisms to acidosis, as e.g. activation of the Na^+/H^+-antiporter or HCO^{3-}/Cl^--exchanger originally thought to defend the intracellular pH.

Cell Swelling in the Brain from Glutamate

The properties of excitatory amino acids particularly of glutamate to induce cell swelling in the brain are known for a long time. Pioneering work has been carried out by Van Harreveld [20] in the early seventies with administration of glutamate to the rat brain *in vivo* by iontophoresis leading not only to an extensive swelling of all cellular elements but also to the development of sheer pannecrosis of the brain parenchyma at the site of infusion. Another procedure to administer glutamate to the brain requiring circumvention of blood-brain barrier is ventriculo-cisternal perfusion of the brain. This results among others in a marked rise of the brain tissue impedance suggestive of an intracellular fluid shift from the extracellular compartment [10] together with the development of massive brain edema. Further, shrinking of the inulin space was observed in experiments using brain slices, which were subjected to glutamate exposure demonstrating again the development of cell swelling [6]. A dose-response behaviour of glial swelling from glutamate has been investigated in details by our laboratory. Thereby, glutamate concentrations of as low as 50 mol were found to induce already a significant although transitory swelling of glial cells, which was intensified at higher glutamate levels [17]. The temporary limitation of the cell swelling response to glutamate at low concentrations was interpreted as an indication of an effective clearance of glutamate from the medium by the suspended glial cells. Obviously, the cell volume normalisation was afforded by effective removal of the amino acid from the extracellular medium.

Observations are available that the glutamate-induced cell swelling can be inhibited by blocking of the Na^+/K^+-ATPase (the active Na^+-pump) by ouabain [12]. This indicates that cell swelling from glutamate not only is brought about by opening of Na^+-channels, but – under the experimental conditions, which were studied – preferably by a cotransport of glutamate together with Na^+-ions into the cells on the basis of the steep extra- to- intracellular Na^+-gradient. This gradient is eliminated by inhibition of the active Na^+-ion extrusion with ouabain. Therefore, uptake of glutamate together with Na^+-ions from the extracellular compartment was not possible anymore, explaining inhibition of the glutamate-induced cell swelling by ouabain. The clinical significance of the glutamate-induced cell swelling during ischemia or trauma relates with the overflow of the amino acid into the interstitial space reaching concentration, which induces cell swelling. As known, there is a massive release of glutamate together with K^+-ions during the terminal anoxic depolarization after complete interruption of the cerebral blood flow. Further, leakage of glutamate across damaged cell membranes can be expected in necrotic brain tissue areas, which has been confirmed by microdialysis and other experiments. Glutamate released in this area may be transported into the penumbra of an ischemic or traumatic focus of necrosis. This process is facilitated by the formation of vasogenic edema spreading from the necrotic core into the surrounding vital parenchyma [3].

The current analysis of the acidosis- as well as glutamate-induced cell swelling may serve as an example of the plethora of mechanisms activated in the brain by acute insults as ischemia or trauma. The resulting cell swelling then can be interpreted both as a homeostasis response but also as sequelae of cell injury. In addition to both glutamate and acidosis, a variety of further mediator agents must be considered, such as arachidonic acid and its metabolites, free radicals, which alone or in concert with other mediators have cell swelling inducing properties eventually causing irreversible cell damage.

Acknowledgement. The secretarial assistance of Pia Handke for preparation of the manuscript is gratefully acknowledged.

References

1. Baethmann A (1978) Pathophysiological and pathochemical aspects of cerebral edema. Neurosurg Rev 1:85–100
2. Baethmann A, Sohler K (1975) Electrolyte – and fluid-spaces of rat brain *in situ* after infusion with dinitrophenol. J Neurobiol 6:3–84
3. Baethmann A, Maier-Hauff K, Schürer L, Lange M, Guggenbichler C, Vogt W, Jacob K, Kempski O (1989) Release of glutamate and of free fatty acids in vasogenic edema. J Neurosurg 70:578–591
4. Baethmann A, Van Harreveld A (1973) Water and electrolyte distribution in grey matter rendered edematous with a metabolic inhibitor. J Neuropathol Exp Neurol 32:408–423
5. Baethmann A, Staub F (1997) Cellular Edema. In: Welch, KMA, Caplan LR, Reis DJ, Siejö BK, Weir B (eds). Primer on Cerebrovascular Diseases. Academic Press, pp 153–156
6. Chan PH, Fishman RA, Lee JL (1979) Effects of excitatory neurotransmitter amino acids on swelling of rat brain cortical slices. J Neurochem 33:1309–1315
7. Hossman K-A, Takagi S (1976) Osmolality of brain in cerebral ischemia. Exp Neurol 51:124–131
8. Jiang Q, Zhang RL, Zhang ZG, Ewing JR, Divine GW, Chopp M (1998) Diffusion-, T_2- and perfusion-weighted nuclear magnetic resonance imaging of middle cerebral artery embolic stroke and recombinant tissue plasminogen activator intervention in the rat. J Cereb Blood Flow Metab 18:758–767
9. Kachel V (1976) Basic principles of electrical sizing of cells and particles and their realization in the new instrument "Metricell". J Histochem Cytochem 24:211–230
10. Kempski O (1982) Die Lokalisation des Glutamat-induzierten Hirnödems. Thesis, Ludwig-Maximilians-University of Munich
11. Kempski O, Chaussy L, Gross U, Zimmer M, Baethmann A (1983) Volume regulation and metabolism of suspended C6 glioma cells: An in vitro model to study cytotoxic brain edema. Brain Res 279:217–228
12. Kempski O, Staub F, Schneider G-H, Weigt H, Baethmann A (1992) Swelling of C6 glioma cells and astrocytes from glutamate, high K^+ concentrations or acidosis In: Yu ACH, Hertz L, Norenberg MD, Syková E, Waxman SG (eds). Progress in Brain Res 94:69–75
13. Kraig RP, Petito CK, Plum F, Pulsinelli WA (1987) Hydrogen ions kill brain at concentrations reached in ischemia. J Cereb Blood Flow Metab 7:379–386
14. Macknight ADC, Leaf A (1977) Regulation of cellular volume. Physiol Rev 37:510–573

15. Mellergård PE, Ouyang YB, Siesjö BK (1992) The regulation of intracellular pH in cultured astrocytes and neuroblastoma cells, and its dependence on extracellular pH in a HCO₃-free solution. Can J Physiol Pharmacol 70:S293–S300
16. Plesnila N, Haberstok J, Peters J, Kölbl I, Baethmann A, Staub F (1999) Effect of lactacidosis on cell volume and intracellular pH of astrocytes. J Neurotrauma 16:831–841
17. Schneider GH, Baethmann A, Kempski O (1992) Mechanisms of glial swelling induced by glutamate. Can J Physiol Pharmacol 70:S334–S343
18. Staub F, Baethmann A, Peters J, Weigt H, Kempski O (1990) Effects of lactacidosis on glial cell volume and viability. J Cereb Blood Flow Metab 10:866–876
19. Van Harreveld A, Ochs S (1956) Cerebral impedance changes after circulatory arrest. Am J Physiol 187:180–192
20. Van Harreveld A, Fifkova E (1971) Light- and electronmicroscopic changes in central nervous tissue after electrophoretic injection of glutamate. Exp Mol Pathol 15:61–81
21. Van Harreveld A (1972) The extracellular space in the vertebrate central nervous system. In: Bourne GH, (ed) The Structure and Function of Nervous Tissue, Vol. 4, pp 447–511. Academic Press

Endogenous Protection Against Hypoxia/ Ischemia in the Brain via Erythropoietin

L. Neeb, K. Ruscher, U. Dirnagl, and A. Meisel

Summary. Tolerance to cerebral ischemia can be induced experimentally by a variety of physical or pharmacological stimuli ('ischemic preconditioning'). We propose that Hypoxia inducible factor-1 (HIF-1) is a master switch of the transcriptional response in the induction of ischemic tolerance. We postulate that erythropoietin (EPO), a key target gene of HIF-1, plays a central role in this scenario. We have modeled ischemic preconditioning *in vitro* (oxygen-glucose deprivation, OGD) and *in vivo* (focal cerebral ischemia in mice; induction of tolerance by hypoxia or desferrioxamine) and provide evidence for the following signaling cascade: HIF-1 is rapidly activated by hypoxia in astrocytes. Following HIF-1 activation, astrocytes express and release EPO. EPO activates the neuronal EPO receptor and subsequently JAK2 and thereby PI3K. PI3K deactivates BAD via Akt-mediated phosphorylation, thus inhibiting hypoxia induced apoptosis in neurons. Our results establish EPO as an important paracrine neuroprotective mediator of ischemic preconditioning.

Key words. Hypoxia-inducible factor-1 (HIF-1) – ischemic tolerance – oxygen glucose – deprivation – preconditioning – stroke

Preconditioning

Hypoxia is a potentially life-threatening emergency that must be dealt with both on the cellular and on the systemic level of the organism. The ability to maintain oxygen homeostasis is essential for the survival of aerobic organisms. Complex aerobic organisms regulate intracellular oxygen concentrations within a narrow range. The physiological response to hypoxia or hyperoxia at either end of the regulated range requires sensitive oxygen sensors coupled to a signal transduction system, which, in turn, effectively activates the regulating response. Ischemic preconditioning is an intrinsic adaptive response to a mild ischemic stimulus protecting from an otherwise 'letal' ischemic insult. In addition to mild hypoxia, many preconditioning stimuli have been described in models of cerebral ischemia *in*

Department of Experimental Neurology and Department of Neurology, Humboldt-University, Berlin, Germany
Address for correspondence: Prof. Dr. Ulrich Dirnagl, Department of Neurology, Humboldt-University Berlin, Schumannstraße 20–21, 10098 Berlin, Germany, Tel.: +49-30-45056-0134, Fax: +49-30-45056-0942, E-Mail: ulrich.dirnagl@charite.de

Maturation Phenomenon in Cerebral Ischemia V
A.M. Buchan et al. (Eds.)
© Springer-Verlag Berlin Heidelberg 2004

vivo [23, 29, 44, 45] and *in vitro* [5, 10, 32]. Ischemic tolerance is a triphasic phenomenon, with an early short-lasting phase of protection that develops within minutes from the initial ischemic insult and lasts approximately 2 h [29]. Subsequently, in the latency phase a cascade of signaling events is initiated by the preconditioning stress, establishing a late phase of tolerance. This late phase becomes apparent 24–72 h after the preconditioning event and lasts for at least 3 days [16, 50]. The signaling pathways that initiate the early and the late phase of preconditioning may be similar. However, due to their dynamics, the protective effect of the early phase must be independent of protein synthesis whereas the late phase is likely to require the synthesis of new proteins. We and others have provided that ischemic preconditioning also exists in human brain and heart. Prodromal Transient Ischemic Attacks (TIAs) [25, 49] and prodromal angina [28] do reduce the size of infarctions in brain and in heart tissue, respectively.

General Scheme of the Late Phase of Preconditioning

The late phase of preconditioning is the result of a complex cascade of cellular events representing an archetypal response to stressful stimuli. The delayed dynamics of the late phase strongly suggest the involvement of *de novo* protein synthesis as crucial for tolerance induction. Thus, it is not surprising that the general inhibitor of protein synthesis, cycloheximide, is able to block the induction of ischemic tolerance [30, 50]. The signaling cascades can be subdivided into 3 major components, the *initiators* or *triggers*, the *transducers* and the *effectors*.

Initiators of the Late Phase of Ischemic Preconditioning

Initiators or *triggers* are metabolites or ligands that were generated in response to the preconditioning ischemic event and were responsible for induction of tolerance. Known initiators are glutamate, adenosine, nitric oxide and oxygen-free radicals. These molecules serve as chemical signals that trigger the development of the late phase of preconditioning [9, 10, 14, 35, 50].

Transducers of the Late Phase of Ischemic Preconditioning

Transducers are signaling molecules that are activated by the initiators and culminate in the expression of the effectors of late preconditioning. The transducers are components of a complex signaling cascade. Among these are protein kinases, transcription factors, and cytokines. So far, three protein kinase pathways were suggested as critical components of ischemic preconditioning: the Extracellular Regulated Kinase pathway [9, 38], Janus-Kinase-2 [7, 24] and Phosphoinositide-3 Kinase (PI3K) [24, 51] pathway. Evidence that transcription factors play a causal role in ischemic preconditioning has been provided for the Nuclear Factor κB (NFκB) [4] for the Hypoxia-inducible factor 1 (HIF-1) [32] and for members of the Activator Protein 1 family (AP-1) [1, 13, 15, 43, 52]. A similar activation of protein kinases or transcription factors can be elicited pharmacologically by a wide variety of agents, including naturally occurring molecules (e.g. reactive oxygen species, endotoxins, TNF-α, Erythropoietin), as well as drugs (NO-donors, adenosine) [3, 7, 9, 11, 20, 24].

Effectors of the Late Phase of Ischemic Preconditioning

Effectors are molecules that are expressed or modified in the brain 2 to 5 days after preconditioning ischemia and are responsible for conferring protection against lethal ischemia. Until now, these effectors are largely unknown. However, there is a growing descriptive literature, providing evidence for a causal role of heat shock proteins [18], the anti-apoptotic protein Bcl2 [39, 40], the neuroprotective protein Tissue Inhibitor of Matrix Metalloproteinase 1 (TIMP-1) [48], and the anti-oxidative protein superoxide dismutase (MnSOD) [45].

Hypoxia Inducible Factor-1 (HIF-1) and Responses to Hypoxia

As mentioned above the time pattern for the late phase of ischemic tolerance is suggestive for underlying genetic programs. Gene expression requires at least one transcription factor. Data from our as well as other groups suggests that the transcription factor HIF-1 serves as a master switch in response to hypoxia [2, 36]. In mammals, HIF-1 functions as a global regulator of oxygen homeostasis that facilitates both oxygen delivery and adaptation to oxygen deprivation through the establishment and the utilization of the cardiac, respiratory, vascular and blood systems. HIF-1 is a basic helix-loop-helix (bHLH) member of the PAS family of transcription factors. The family consists of two classes of proteins, class I and II that heterodimerize with each other. The prototype of this family of transcription factors is HIF-1, consisting of HIF-1α and HIF-1β [47]. The HIF-1β subunit is constitutively expressed, whereas the HIF-1α subunit is precisely regulated by oxygen-dependent Prolyl Hydroxylase domain containing proteins (PHD). Under normoxic conditions PHD prolyl hydroxylates HIF-1α. This hydroxylation is required for interaction of HIF-1α with VHL which leads to degradation of the protein. Low oxygen concentrations are rate limiting for PHD. Thus, under hypoxic conditions HIF-1α is not prolyl hydroxylated and can therefore heterodimerize with HIF-1β [8, 12]. This leads to a transcription initiation complex which initiates the mRNA expression of the HIF-1 target genes. These gene products are involved in cell proliferation and survival, in erythropoiesis, in angiogenesis and vascular remodeling, in energy metabolism and vasomotor response [reviewed in 37]. Consequently these genes play a main role in oxygen homeostasis.

Erythropoietin as a Paracrine Mediator of Ischemic Tolerance in the Brain

From the more than 40 known HIF-1 target genes, we postulated in a candidate approach a central role for erythropoietin (EPO) in mediating ischemic tolerance.

EPO is a glycoprotein hormone that regulates hematopoiesis by promoting cell survival, proliferation and differentiation of immature erythroid cells. EPO is mainly expressed in adult kidney and fetal liver tissue. EPO expression is regulated by transcriptional activation, where HIF-1 is of essential functional importance [36]. Only recently astrocytes have been found to produce EPO, while neurons express the EPO receptor (EpoR) [3, 6, 19, 21, 22, 26]. This may have implications for IP, since EPO has potent neuroprotective properties [26, 34, 41]. EPO-induced neuroprotection seems largely mediated by anti-apoptotic signaling cascades, which are well established for the role of EPO in hematopoiesis [17, 42]. Therefore, we tested the following hypotheses:

1) In astrocytes hypoxia induces EPO synthesis in a HIF-1-dependent fashion.
2) EPO released into the extracellular space acts as a paracrine endogenous neuroprotectant.
3) Neuroprotection occurs via a cascade of protein phosphorylation that counteracts hypoxia-induced apoptosis.

Astrocytes Produce EPO mRNA HIF-1 Mediated after Oxygen Glucose Deprivation

In response to OGD astrocytes very rapidly express EPO mRNA. Already 60 min after OGD, expression reached a maximum with a 12-fold induction over baseline. This was accompanied by a strong induction of EPO protein. Since EPO expression is mainly controlled by the transcription factor HIF-1 we investigated whether activation of HIF-1 precedes EPO expression. Using a fluorescent based electrophoretic mobility shift assay [33] we observed marked HIF-1 DNA binding activity in nuclear extracts from astrocytes after OGD-stimulation for 180 min.

Recombinant human EPO (RhEPO) Time- and Dose-Dependently Protects Neurons from OGD

Cultured rat cortical neurons expressed EpoR Recombinant human (rh) EPO induced tolerance against OGD in a dose-dependent manner. Primary cortical neurons were pretreated with different concentrations of rhEPO for 48 h. Only the highest concentration of rhEPO afforded statistically significant neuronal survival after 120 min of OGD, as assessed 24 h later by LDH assay (75% protection). Even a 5 min incubation period with 100 U/l rhEPO to 72 h before OGD already afforded tolerance against OGD in neurons. The highest levels of protection were achieved after 48 h of incubation with rhEPO. To test whether neuroprotection is mediated by the specific interaction of EPO with its cognate receptor, we coapplied 100 U/L rhEPO either with a soluble EPO-receptor (sEpoR) or with an antibody against the EPO-receptor (antiEpoR). Both, sEpoR or antiEpoR blocked EPO-induced ischemic tolerance. In phase-contrast micrographs of OGD-challenged neurons rhEPO blocked OGD-induced neuronal apoptosis. To further substantiate the antiapoptotic effect of EPO we analyzed its effect on DNA-fragmentation as a hallmark of apoptosis. Preincubation with 100 U/L EPO significantly reduced

DNA-fragmentation induced by OGD, SEpoR as well as antiEpoR were able to block the antiapoptotic effect of EPO.

Induction of Tolerance in Cortical Neurons by Astrocytic OGD Preconditioned Medium is Mediated by Erythropoietin

We next investigated whether OGD-treated astrocytes release a factor capable of protecting neurons from OGD, and whether this factor is EPO. Astrocytes were subjected to 180 min of OGD, after which they were allowed to condition medium for 24 h. Subsequently, conditioned medium was transferred to neuron-enriched cultures and neurons were exposed to this medium for 48 h. Medium, OGD-conditioned by astrocytes, protected neurons from OGD. LDH assay 24 h following OGD demonstrated a 75% increase in neuronal viability compared to "control-conditioned" medium. Medium transfer-induced neuronal protection was completely blocked by sEpoR or antiEpoR, indicating that EPO is the neuroprotective factor released by OGD-stimulated astrocytes.

Neuron to Neuron Transfer of Protection is not Mediated by EPO

Is EPO also involved in OGD preconditioning of enriched neurons in culture? Neurons were subjected to a preconditioning interval of OGD (90 min), after which they were allowed to condition medium for 24 h. Transfer of this medium to naive neuron-enriched cultures and subsequent exposure for 48 h conferred protection to neurons against a normally lethal OGD (120 min). Thus, neurons, like astrocytes, release protective factor(s). However, in this case EPO is not involved, since neither sEpoR nor antiEpoR affected the protective effect. In line with this finding we observed no induction of EPO-mRNA in neurons following a preconditioning OGD-period of 60 min.

Neuroprotection of EPO *in vivo*

Evidence from our *in vitro* model suggests an essential role for EPO in ischemic tolerance in the brain *in vivo*. To test this hypothesis we used our *in vivo* model of focal cerebral ischemia of the mouse in a hypoxic preconditioning paradigm. Three hundred min of hypoxia with 8% oxygen induced robust neuroprotection against experimental stoke elicited 72 h later. When we applied soluble EpoR intracerebroventricularly (i.c.v.) we partially but significantly blocked the neuroprotective effect of EPO. We started i.c.v. instillation immediately after preconditioning hypoxia and stopped 6 h before ischemia. As control we used a heat-inactivated sEpoR. Because of technical limitations of this approach we can only speculate whether the partial inhibition is due to an involvement of other mechanisms, or a result of an insufficient blockade of the inhibiting sEpoR (Prass et al., manuscript in preparation).

EPO-Signaling Pathways – EPO-Induced Neuroprotection is JAK2-Dependent

Having demonstrated that EPO is a neuroprotective paracrine mediator from astrocytes to neurons in ischemic preconditioning we further studied EpoR-mediated signaling pathways. Several studies have shown that EPO mediates its effects by tyrosine phosphorylation. Rather than containing an intrinsic protein tyrosine kinase activity, the relatively short cytoplasmatic domain of the EpoR interacts with non-receptor tyrosine kinases. A key kinase in this cascade is JAK2, which is constitutively associated with the EpoR. The first step of this intracellular signaling is EPO-induced homodimerization of EpoR. Homodimerization of EpoR induces a conformational change of the whole protein complex, which places JAK2 in close proximity to EpoR, thus promoting trans-tyrosine phosphorylation and activation of JAK2. JAK2 induces phosphorylation of cytosolic tyrosines in the EpoR. This allows docking to the EpoR cytosolic domain of several signaling proteins, including PI3K and STAT1, STAT3 and STAT5. These proteins, in turn, become tyrosine phosphorylated and thus activated [27]. NF-kappaB is activated by phosphorylation of its inhibitor subunit IkappaB.

Since, JAK2 is the key kinase in the signal transduction pathways activated by the EpoR, we first explored the role of JAK2-dependent pathways for EPO's anti-apoptotic effect on OGD-induced apoptosis in neurons. We found that rhEPO-induced neuroprotection is significantly attenuated in the presence of the specific JAK2 inhibitor AG490. AG490 alone had no effect on neuronal viability. Our finding is in agreement with recent results by Digicaylioglu and Lipton (2001) [7], who also provide strong evidence for an essential role of EpoR-mediated JAK2-activation in neuroprotection.

JAK2/STAT Pathway

The JAK2-STAT pathway plays a key role in the anti-apoptotic signaling of EPO in hematopoiesis. In contrast, we found no evidence for induction of STAT5, STAT3 or STAT1 in EPO-induced neuroprotection. Furthermore, we analyzed anti-apoptotic target genes of these STATs, which are known to be protective in erythroid progenitors. However, EPO did not induce transcription of the antiapoptotic bcl-2 and bcl-XL genes in cortical neurons, neither after 12 nor after 48 h of EPO application. We therefore conclude that the JAK/STAT pathway does not play a significant role in EPO-induced neuroprotection in our system.

JAK2/PI3K Pathway

The ability of trophic factors to promote survival has been attributed, at least in part, to PI3K and its molecular targets. In particular, in erythroid precursors EPO prevents apoptotic cell death via a PI3K-dependent pathway [46]. We treated cortical neurons with 100 U/l rhEPO in the presence of a specific inhibitor of PI3K and exposed them to a lethal OGD interval either 1 or 24 h after pretreatment. Our data showed that PI3K inhibition significantly diminished EpoR-mediated neuroprotection in the short protection window as in the long protection windows. PI3K suppresses apoptotic cell death through its downstream effector protein kinase B (PKB)/Akt. Activated Akt kinase plays a central role in suppressing apoptosis by modulating the activities of Bcl2-family proteins and caspase 9. PI3K specifically induces phosphorylation at Thr308 and Ser473, which play central roles in Akt activation. We found EPO-induced and PI3K-mediated activation of Akt kinase activity in cultured cortical neurons. In EPO-stimulated cortical neurons we found an activation of Akt kinase which is blocked by sEpoR and PI3K inhibition through LY294002. In accordance with this finding, Sirén et al. (2001) reported EPO-induced phosphorylation of Akt at Ser473 by hypoxia in hippocampal neurons. Activated Akt, in turn, can specifically phosphorylate Bad at Ser136 and can thus inactivate Bad. Phosphorylated Bad prevents apoptosis, since unphosphorylated Bad is capable of forming heterodimers with the antiapoptotic proteins Bcl-X$_L$ or Bcl-2 and thereby antagonizes their antiapoptotic function. To investigate whether EPO can induce phosphorylation of Bad we measured Bad phosphorylation by using a phosphospecific anti-Bad antibody. Following pretreatment of cortical neurons with EPO for 1 h Bad phosphorylation increased significantly, while it was significantly diminished after blockade of EpoR by sEpoR as well as after specific inhibition of PI3K. Taken together, we can describe a correlation between EPO-induced and PI3K-mediated phosphorylation of Bad at Ser136 and neuroprotection. These results suggest that Bad phosphorylation plays a role in EPO-induced prevention of neuronal apoptosis.

JAK2/NFκB Pathway

Recently, the activation of the transcription factor NF-κB has been reported as an essential step in EPO-induced neuroprotection [7]. This pathway is JAK2-dependent but PI3K-independent and may thus explain why in our study EPO-induced neuroprotection is only partially blocked by the specific PI3-kinase inhibitor LY294002, although this compound inhibited Akt kinase completely and Bad-phosphorylation nearly completely to control level. Using Western blotting, Digicaylioglu and Lipton (2001) showed that EPO induces the expression of the Inhibitor of Apoptosis Proteins cIAP2 and XIAP, which correlates with EPO-induced neuroprotection. These data suggest an involvement of IAPs in the anti-apoptotic signaling of EPO. Remarkably, the activation of NF-κB is a key event for IP, as recently demonstrated in an in-vivo model of global ischemia [4, 31].

Our current understanding of EPO-mediated neuroprotection is summarized in Fig. 1. Following EPOs binding to EpoR JAK2 is activated. The JAK/STAT pathway is not involved in EPO-mediated neuroprotection. However, we provide evidence for an involvement of the PI3K/Akt pathway. Activation of this pathway cumulates in the phosphorylation and thereby inactivation of the proapoptotic Bcl-2 protein Bad. The transcriptional repression of Bax, another proapoptotic Bcl-2 family protein does not play a role in Epo-mediated neuroprotection. Lipton and coworkers provide strong evidence for an essential role of NF-kappaB in EPO-mediated neuroprotection, possibly by transcriptional activation of Inhibitors of Apoptosis Proteins.

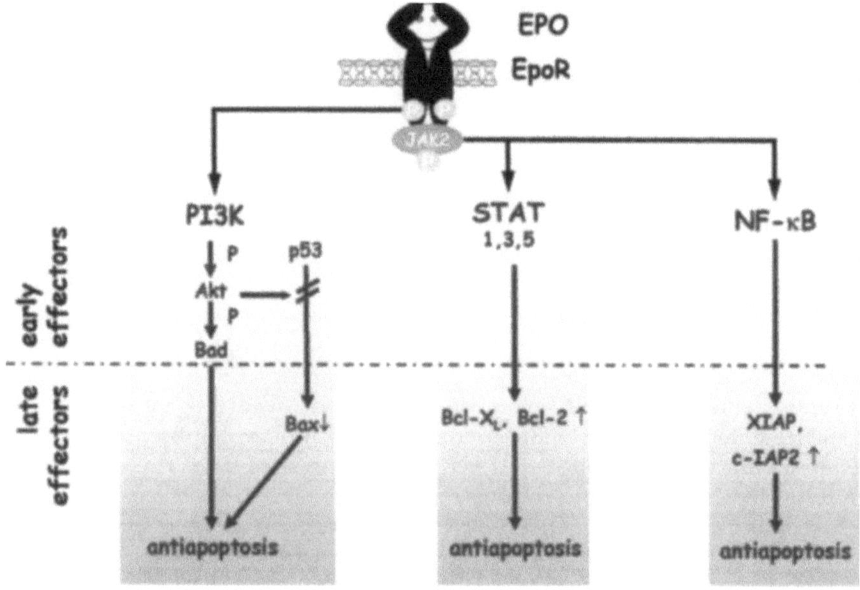

Fig. 1. Downstream pathways to Erythropoietin inhibition of apoptosis

Conclusions

In the brain as well as other organs a long lasting tolerant state against hypoxia or focal cerebral ischemia can be induced. There is strong evidence for gene expression programs as underlying mechanisms of this protection. We identified a paracrine signal transduction pathway from astrocytes to neurons. The key mediator of this paracrine neuroprotection from astrocytes to neurons but not from neurons to neurons was EPO, which acts via its cognate receptor. EPO-induced neuroprotection is effective within minutes and sustained for many hours with continued EPO exposure. The very fast neuroprotection induced by EPO is not consistent with the induction of a protective gene expression by EPO. For such fast cellular responses, posttranslational regulation, for example protein phosphorylation, is more likely the responsible mechanism. In neurons, exogenously applied EPO induces fast neuroprotection. However, since the neuroprotective effect is sustained for more than 24 h, an additional induction of gene expression programs seems to be likely. The JAK2/NF-kappaB-mediated pathway may account for this long lasting neuroprotection by EPO. Our data provide evidence for a new, endogenous signaling cascade inducing tolerance against ischemia in the brain. Our results, together with those of other authors using different models to simulate IP, demonstrate that IP in the brain is the result of complex signaling programs involving various cellular elements of the brain. By a combination of transcriptional as well as posttranslational mechanisms the brain can protect itself against the deprivation of substrate. Understanding these mechanisms may allow us in the future to induce or boost endogenous protection in patients as a novel strategy to safeguard the brain against hypoxia/ischemia. The drug DFO as an inducer of IP could become a promising tool in patients when ischemia can be anticipated, e.g. in neurosurgery. Clinical stroke trials evaluating the effectivness of EPO are already under way.

Acknowledgements. Supported by the Deutsche Forschungsgemeinschaft (DFG) and the Herman and Lilly Schilling Foundation.

References

1. Barone FC, White RF, Spera PA, Ellison J, Currie RW, Wang X, Feuerstein GZ (1998) Ischemic preconditioning and brain tolerance: temporal histological and functional outcomes, protein synthesis requirement, and interleukin-1 receptor antagonist and early gene expression. Stroke 29:1937–1950
2. Bergeron M, Gidday JM, Yu AY, Semenza GL, Ferriero DM, Sharp FR (2000) Role of hypoxia-inducible factor-1 in hypoxia-induced ischemic tolerance in neonatal rat brain. Ann Neurol 48:285–296
3. Bernaudin M, Bellail A, Marti HH, Yvon A, Vivien D, Duchatelle I, Mackenzie ET, Petit E (2000) Neurons and astrocytes express EPO mRNA: oxygen-sensing mechanisms that involve the redox-state of the brain. Glia 30:271–278
4. Blondeau N, Widmann C, Lazdunski M (2001) Activation of the Nuclear Factor-κB is a key event in brain tolerance. J Neurosci 21:4668–4677
5. Bruer U, Weih M, Isaev N, Meisel A, Ruscher K, Bergk A, Trendelenburg G, Wiegand F, Victorov I, Dirnagl U (1997) Induction of tolerance in rat cortical neurons: hypoxic preconditioning. FEBS Lett 414:117–121. Digicaylioglu M, Bichet S, Marti HH, Wenger RH, Rivas LA, Bauer C, Gassmann M (1995) Localization of specific erythropoietin binding sites in defined areas of the mouse brain. Proc Natl Acad Sci USA 92:3717–3720
6. Digicaylioglu M, Lipton SA (2001) Erythropoietin-mediated neuroprotection involves cross-talk between Jak2 and NF-kappaB signalling cascades. Nature 412:641–647
7. Epstein AC, Gleadle JM, McNeill LA, Hewitson KS, O'Rourke J, Mole DR, Mukherji M, Metzen E, Wilson MI, Dhanda A, Tian YM, Masson N, Hamilton DL, Jaakkola P, Barstead R, Hodgkin J, Maxwell PH, Pugh CW, Schofield CJ, Ratcliffe PJ (2001) C. elegans EGL-9 and mammalian homologs define a family of dioxygenases that regulate HIF by prolyl hydroxylation. Cell 107:43–54
8. Gonzalez-Zulueta M, Feldman AB, Klesse LJ, Kalb RG, Dillman JF, Parada LF, Dawson TM, Dawson VL (2000) Requirement for nitric oxide activation of p21(ras)/extracellular regulated kinase in neuronal ischemic preconditioning. Proc Natl Acad Sci USA 97:436–441
9. Grabb MC, Choi DW (1999) Ischemic tolerance in murine cortical cell culture: Critical role for NMDA Receptors. J Neurosci 19:1657–1662
10. Heurteaux C, Lauritzen I, Widmann C, Lazdunski M (1995) Essential role of adenosine, adenosine A1 receptors, and ATP-sensitive K+ channels in cerebral ischemic preconditioning. Proc Natl Acad Sci USA 92:4666–4670
11. Jaakkola P, Mole DR, Tian YM, Wilson MI, Gielbert J, Gaskell SJ, Kriegsheim AV, Hebestreit HF, Mukherji M, Schofield CJ, Maxwell PH, Pugh CW, Ratcliffe PJ (2001) Targeting of HIF-alpha to the von Hippel-Lindau ubiquitylation complex by O_2-regulated prolyl hydroxylation. Science 292:468–472
12. Kapinya K, Penzel R, Sommer C, Kiessling M (2000) Temporary changes of the AP-1 transcription factor binding activity in the gerbil hippocampus after transient global ischemis and ischemic preconditioning. Brain Res 872:283–293
13. Kato H, Liu Y, Araki T, Kogure K (1992) MK-801, but not anisomycin, inhibits the induction of tolerance to ischemia in the gerbil hippocampus. Neurosci Lett 139(1):118–121
14. Kato H, Kogure K, Araki T, Itoyama Y (1995) Induction of Jun-like immunoreactivity in astrocytes in gerbil hippocampus with ischemic tolerance. Neurosci Lett 189:13–16
15. Kitagawa K, Matsumoto M, Tagaya M, Hata R, Ueda H, Niinobe M, Handa N, Fukunaga R, Kimura K, Mikoshiba K (1990) 'Ischemic tolerance' phenomenon found in the brain. Brain Res 528:21–24

16. Lawson AE, Bao H, Wickrema A, Jacobs-Helber SM, Sawyer ST (2000) Phosphatase inhibition promotes antiapoptotic but not proliferative signaling pathways in erythropoietin-dependent HCD57 cells. Blood 96:2084–2092
17. Liu Y, Kato H, Nakata N, Kogure K (1993) Temporal profile of heat shock protein 70 synthesis in ischemic tolerance induced by preconditioning ischemia in rat hippocampus. Neuroscience 56:921–927
18. Liu C, Shen K, Liu Z, Noguchi CT (1997) Regulated human erythropoietin receptor expression in mouse brain. J Biol Chem 272:32395–32400
19. Liu J, Ginis I, Spatz M, Hallenbeck JM (2000) Hypoxic preconditioning protects cultured neurons against hypoxic stress via TNF-alpha and ceramide. Am J Physiol Cell Physiol 278:C144–C153
20. Marti HH, Wenger RH, Rivas LA, Straumann U, Digicaylioglu M, Henn V, Yonekawa Y, Bauer C, Gassmann M (1998) Erythropoietin gene expression in human, monkey and murine brain. Eur J Neurosci 8:666–676
21. Masuda S, Okano M, Yamagishi K, Nagao M, Ueda M, Sasaki R (1994) A novel site of erythropoietin production. Oxygen-dependent production in cultured rat astrocytes. J Biol Chem 269:19488–19493
22. Matsushima K, Hakim AM (1995) Transient forebrain ischemia protects against subsequent focal cerebral ischemia without changing cerebral perfusion. Stroke 26:1047–1052
23. Meisel A, Ruscher K, Freyer D, Karsch M, Isaev N, Sawitzki N, Priller J, Volk HD, Dirnagl U (2001):Endogenous tolerance against hypoxia/ischemia in the brain by astroglial-neuronal signaling via erythropoietin. J Neurosci (submitted)
24. Moncayo J, de Freitas GR, Bogousslavsky J, Altieri M, van Melle G (2000) Do transient ischemic attacks have a neuroprotective effect? Neurology 54:2089–2094
25. Morishita E, Masuda S, Nagao M, Yasuda Y, Sasaki R (1997) Erythropoietin receptor is expressed in rat hippocampal and cerebral cortical neurons, and erythropoietin prevents in vitro glutamate-induced neuronal death. Neuroscience 76:105–116
26. Oda A, Sawada K, Druker BJ, Ozaki K, Takano H, Koizumi K, Fukada Y, Handa M, Koike T, Ikeda Y (1998) Erythropoietin induces tyrosine phosphorylation of Jak2, STAT5A, and STAT5B in primary cultured human erythroid precursors. Blood 92:443–451
27. Ottani F, Galvani M, Ferrini D, Sorbello F, Limonetti P, Pantoli D, Rusticali F (1995) Prodromal angina limits infarct size. A role for ischemic preconditioning. Circulation 91:291–297
28. Perez-Pinzon MA, Xu GP, Dietrich WD, Rosenthal M, Sick TJ (1997) Rapid preconditioning protects rats against ischemic neuronal damage after 3 but not 7 days of reperfusion following global cerebral ischemia. J Cereb Blood Flow Metab 17:175–182
29. Prass K, Schumann P, Wiegand F, Dirnagl U (1998) Induced tolerance to ischemia: From heart to brain. In: Kriegelstein J (ed) Pharmacology of cerebral ischemia 1998. medpharm Scientific Publishers, Stuttgart, pp 69–83
30. Ravati A, Ahlemeyer B, Becker A, Klumpp S, Krieglstein J (2001) Preconditioning-induced neuroprotection is mediated by reactive oxygen species and activation of the transcription factor nuclear factor-kappaB. J Neurochem 78:909–919
31. Ruscher K, Isaev N, Trendelenburg G, Weih M, Iurato L, Meisel A, Dirnagl U (1998) Induction of hypoxia inducible factor 1 by oxygen glucose deprivation is attenuated by hypoxic preconditioning in rat cultured neurons. Neurosci Lett 254:1171–1220
32. Ruscher K, Reuter M, Kupper D, Trendelenburg G, Dirnagl U, Meisel A (2000) A fluorescence based non-radioactive electrophoretic mobility shift assay. J Biotechnol 78:163–170
33. Sakanaka M, Wen TC, Matsuda S, Masuda S, Morishita E, Nagao M, Sasaki R (1998) In vivo evidence that erythropoietin protects neurons from ischemic damage. Proc Natl Acad Sci USA 95:4635–4640
34. Schumann P, Prass K, Wiegand F, Ahrens M, Megow D, Dirnagl U (1998) Oxygen free radicals and ischaemic preconditioning in the brain: Preliminary data and a hypothesis. In: Ito U, Klatzo I (eds) Maturation phenomenon in cerebral ischemia III. Springer, Heidelberg New York, pp 95–103
35. Semenza GL (2000), HIF-1: mediator of physiological and pathophysiological responses to hypoxia. J Appl Physiol 88:1474–1480
36. Semenza GL (2000b) Hypoxia, clonal selection, and the role of HIF-1 in tumor progression. Crit Rev Biochem Mol Biol 35:71–103
37. Shamloo M, Rytter A, Wieloch T (1999) Activation of the extracellular signal-regulated protein kinase cascade in the hippocampal CA1 region in a rat model of global cerebral ischemic preconditioning. Neuroscience 93:81–88
38. Shimazaki K, Ishida A, Kawai N (1994) Increase in bcl-2 oncoprotein and the tolerance to ischemia-induced neuronal death in the gerbil hippocampus. Neurosci Res 20:95–99

39. Shimizu S, Nagayama T, Jin KL, Zhu L, Loeffert JE, Watkins SC, Graham SH, Simon RP (2001) Bcl-2 antisense treatment prevents induction of tolerance to ischemia in the rat brain. J Cereb Blood Flow Metab 21:233–243
40. Siren AL, Fratelli M, Brines M, Goemans C, Casagrande S, Lewczuk P, Keenan S, Gleiter C, Pasquali C, Capobianco A, Mennini T, Heumann R, Cerami A, Ehrenreich H, Ghezzi P (2001) Erythropoietin prevents neuronal apoptosis after cerebral ischemia and metabolic stress. Proc Natl Acad Sci USA 98:4044–4049
41. Socolovsky M, Fallon AE, Wang S, Brugnara C, Lodish HF (1999) Fetal anemia and apoptosis of red cell progenitors in Stat5a-/-5b-/- mice: a direct role for Stat5 in Bcl-X(L) induction. Cell 98:181–191
42. Sommer C, Gass P, Kiessling M (1995) Selective c-JUN expression in CA1 neurons of the gerbil hippocampus during and after aquisition of an ischemia-tolerant state. Brain Pathol 5:135–144
43. Stagliano NE, Perez-Pinzon MA, Moskowitz MA, Huang PL (1999) Focal ischemic preconditioning induces rapid tolerance to middle cerebral artery occlusion in mice. J Cereb Blood Flow Metab 19:757–761
44. Toyoda T, Kassell NF, Lee KS (1997) Induction of ischemic tolerance and antioxidant activity by brief focal ischemia. NeuroReport 8:847–851
45. Uddin S, Kottegoda S, Stigger D, Platanias LC, Wickrema A (2000) Activation of the Akt/FKHRL1 pathway mediates the antiapoptotic effects of erythropoietin in primary human erythroid progenitors. Biochem Biophys Res Commun 275:16–19
46. Wang GL, Semenza GL (1995) Purification and characterization of hypoxia-inducible factor 1. J Biol Chem 270:1230–1237
47. Wang X, Yaish-Ohad S, Li X, Barone FC, Feuerstein GZ (1998) Use of suppression subtractive hybridization strategy for discovery of increased tissue inhibitor of matrix metalloproteinase-1 gene expression in brain ischemic tolerance. J Cereb Blood Flow Metab 18:1173–1177
48. Weih M, Harms L, Dirnagl U, Kallenberg K, Einhaupl KM (1999) Attenuated stroke severity after prodromal TIA: Is it ischemic tolerance? Stroke 30:1851–1854
49. Wiegand F, Liao W, Busch C, Castell S, Knapp F, Lindauer U, Megow D, Meisel A, Redetzky A, Ruscher K, Trendelenburg G, Victorov I, Riepe M, Diener HC, Dirnagl U (1999) Respiratory chain inhibition induces tolerance to focal cerebral ischemia. J Cereb Blood Flow Metab 19:1229–1237
50. Yano S, Morioka M, Fukunaga K, Kawano T, Hara T, Kai Y, Miyamoto E, Ushio Y (2001) Activation of Akt/protein kinase B contributes to induction of ischemic tolerance in the CA1 subfield of gerbil hippocampus. J Cereb Blood Flow Metab 21:351–360
51. Yoneda Y, Kuramoto N, Azuma Y, Ogita K, Mitani A, Zhang L, Yanase H, Masuda S, Kataoka K (1998) Possible involvement of activator protein-1 DNA binding in mechanisms underlying ischemic tolerance in the CA1 subfield of gerbil hippocampus. Neuroscience 86:79–89

Ultrastructural Temporal Profile of the Dying Neuron and Surrounding Astrocytes in the Ischemic Penumbra: Apoptosis or Necrosis?

U. Ito, T. Kuroiwa, S. Hanyu, Y. Hakamata, E. Kawakami, I. Nakano, and K. Oyanagi

Summary. We investigated the temporal profile of isolated dying neurons (disseminated selective neuronal necrosis: DSNN) and the behaviors of astrocyte surrounding these dying neurons, in the ischemic penumbra of the cerebral cortex. In the ischemic penumbra, DSNN progressed slowly until 3 weeks after the ischemic insult. Cell bodies, cell processes, and end-feet of living astrocytes became swollen, with an increase in the number and in the volume of the mitochondria and accumulation of glycogen granules. The DSNN started 15 min after the ischemic insult, and progressed with increasing numbers of dark neurons having various degrees of electron density during 5 to 24 h. The isolated dark neurons showed homogeneous condensation of their cytosol, organelles, and nucleus, in which small loosely aggregated chromatin condensates were observed in the nuclear matrix and along the margin of the nuclear membrane. These chromatin condensations were positive for TUNEL staining. The swollen astrocytic cell processes surrounded the dark neurons. Astrocytic swelling was most prominent near the dendritic synapses. Finally, the isolated dark neurons became completely shrunken with very high electron density of the entire cell containing degenerated mitochondria having swollen matrices with occasional woolly densities. The shrunken neuron was fragmented into electron-dense debris by invading astrocytic cell processes. Some of the debris was phagocytized by astrocytes, and others moved into the extracellular space and were phagocytized by the perivascular microglia. Macrophages and other inflammatory cell were not observed in the penumbra. The ultrastructural characteristics of DSNN, in the present study, suggested necrotic neuronal death instead of apoptosis. Condensation of the isolated neuron was induced by swelling of astrocytic cell processes surrounding the dark neuron.

Key words. Apoptosis vs. necrosis – astrocytic swelling – disseminated selective neuronal necrosis – ischemic penumbra – maturation phenomenon of ischemic injuries – neuronal death

Umeo Ito[1,3,4], Toshihiko Kuroiwa[2], Shuji Hanyu[3], Youji Hakamata[3], Emiko Kawakami[4], Imaharu Nakano[3], Kiyomitsu Oyanagi[4]
[1] Department of Neurosurgery, Musashino Red Cross Hospital, Tokyo
[2] Department of Neuropathology, Medical Research Institute, Tokyo Medical and Dental University, Tokyo
[3] Department of Neurology, Jichi Medical School, Tochigi
[4] Tokyo Metropolitan Institute of Neuroscience, Tokyo
Correspondence to: Umeo Ito, MD, PhD, 4-22-24, Zenpukuji, Suginami-ku, Tokyo 167-0041, Japan, Tel.: +81-3-3390-2329, Fax: +81-3-3301-5600, E-Mail: umeo-ito@nn.iij4u.or.jp

Maturation Phenomenon in Cerebral Ischemia V
A.M. Buchan et al. (Eds.)
© Springer-Verlag Berlin Heidelberg 2004

Introduction

Recently, the topic of apoptosis vs. necrosis of dying neurons after ischemic insult has been a matter of controversy [1, 4, 23]. We report our findings herein as well as discuss apoptosis vs. necrosis as the cause of this death.

Cerebral infarction develops rapidly after a large ischemic insult has occurred. We developed a model to induce a large ischemic penumbra around a small focal infarction in the cerebral cortex of Mongolian gerbils [7, 9] by giving a threshold amount of ischemic insult to induce cerebral infarction. The histopathology of this model revealed disseminated eosinophilic ischemic neurons (disseminated selective neuronal necrosis: DSNN) that increased in number in a large area of the cerebral cortex after revascularization, and a focal infarction developed only in the frontal lobe by 24 h after the start of recirculation [10]. Electron-microscopically, these disseminated eosinophilic ischemic neurons were observed as dark neurons with increased cytosolic electron-density. These dark neurons increased in numbers until day 4, and new one were still appearing 3 weeks after the start of recirculation. This observation corresponds to the maturation phenomenon of ischemic injuries [11], the original concept of the delayed neuronal death described in CA1 neurons [17].

Using this model, in this present study, we examined the ultrastructural temporal profile of these dying dark neurons in the ischemic penumbra of the parietal cortex with special attention given to the behavior of the astrocytes surrounding the dying neurons.

Materials and Methods

Under 2% halothane, 70% nitrous oxide, and 30% oxygen anesthesia, the left carotid artery of adult Mongolian gerbils was twice occluded for 10 min each time, with a 5 h interval between the 2 occlusions [8]. After each cervical surgery, animals soon recovered from the anesthesia and moved spontaneously. Ischemia-positive animals were selected based on the stroke index score determined after the first occlusion [26].

The gerbils were sacrificed at various times, i.e., at 15 min, at 5, 12, 24 h, at 4 days, and at 1, 2, 3 weeks following the second ischemic insult by intracardiac perfusion with glutaraldehyde fixative for electron microscopy and phosphate-buffered formaldehyde fixative for light microscopy.

Ultrathin sections including the 3rd~5th cortical layers were prepared from the parietal lobe of the left ischemic cerebral hemisphere at the mid-point between the interhemispheric and rhinal fissures as coronal sections at the level of the infundibulum. Alternative sections were double stained by uranyl acetate and lead solution, and observed with a Hitachi electron microscope. Paraffin sections were separately stained with hematoxylin-eosin (HE), periodic acid fuchsin Schiff (PAS), and TUNEL reagents (ApopTag: Intergen).

Results

Swelling of astrocytic cell processes, perivascular end-feet, and cell body had already begun 15 min after the ischemic insult, without accumulation of glycogen granules [12]. At this stage, isolated dark neurons with diffuse, increased electron density were found disseminated among the almost normal-looking neurons. Some of the normal-appearing neurons showed disaggregated ribosomes and slight swelling of the rough endoplasmic reticulum (rER) and Golgi apparatus. The swollen astrocytic cell processes surrounded these dark neurons. Small dots of chromatin condensation were observed scattered in the nuclear matrix as well as along the nuclear membrane of the dark neurons. However, no swelling was observed in mitochondria and the other cytosolic organelles. These dark neurons were never found in the control animals after the same procedures for fixation and preparation as used for the postischemic animals.

From 5 to 24 h, isolated dark neurons with different grades of high electron density increased in number among the almost normal looking neurons (Fig. 1 d), some of which showed disaggregation of ribosomes and slight swelling of their rER and Golgi apparatuses as seen at 15 min. These dark neurons still newly appeared even 3 weeks after the start of recirculation. In the swollen astrocytic cell processes, enlarged mitochondria having slightly swollen cristae and boosted electron density of the matrices had increased in number, together with increased accumulation of glycogen granules [12, 13]. These dark neurons were compatible with ischemic neuronal change seen by HE staining. These ischemic neurons were scattered among the normal-looking neurons in the ischemic penumbra of the cerebral cortex, and had a homogeneous eosinophilic cytosol containing a pycnotic and/or karyorrhectic nucleus (Fig. 2 a) [10]. In this stage, TUNEL-positive nuclei showing strong positively on the dotted chromatin condensations of the neuron were also observed scattered in the ischemic penumbra of the cerebral cortex (Fig. 2 b). The dark neurons were surrounded by remarkably swollen astrocytic cell processes, especially prominent nearby the dendritic synapses of the dark neurons [31] (Fig. 1 c), and showed darkened nuclei in which dots of chromatin condensates were observed (Fig. 1 b). Later than 12 h after the ischemic insult, some mitochondria of these dark neurons showed partial swelling of their matrices and disintegration of cristae with woolly densities (Fig. 1 a). Among these dark neurons, completely shrunken neurons surrounded by remarkably swollen astrocytic cell processes (Fig. 1 c) were increased in number from 12 to 24 h after the ischemic insult. These shrunken neurons often contained mitochondria with swollen matrices having woolly densities.

By 4 days after the ischemic insult, these shrunken neurons were fragmented by astrocytic cell processes that had invaded and separated the shrunken neuronal cytosol and nucleus. Up to this stage, no inflammatory cells and macrophages appeared in the ischemic penumbra. However, phagocytic activity of the perivascular microglia was observed.

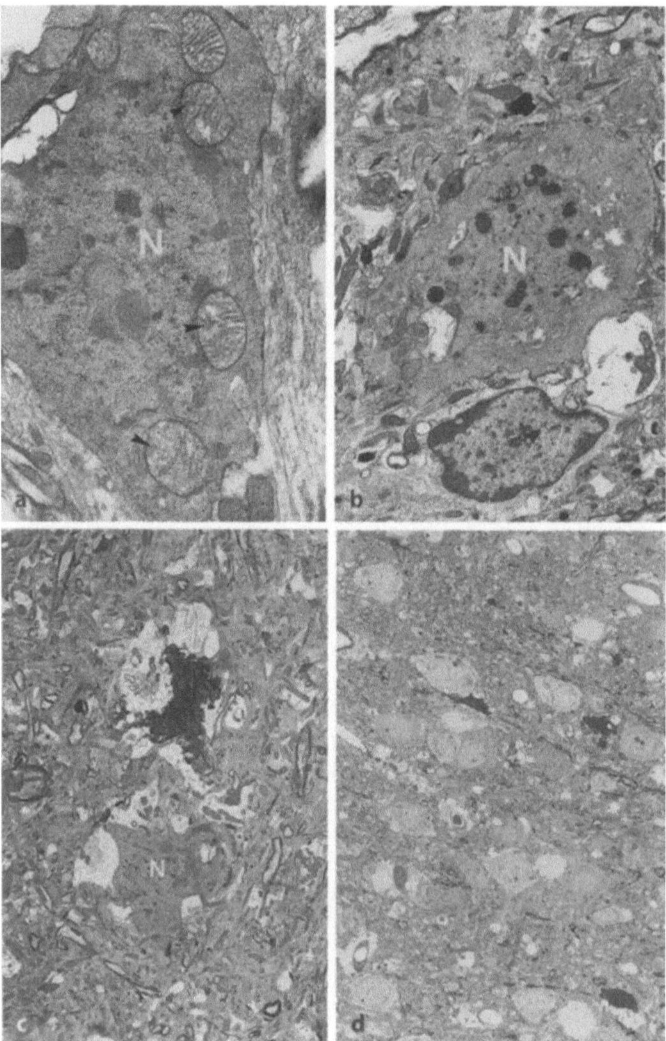

Fig. 1a–d. Electron microscopy of electron dense dying neurons: **a** Later than 12 h after the ischemic insult, some mitochondria of these dark neurons shows partial swelling of their matrices and
disintegration of cristae with woolly densities (arrows) (×6000); **b** the dark neuron shows darkened nucleus (N) in which dots of chromatin condensates are observed. Nuclear membrane is
swollen and disintegrated (×4000); **c** two dark neurons with different grade of electron density are
surrounded by remarkably swollen astrocytic cell processes, especially prominent nearby the
dendritic synapses of the dark neurons. N: nucleus (×1500); **d** twelve hours after ischemic insult,
isolated dark neurons with different grades of high electron density increased in number among
the almost normal looking neurons (×500)

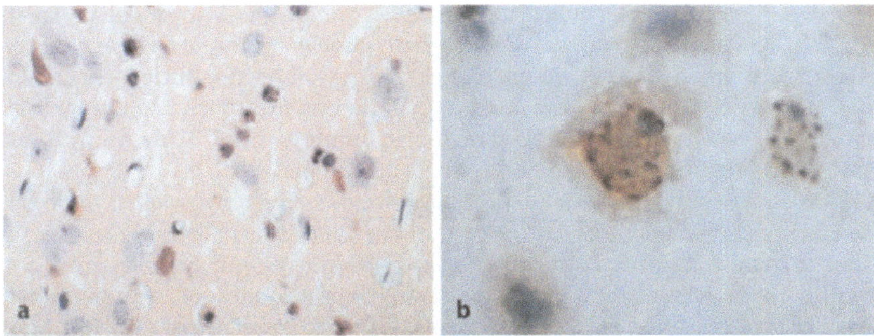

Fig. 2a,b. Light microscopy of neurons showing eosinophilic ischemic change: **a** the ischemic neurons are scattered among the normal-looking neurons in the ischemic penumbra of the cerebral cortex, and have a homogeneous eosinophilic cytosol containing a pycnotic and/or karyorrhectic nucleus (H.E. ×400); **b** TUNEL-positive nuclei showing strong positivity on the dotted chromatin condensations of the neuron are observed scattered in the ischemic penumbra of the cerebral cortex (TUNEL ×800)

Discussion

Kerr and Wyllie (1972) [15] found apoptosis after a mild ischemic insult to the liver cells; i.e., as the isolated cell died, the nucleus became condensed with large crescent-shaped chromatin condensates and fragmented into separate pieces, the condensed cytosol was also fragmented into separate apoptotic bodies containing fragmented nuclear material and normal cytosolic organelles; and these apoptotic bodies were phagocytized by macrophages. While, in necrosis, not isolated but a massive number of cells die with swollen cytosol and organelles; and the nucleus is also swollen with small chromatin condensates. Finally all cellular and nuclear membranes are ruptured. As some neurons die isolatedly with cellular condensation after a mild ischemic insult to the brain, the possibility of apoptotic neuronal death has been discussed.

The present study was aimed at elucidating time course of the ultrastructural morphological changes occurring in isolated cells undergoing neuronal death in the cerebral cortex with special regard to the behavior of the surrounding astrocytic cell processes after a mild ischemic insult, in order to discuss better the topic of apoptosis vs. necrosis for the neuronal death.

In the present model, post ischemic injuries mature slowly in the cerebral cortex. The temporal profile of histopathology revealed the development of disseminated selective neuronal necrosis (DSNN) only in the penumbra area of the cerebral cortex after the ischemic insult. However, in the infarcted focus of the frontal lobe, following temporary development of DSNN, all neurons and astrocytes except for the capillary wall itself undergo massive necrosis with marked swelling and destruction of the entire membranous system of the cells [10–12, 22].

In the penumbra, under EMS observation, the DSNN appeared from the early post ischemic stage at 15 min after the start of recirculation, as isolated dark neurons with different grades of high electron density of their cytosol and nucleus. In these dark neurons, no swelling of mitochondria and other cytosolic organelles was observed, except in shrunken neurons with very high electron density of the entire cell. As the dark neuron showed small loosely aggregated chromatin condensates scattered in the nuclear matrix as well as along the nuclear membrane, cellular activity had probably decreased [22]. These isolated dark neurons with different grades of electron density increased in number from 5 min to 24 h, and were still newly appeared 3 weeks after the ischemic insult.

The astrocytic cell bodies and processes were remarkably swollen, especially those nearby the dendritic synapses of the dark neurons; their mitochondria increased in number and size; and they accumulated glycogen granules from 5 to 24 h after the ischemic insult. These astrocytic mitochondria showed moderately swollen cristae and increased electron densities of their matrices [12, 13]. These findings highly suggest activated astrocytic energy metabolism, generating lactate as a neuronal fuel [2, 6], scavenging potassium [25], neurotransmitters [19, 24], and other metabolites from the neuron and dendritic synapses, and also promoting survival of the astrocytes themselves [18]. As the neuronal fuel would not be transferred smoothly to neurons due to deranged neuronal energy metabolism, glycogen would accumulate in the astrocytic cell processes via gluco-neogenesis from lactate followed by glycogenesis [3, 6, 28–30].

Among the dark neurons with increased electron density, completely shrunken neurons with very high electron density of the entire cell body and axons increased in number from 5 to 24 h after the ischemic insult. Later than 12 h after the ischemic insult, some mitochondria of the isolated dark neurons showed partial swelling of their matrices, and disintegrated cristae with woolly densities. However, many of the mitochondria in the shrunken neurons showed swollen matrices with occasional woolly densities and disintegrated cristae; and such cells were considered to irreversibly damaged and unable to survive [5]. From our findings, not every dark neuron died to become a shrunken neuron, but some of them survived. Further quantitative study of the fate of these isolated dark neuron is necessary.

Four days after the ischemic insult, these condensed neuronal bodies became separated and fragmented by invasion of tiny astrocytic cell processes. Some of these fragments were phagocytized by astrocytic cell processes, and others were seen moving in the intercellular spaces. No inflammatory cells or phagocytes were seen in this study. Only the perivascular microglia showed phagocytic activities. Our present study suggests that the astrocytic cell processes around the dark neuron reacted to rescue the injured neuron by generating fuels, scavenging neuronal and synaptic metabolites [14, 16, 19, 25, 27, 30] and disposed dead neurons by shrinking, smashing and phagocitizing them.

Necrosis versus apoptosis in ischemic neuronal death has been controversial. In the present study, some of the morphological findings on the dark neurons indicated classical necrosis: i.e., on eosinophilic ghosting of cells in terms of histology; small loose aggregates of nuclear chromatin in the nuclear matrix and margin, instead of the marginates of condensed coarse granular aggregates seen in apoptosis; and a lack of apoptotic bodies. Others corresponded to apoptosis: scattered individual cells affected in terms of histology (shrinkage necrosis); lack of exudative inflammation, and condensation of the cytosol with structurally intact mitochondria and other organelles, instead of swelling of all cell component followed by rupture of cell membrane and destruction of cytosolic organelles as seen in necrosis [21, 33]. But, sometimes dead cells are swollen and sometimes they are shrunken depending on the cellular environment [21]. In the ischemic penumbra, swollen perineuronal astrocytes seemed to cause condensation of the necrotic neuron, resulting in shrinkage. While, when astrocytes died in the infarcted area, all neurons and astrocytes swelled and all of their membrane systems were ruptured. The macrophage seemed not to be able to enter the intact neuropils like those in the penumbra, but could enter the infarct focus where neuropils have been disrupted.

Conclusion

In conclusion: 1) The ultrastructural characteristics of DSNN, in the present study, suggested necrotic neuronal death instead of apoptosis. Condensation of the isolated neuron was induced by swelling of astrocytic cell processes surrounding the dark neuron. 2) Nuclear chromatin condensations observed from the early stage in appearance of the dark neurons indicated an initiating factor of the neuronal derangement. As the TUNEL procedure resulted in staining at the nuclear chromatin condensates of the dark neuron, in the present study, some DNA derangement seemed to be involved in this process [20, 32]. Further ultrastructural TUNEL staining is necessary to determine the ultrastructural characteristics of the TUNEL-positive neurons. 3) Not all dark neurons seems to die, but some of them to survive. Mitochondrial dysfunction seemed to be a determining factor of the irreversibility.

References

1. Colbourne F, Sutherland GR, Auer RN (1999) Electron microscopic evidence against apoptosis as the mechanism of neuronal death in global ischemia. J Neurosci 19(11):4200–4210
2. Coles JA (1995) Glial cells and the supply of substrates of energy metabolism to neurons. In: Kettenmann H, Ransom B (eds) Neuroglia. Oxford University Press, New York Oxford, pp 793–804
3. Dringen R, Schmoll D, Cesar M, Hamprecht B (1993) Incorporation of radioactivity from [14C] lactate into the glycogen of cultured mouse astroglial cells. Evidence for gluconeogenesis in brain cells. Biol Chem Hoppe Seyler 374(5):343–347
4. Dux E, Oschlies U, Uto A, Kusumoto M, Hossmann KA (1996) Early ultrastructural changes after brief histotoxic hypoxia in cultured cortical and hippocampal CA1 neurons. Acta Neuropathol (Berl) 92(6):541–544
5. Ghadially FN (1997) Ultrastructural Pathology of the Cell and Matrix, 4th edn. Butterworth-Heinemann 18–29:246–258
6. Hamprecht B, Dringen R (1995) Energy metabolism. In: Kettenmann H, Ransom B (eds) Neuroglia. Oxford University Press, New York Oxford, pp 473–487
7. Hanyu S, Ito U, Hakamata Y, Yoshida M (1995) Transition from ischemic neuronal necrosis to infarction in repeated ischemia. Brain Research 686:44–48
8. Hanyu S, Ito U, Hakamata Y, Nakano I (1997) Topographical analysis of cortical neuronal loss associated with disseminated selective neuronal necrosis and infarction after repeated ischemia. Brain Res 767(1):154–157
9. Ito U, Hanyu S, Hakamata Y, Kuroiwa T, Yoshida M (1997) Features and threshold of infarct development in ischemic maturation phenomenon. In: Ito U, Kirino T, Kuroiwa T, Klatzo I (eds) Maturation Phenomenon in Cerebral Ischemia II. Springer, Berlin Heidelberg, pp 115–121
10. Ito U, Hanyu S, Hakamata Y, Arima K, Oyanagi K, Kuroiwa T, Nakano I (1999) Temporal profile of cortical injury following ischemic insult just-below and at the threshold level for induc-

tion of infarction: light and electron microscopic study. In: Ito U, Orzi F, Kuroiwa T, Fieschi C, Klatzo I (eds) Maturation Phenomenon in Cerebral Ischemia III. Springer, Berlin Heidelberg, pp 227–235

11. Ito U, Spatz M, Walker J Jr, Klatzo I (1975) Cerebral ischemia in mongolian gerbils. I. Light microscopic observations. Acta Neuropathol Berl 32(3):209–223

12. Ito U, Kuroiwa T, Hanyu S, Hakamata Y, Arima K, Oyanagi K, Nakano I (1999) Ultrastructural features of astroglial mitochondria following temporary ischemia at threshold level to induce infarction. In: Krieglstein J (ed) Pharmacology of cerebral ischemia. medpharm Science Publication, Stuttgart, pp 113–117

13. Ito U, Kuroiwa T, Hanyu S, Hakamata Y, Arima K, Oyanagi K, Nakano I (2000) Ultrastructure and morphometry of astroglial mitochondria following temporary threshold ischemia to induce focal infarction. In: Bazan NG, Ito U, Kuroiwa T, Klatzo I (eds) Maturation Phenomenon in Cerebral Ischemia IV. Springer, Berlin Heidelberg, pp 253–259

14. Kempski O, Volk C (1997) Glial protection against neuronal damage. In: Ito U, Kirino T, Kuroiwa T, Klatzo I (eds) Maturation Phenomenon in Cerebral Ischemia II. Springer, Berlin Heidelberg, pp 115–121

15. Kerr JF, Wyllie AH, Currie AR (1972) Apoptosis: a basic biological phenomenon with wide-ranging implications in tissue kinetics. Br J Cancer 26(4):239–257

16. Kimelberg HK, Rutledge E, Goderie S, Charniga C (1995) Astrocytic swelling due to hypotonic or high K+ medium causes inhibition of glutamate and aspartate uptake and increases their release. J Cereb Blood Flow Metab 15(3):409–416

17. Kirino T (1982) Delayed neuronal death in the gerbil hippocampus following ischemia. Brain Res 239(1):57–69

18. Kraig RP, Lascola CD, Caggiano A (1995) Glial response to brain ischemia. In: Kettenmann H, Ransom B (eds) Neuroglia. Oxford University Press, New York Oxford, pp 964–976

19. Levi G, Gallo V (1995) Release of neuroactive amino acids from glia. In: Kettenmann H, Ransom B (eds) Neuroglia. Oxford University Press, New York Oxford, pp 815–826

20. Li PA, Rasquinha I, He QP, Siesjo BK, Csiszar K, Boyd CD, MacManus JP (2001) Hyperglycemia enhances DNA fragmentation after transient cerebral ischemia. J Cereb Blood Flow Metab 21(5):568–576

21. Majno G, Joris I (1995) Apoptosis, oncosis, and necrosis. An overview of cell death. Am J Pathol 146(1):3–15

22. Marcoux FW, Morawetz RB, Crowell RM, DeGirolami U, Halsey JH Jr (1982) Differential regional vulnerability in transient focal cerebral ischemia. Stroke 13(3):339–346

23. Martin LJ (2001) Neuronal cell death in nervous system development, disease, and injury (Review). Int J Mol Med 7(5):455–478

24. Martin DL (1995) The role of glia in the inactivation of neurotransmitters. In: Kettenmann H, Ransom B (eds) Neuroglia. Oxford University Press, New York Oxford, pp 732–745

25. Newman EA (1995) Glial cell regulation of extracellular potassium. In: Kettenmann H, Ransom B (eds) Neuroglia. Oxford University Press, New York Oxford, pp 717–731

26. Ohno K, Ito U, Inaba Y (1984) Regional cerebral blood flow and stroke index after left carotid artery ligation in the conscious gerbil. Brain Res 297(1):151–157

27. Rosenberg GA, Aizeman E (1989) Hundred-fold increase in neuronal vulnerability to glutamate toxicity in astrocyte-poor cultures of rat cerebral cortex. Neuroscience Letters 103:162–168

28. Sokoloff L (1992) Energy metabolism and effects of energy depletion or exposure to glutamate. Can J Physiol Pharmacol 70(12):S107–112

29. Sokoloff L, Gotoh J, Law MJ, Takahashi S (1996) Functional activation of energy metabolism in nervous tissue: Roles of neurons and astroglia. In: Krieglstein J (ed) Pharmacology of cerebral ischemia. medpharm Scientific Publ, Stuttgart, pp 259–270

30. Swanson RA, Choi DW (1993) Glial glycogen stores affect neuronal survival during glucose deprivation in vitro. J Cereb Blood Flow Metab 13(1):162–169

31. von-Lubitz DK, Diemer NH (1983) Cerebral ischemia in the rat: ultrastructural and morphometric analysis of synapses in stratum radiatum of the hippocampal CA-1 region. Acta Neuropathol Berl 61(1):52–60

32. Wijsman JH, Jonker RR, Keijzer R, van de Velde CJ, Cornelisse CJ, van Dierendonck JH (1993) A new method to detect apoptosis in paraffin sections: in situ end-labeling of fragmented DNA. J Histochem Cytochem 41(1):7–12

33. Wyllie AH, Kerr JF, Currie AR (1980) Cell death: the significance of apoptosis. Int Rev Cytol 68:251–306

The Maturation Phenomenon as an Expression of C. & O. Vogt's Theory of Pathoclisis

I. Klatzo

Summary. This presentation is aimed at spotlighting from obscurity two brain researchers – the couple of Cécile and Oskar Vogt – who, about one hundred years ago, in a visionary way, laid the foundations for the most spectacular progress which is now unfolding in Neuroscience. Although coming from vastly different backgrounds and environments, they were both endowed with brilliant, searching minds, and became, from their early youth, independently obsessed with the elucidation of the basic working of the human mind, and particularly with finding the real forces which regulate the *body-mind interrelationship*.

Arriving independently at the conclusion that functional and psychic events must be anchored in specific, regional brain structures, they conceived as their major objective the *correlation of functional brain phenomena with particular cortical areas and various brain structures*. This eventually led to the Vogts' monumental cytoarchitectonic studies on the cerebral cortex which, in collaboration with K. Brodmann, reached their peak in distinguishing about 200 distinct cortical areas [3].

Key words. C&O Vogt biography – pathoclisis

Introduction

In an effort to establish a correlation between anatomy and function, the Vogts used *cortical electrostimulation in primates*, and were able to demonstrate in many instances a synchronous relation between evoked functional responses and distinct cytoarchitectonic areas. The Vogts' data on coupling anatomy with function were strongly supported by observations of the neurosurgeons O. Foerster and W. Penfield, who had the unique opportunity to stimulate the exposed human cortex during surgical operations [2, 3, 12]. At the present time, the cytoarchitectonic maps of the Vogts and Brodmann are used as a reference in the mapping of the cerebral cortex and of the subcortical brain centers by sophisticated methods, such as by PET, MRI and various markers for neurotransmitters.

Correspondence to: Dr. Igor Klatzo, 19022 Canadian Court, Gaithersburg, MD, 20886, USA, Tel.: 1-301-869-5502, E-Mail: klatzo@aol.com

Maturation Phenomenon in Cerebral Ischemia V
A. M. Buchan et al. (Eds.)
© Springer-Verlag Berlin Heidelberg 2004

The monumental work of the Vogts on brain cytoarchitectonics, which is so relevant for identifying functional brain areas, was honored by the First Vogt-Brodmann Symposium on Perspectives of Architectonic Brain Mapping, held in Düsseldorf-Jülich, on June 20–21, 1999. In the Welcoming Address of this symposium, Cécile and Oskar Vogt were recognized as "key figures in establishing modern brain research", and their work with Brodmann as of "fundamental importance to neuropsychology, neurophysiology and brain mapping".

The greatest of the Vogts' contribution to the field of Neuroscience, however, is the theory of *pathoclisis*. This evolved from Oskar Vogt's persistent, almost obsessive interest in *molecular genetics* [11].

As a youngster with an inquisitive mind, Oskar became fascinated with the *variability* within species of various insects and creatures which he was collecting in the marshes surrounding his birth town of Husum in Schleswig Holstein. Under the influence of Darwin's theory of evolution, Oskar's interpretation of variability tended to assume the involvement of *genetic mutations* which could be induced by *environmental changes*. During his medical studies, he was impressed with the variability observed in various human brain disorders, and this led him to consider whether *mechanisms of variability, observed in insects such as the bumble-bee, could not also be applicable to the human brain.*

With such considerations in the background and after securing the possibility for independent brain research in Berlin, the Vogts arrived, on the basis of their cytoarchitectonic observations, at the concept of pathoclisis.

According to the Vogts' definition, *pathoclisis* (from the Greek: abnormal predisposition) *represents a genetically determined excessive variability, reaching in intensity a pathological change, and which is based on the modulation of the molecular composition of specific neuronal constituents in particular brain regions.*

As Copernicus' celestial theory stated that the earth is simply a planet rotating around the sun due to *physical gravitational forces*, so in the Vogts' pathoclisis theory, *molecular genetics* appears to regulate our human individual development, as well as significantly determining when and in which way we become ill.

The concept of pathoclisis has many ramifications. First of all, it can be assumed that the *selective vulnerability* of various neuronal regions to a given noxious factor is an expression of pathoclisis, in which various neuronal populations, endowed with different molecular configurations, are expected to react differently.

The Vogts were the first to recognize, what is now accepted, that *sublethal cell injury* (such as in areas of the penumbra) *activates genetic molecular mechanisms, accelerating programmed apoptotic cell death, as well as inducing a variety of defensive mechanisms which ameliorate damage and stimulate cell recovery.* The confrontation of forces, promoting either apoptotic death or cell recovery, resembles a *battle between the forces of good and evil, determining whether a cell will live or die, with the outcome of the struggle depending on the balance of genomic responses characteristic of each topistic unit.*

Influenced by previous observations on trans-neuronal or trans-synaptic degeneration, which affects previously healthy neurons when they are deprived of synaptic stimulation from interconnected, severely damaged or destroyed neuronal centers, the Vogts arrived in their *studies on aging* to the general conclusion that "inactivity accelerates and activity delays the aging of nerve cells". This conclusion was supported by the classical U.V. spectroscopic studies of Hydén [4], showing the correlation of high neuronal activity with stimulation of the RNA and protein-producing "nucleolar apparatus", whereas inactivity was associated with a fading replenishment of structural proteins, the latter essential for normal function.

In agreement with the theory of pathoclisis, the Vogts found that the cell populations in various *topistic regions had different, genetically-regulated aging schedules*. In the "elite" brains of their collection, the areas related to some outstanding intellectual ability revealed conspicuously delayed aging, whereas in the athétose double, the "hard" working neurons in subcortical centers, responsible for increased mobility, tended to become hypertrophic [2, 9, 10].

In other studies on aging, using the term *involution* instead of *apoptosis*, the Vogts can be credited with the first microscopic description of apoptotic changes, such as the clumping and extrusion of nuclear chromatin, fading of the nucleolus and cell pyknosis, all currently considered as characteristic markers for apoptosis.

In cerebral ischemia, the involvement of pathoclisis in the Maturation Phenomenon (M.Ph.) appears in *the variability of continuing fluctuating genomic expressions affecting the demise or survival of particular topistic units*, and this has been especially apparent in my studies on cerebral ischemia in which I was engaged about 4 years ago at Prof. N.G. Bazan's Neuroscience Center of Excellence in New Orleans.

The main focus of our studies in New Orleans [1] was centered on manifestations of *apoptosis* with regard to neuronal survival in gerbils subjected either to repeated, separated by various time intervals, or to single ischemic exposures.

In the *repetitive ischemia*, all groups of gerbils were sacrificed 3 days after the second 6 min. bilateral carotid occlusion, whereas the preceding neuroprotective 2 min ischemia was carried out at 15 min, 1 h, and 3 days before the second insult.

The observations revealed that the *patterns of ischemic injury* markedly *differed*, depending on *the time interval* between the occlusions (Fig. 1 a).

As to be expected, the gerbils with a *3-day* interval between occlusions indicated a clear neuroprotection with regard to the CA1 pyramidal neurons, which appeared relatively well-preserved and showed TUNEL staining in less than 20% of the neurons. This pattern differed greatly from that in gerbils in which the carotid occlusions were separated by *one hour*. Animals in this group revealed a striking amount of damage in the thalamus and parietal cortex (Fig. 2 a, b), demonstrable with Nissl and TUNEL staining, the latter showing numerous TUNEL-positive neurons, particularly in the ventral thalamic nuclei (Fig. 2 a), whereas no significant cortical or thalamic damage was observed in a single 6 min occlusion. Otherwise the gerbils with a one hour interval revealed no significant evidence of protection concerning the CA1 sector, the great majority of pyramidal neurons demonstrating strong TUNEL staining in the nuclei and cytoplasm (Fig. 2 d), and the Nissl preparations revealing severe neuronal damage. These observations support the assumption that the first ischemic exposure induces a rapidly changing panorama of genomic expressions, which, *depending on the timing* of the second insult, may affect differently the pattern of neuronal reactivity and the resulting injury.

For the interpretation of the striking involvement of the thalamus in animals with a one hour interval between the occlusions, one of the Vogts' postulates might be relevant, namely that the theory of *pathoclisis extends, besides neurons, to other components of the brain tissue, that is to glia and to elements of the vascu-*

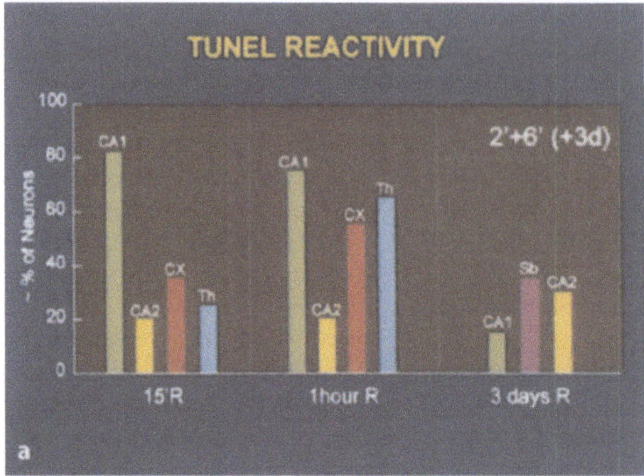

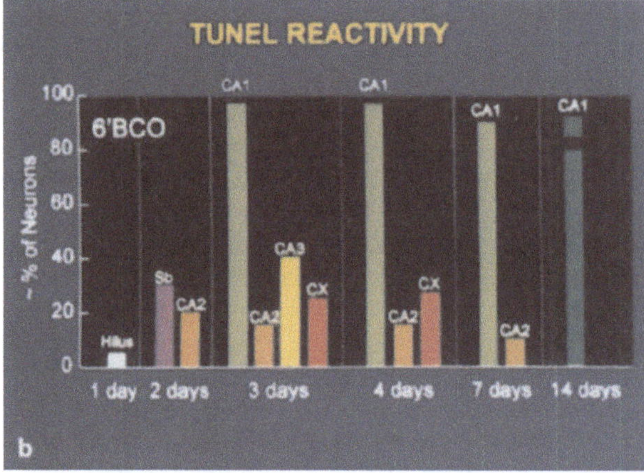

Fig. 1. a TUNEL reactivity (%) in gerbils subjected to repetitive ischemia, sacrificed 3 days after the second 6 min bilateral carotid occlusion, whereas the first 2 min occlusion was carried out at 15 min, 1 hour and 3 days before the second insult. **b** TUNEL reactivity in animals subjected to a single 6 min occlusion

lature. Since our previous investigations, carried out in Bethesda [5], had shown that a one hour interval between occlusions can be associated with striking changes in the BBB permeability affecting selective thalamic regions, it can be speculated that the intensity of ischemic injury in the thalamus may be related to genomic reactivity of the vasculature in selected thalamic regions.

In the single 6 min occlusion ischemia our observations indicated that *selective vulnerability*, which reflects the *intensity* of injury in selective neuronal populations, differs from *selective sensitivity*, characterized by the induction of genomic expressions of mostly a transitory nature in areas not associated with lethal neuronal injury.

Thus, the CA 3 segment, three days after a single 6 min occlusion, revealed TUNEL staining in about 30% of the neurons, which lasted for only one day, and was confined

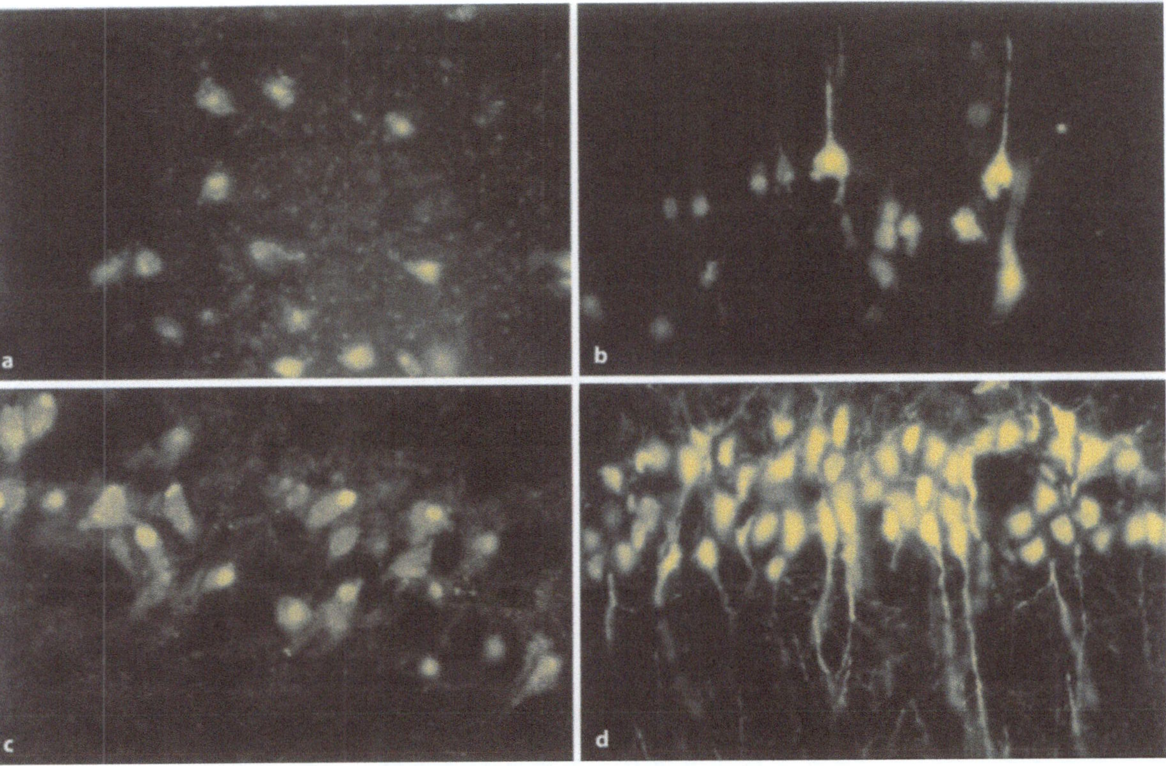

Fig. 2. a TUNEL staining in the ventral thalamic nucleus of a gerbil sacrificed 3 days after the second 6 min occlusion, which was preceded by one hour by a 2 min occlusion. **b** Parietal cortex of the animal in **a** showing numerous TUNEL-positive neurons. **c** CA3 sector of hippocampus in a gerbil sacrificed 3 days after a single 6 min occlusion showing TUNEL labeling confined to mostly eccentrically positioned nuclei. **d** CA1 sector of gerbil shown in **c** with TUNEL labeling displaying intense positive staining, including the cytoplasm and processes of pyramidal neurons

to peripherally-translocated neuronal nuclei (Fig. 2c), whereas a quantitative assay in animals 7 days after the ischemic insult showed no neuronal loss in the CA 3 sector. CA 1, the sector most vulnerable to injury, remained invariably negative up to the third day in all 6 min occlusions. At which time the intense TUNEL staining of pyramidal neurons involved the cytoplasm, in addition to the nuclei (Fig. 2d), and was associated with morphological evidence of severe neuronal damage.

Discussion

In general our observations both in repeated and single ischemia, demonstrating that *neurons with TUNEL-positive staining confined to the nuclei may be capable of recovery,* suggest the possibility that apoptotic *fragmentation of DNA within the nucleus, with preserved integrity of the nuclear membrane, does not represent the point of no return as long as cytoplasmic mitochondrial DNA can still be active.*

As visionaries, preoccupied with the importance of genetic mechanisms in brain pathology, the Vogts predicted that in the future these mechanisms *could be manipulated to enhance certain positive genetic expressions, or to suppress harmful ones* [7]. By this they envisaged what is currently one of the main subjects in Neuroscience, mainly an effort to *conquer by genetic manipulations* various neurodegenerative diseases, such as Friedreich's ataxia, amyotrophic sclerosis, Alzheimer's and Parkinson's disease [6]. In this respect, the insertion of corresponding disease genes into mice has produced a number of mouse disease models, which are currently studied in many laboratories for identifying the precise changes in molecules and cells that underlie various neurodegenerations.

I would like to mention that the greatness of the Vogts is based both on their penetrating vision of the future for brain research and on the sterling quality of their character, which sustained a "test of fire" during the terror-stricken time of the Nazis when they risked their lives to save some of their friends who happened to be of the Jewish faith. Sailing through the rough seas of intolerance and terror, the Vogts maintained their integrity and unshaken beliefs. Their lives and contributions to brain research should serve as an inspiration to a new generation of neuroscientists.

References

1. Gordon WC, Colangelo V, Bazan NG, Klatzo I (1999) Aspects of the Maturation Phenomenon observed by the TUNEL method. In Maturation Phenomenon in Cerebral Ischemia III. Springer, Berlin Heidelberg, pp 15–23
2. Hassler R (1959) Cécile und Oskar Vogt. In: Kolle K (ed) Grosse Nervenärzte. Georg Thieme, Stuttgart, pp 45–65
3. Haymaker W (1951) Cécile and Oskar Vogt. Neurology 1:179–204
4. Hydén H (1943) Protein metabolism in the nerve cell during growth and function. Acta Physiol Scand 6 (suppl XVII):1–136
5. Nagashima G, Nowak TS Jr, Joo F, Ikeda J, Rützler C, Lohr J, Klatzo I (2000) The role of the blood-brain barrier in ischemic brain lesions. In: Johansson BB, Owman C, Widner H (eds) Pathophysiology of the Blood-Brain Barrier Fernström Foundation Series,Elsevier Publ
6. Neuroscience Newsletter, May–June 2000, p 7

7. Vogt C & O (1919) Allgemeine Ergebnisse unserer Hirnforschung. J f Psychol u Neurol.25:227–461
8. Vogt C & O (1922) Erkrankungen der Grossnirnrinde im Lichte der Topistic, Pathoclise und Pathoarchitektonik. J f Psychol u Neurol 28:1–171
9. Vogt C & O (1947) Über Wesen und Ursache des Alterns der Hirnzellen. Forschungen und Fortschritte 21/23:4–6
10. Vogt C & O (1946) Aging of nerve cells. Nature 157:304
11. Vogt C & O (1929) Hirnforschung und Genetik. J f Psychol u Neurol 39:438–446
12. Vogt C & O (1926) Die vergleichend-architektonische und die vergleichend reizphysiologische Felderung der Grosshirnrinde unter besonderer Berücksichtigung der menschlichen. Naturwissensch 14:1190–1228

Protection by Apomorphine of Dopaminergic Neurons Following Acute Inhibition of Oxidative Metabolism in Rodents

F. Orzi, G. Battaglia, F. Nicoletti, F. Girardi, C. Busceti, and F. Fornai

Key words. Apomorphine – neuroprotection – oxidative metabolism – methyl-phenyl-tetra-hydropyridine (MPTP) – methamphetamine – ischemia, mice

Introduction

A vast body of research indicates that a number of drugs or procedures aimed to reduce excitatory neurotransmission, reactive oxygen species (ROS) production or inflammation do ameliorate ischemic damage, as assessed by histopathology, in laboratory animals. The optimism associated with the promising results in laboratory animals, however, contrasts with the negative findings of several clinical trials, as a number of compounds that reduce infarct size in animal models of stroke were not successful in clinical trials. Reasons for the apparent discrepancy are various, and probably include heterogeneity of the stroke population or the long time interval between onset of stroke and beginning of the treatment. Two additional aspects, potentially relevant to the discrepancy, refer to the clinical criteria for evaluation of the efficacy, and to the plasma concentration and side effects of a drug. Histopathological assessment of the size of the ischemic lesion may well not account for the complex, mostly unclear, relationship between histopathological damage and functional consequences [21]. Moreover, most of the drugs proven effective in reducing the infarct size in rodents were used at doses which would cause unacceptable adverse effects in humans [8].

In the present study we sought to determine whether apomorphine, a drug used in persons with movement disorders, is neuroprotective following acute inhibition of oxidative metabolism by systemic administration of MPTP, in the mouse. The drug was used by continuous infusion, at a dosage comparable to the one used in the treatment of severe on-off fluctuations in Parkinson's disease [44].

Data have accumulated in the last few years suggesting that dopamine agonist drugs are neuroprotective. Bromocriptine [24, 31, 34, 35], pergolide [10, 38, 39],

Francesco Orzi[1,2], Guiseppe Battaglia[2], Ferdinandio Nicoletti[4,2], Francesca Girardi[1], Carla Busceti[2] and Francesco Fornai[3]
[1] Dep. Neurological Sciences, University of Roma "La Sapienza"
[2] Neuromed Institute, Pozzilli (IS)
[3] Dep. Human Morphology and Applied Biology, University of Pisa
[4] Dep. Human Physiology and Pharmacology, University of Roma "La Sapienza", Italy

Correspondence to: Prof. Francesco Orzi, Dipartimento di Scienze Neurologiche, Universita di Roma "La Sapienza", Il Facolta Di Medicina, Policlinico Sant Andrea, Via di Grottarossa, Roma, Italy, Tel.: 39 06 4991 4708, Fax: 39 06 495 8344, E-mail: francesco.orzi@uniromal.it

Maturation Phenomenon in Cerebral Ischemia V
A. M. Buchan et al. (Eds.)
© Springer-Verlag Berlin Heidelberg 2004

ropinirole [7, 22], pramipexole [5, 18, 30], lisuride [4], piribedil [9], or apomorphine [11, 16, 17] have all been shown to protect neurons or brain tissue under a variety of noxious conditions [6]. Most of the studies were carried out in cultured cells or in vitro, in a few cases the protection was observed in animal models of Parkinson's disease [15] or during brain ischemia [3, 4, 18, 31, 38].

The mechanisms by which protection occurs, however, are uncertain. Increased dopamine metabolism causes excessive production of ROS, and there is substantial evidence that oxidative stress is associated with both acute and chronic neurodegeneration. Excessive dopamine turnover occurs in Parkinson's disease during levodopa therapy or because of the increased activity of the surviving neurons [1, 33], and in ischemia because of the depolarization-induced striatal dopamine release [14, 28]. A hypothesis, therefore, poses that the efficacy of dopamine agonists in reducing the damage in animal models of Parkinson's disease or brain ischemia is mediated by the agonist-induced reduction of the dopamine turnover, with secondary reduction of the associated ROS production. This hypothesis has a number of flows. For instance, although dopamine agonists decrease the ischemia-induced striatal dopamine release, the distribution of the protection, in models of global ischemia, involves both dopaminergic (nigrostriatal) and non-dopaminergic (hippocampal) neurons.

Alternative hypotheses ascribe the protection to intrinsic antioxidant properties of the agonists [37] or to production of neurotrophic factors [29]. In the studies in which the protection was observed in vivo, in animal models of Parkinson's disease [15, 17], or brain ischemia [4, 18, 31, 38], mechanisms associated with systemic administration of the drugs, including changes in cerebral blood flow, and body temperature, cannot be ruled out. In the same studies, drugs were used at doses exceeding those currently used in humans for therapy of neurological diseases.

In the monkey, MPTP induces the loss of pigmented neurons of the SNpc [19], with a topology similar to that observed in humans [19, 42], and with depletion of striatal dopaminergic markers [20]. MPTP readily crosses the blood-brain barrier and in glial cells it is converted in MPP^+ that is taken up by the dopamine transporters and accumulates in mitochondria. The best defined MPTP effect is the inhibition of complex I, with secondary ATP depletion, loss of mitochondria membrane potential, changes in calcium homeostasis and radical formation [2]. By inhibiting oxidative metabolism, MPTP causes ATP depletion. The typical neuronal loss, mostly within the substantia nigra compacta, is associated with oxidative stress due to increased production of ROS. A hypothesis poses that the initial loss of dopaminergic neurons induces compensatory increases in dopamine (DA) production with increased oxidative metabolism of DA and ensuing production of ROS. MPTP poisoning is accompanied in the experimental animal by glial reaction, astrogliosis, expression of MHC class I and II antigens and lymphocyte infiltration [23, 25, 26]. Recently similar inflammatory changes were found in the brain of subjects who became parkinsonian as a result of exposure to MPTP [27], clearly showing that the inflammatory response occurs both in rodents and human and non-human primates [32]. Since 1996 several studies have stressed the potential involvement of apoptosis in Parkinson disease (PD) [2]. Altogether these findings point to remarkable similarities between neurodegeneration associated with ischemia and with MPP+.

In the present study we show in the mouse that apomorphine, administered after the MPTP-induced striatal damage has occurred, enhances the spontaneous recovery. The drug was administered continuously, at doses similar to those used in persons with movement disorders.

Neuroprotection by pretreatment with apomorphine towards acute MPTP toxicity had been previously shown [17]. In order to clarify the mechanism of the neuroprotection, the effects of apomorphine were evaluated also in animals administered methamphetamine. The drug is selectively toxic for dopamine (DA) and 5-hydroxy-tryptamine nerve terminals and produces a striatal damage comparable to the one obtained by MPTP, but with different mechanisms. We show that pre-treatment with apomorphine dose-dependently protects striatum from methamphetamine-induced DA and tyrosine hydroxylase (TH) reduction.

Materials and Methods

Experiments were carried out in 10 week old C57 black mice. On the first day of the treatment (day 1) all the animals received intraperitoneal injection of either saline, or MPTP (30 mg/kg). The animals were then administered either apomorphine or saline, continuously for 28 days by means of an osmotic pump (Alzet) implanted subcutaneously on day 1. The nominal delivery of the pump is 0.25 µL/h, corresponding to a cumulative dose of 3.15 mg/kg/day of apomorphine. Control animals received an equivalent volume of saline. There were, therefore, 4 animal groups (n = 8–10): MPTP lesioned or sham animals administered either apomorphine or saline for 28 days. Apomorphine/saline delivery by the pumps was effective at 40 h after MPTP administration. On day 30, animals were sacrificed, brains removed and dissected. The striatum from one half was immediately processed for assay of the catecholamines by means of HPLC. The other half was snap-frozen in ice/isopentane and stored until processing for TH immunostaining which was carried out in 4 striatal consecutive sections (10 µm).

In a different set of experiments, three groups of animals (n = 10, for each group) were administered intraperitoneally 1, 5, or 10 mg/kg of R-apomorphine 15 min before each intraperitoneal administration of methamphetamine (5 mg/kg×3, 2 h apart). Five control groups (n=10, for each group) received saline, methamphetamine, or 1, 5, or 10 mg/kg of R-apomorphine 15 min before saline. Five days following the treatment animals were sacrificed and the striatum processed for monoamine levels and tyrosine hydroxylase activity. In addition, 8 animal goups (n=7) were used to assess the effect of apomorphine + saline or apomorphine + methamphetamine on body temperature.

Results

A 50–60% reduction of striatal dopamine levels was observed in the MPTP-treated animals. A reduction of a comparable intensity was observed in the animals administered methamphetamine.

The treatment with the continuous apomorphine infusion (which started 40 h following the acute administration of MPTP) significantly enhanced the spontaneous recovery that follows the dopaminergic loss induced by the toxin (Fig. 1, 2).

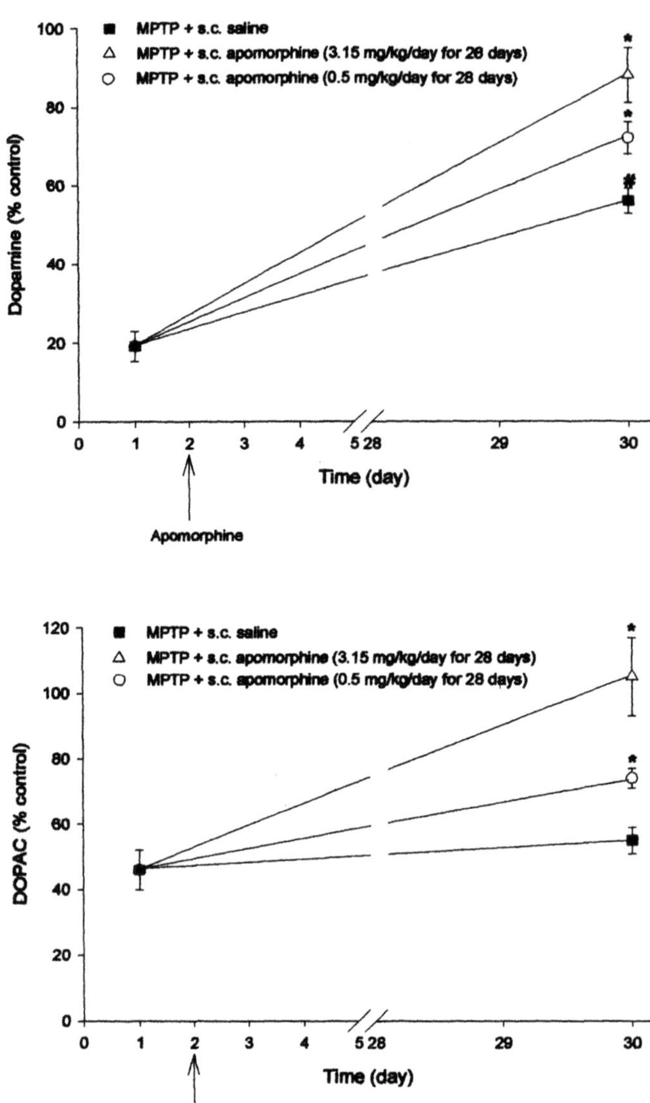

Fig. 1. Effects of subcutaneous, continuous infusion of apomorphine on the recovery of striatal DA and DOPAC levels following a single injection of MPTP. Apomorphine or saline injection started 40 hours following MPTP injection, values measured on day 1 represent, therefore, the intensity of the tissue damage induced by MPTP per se. On day 30, recovery is significantly (one-way Anova and Fisher PLSD) enhanced in animals receiving apomorphine, in a dose-dependent way. Values (means ± SEM) are expressed as percentage of control values

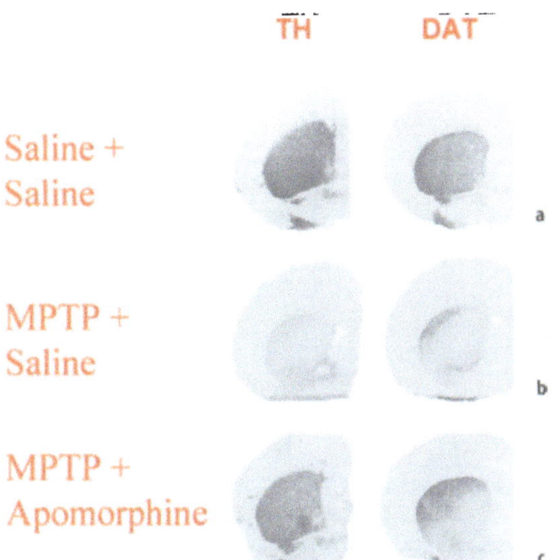

Fig. 2. Representative tyrosine hydroxylase (TH) and dopamine transporter (DAT) staining of the striatum from (**a**) control mice (saline+saline), (**b**) animals administered MPTP+saline, or (**c**) MPTP+subcutaneous, continuous apomorphine (3.15 mg/kg/day). There is partial, significant rescuing by apomorphine of the MPTP-induced reduction of TH and DAT staining

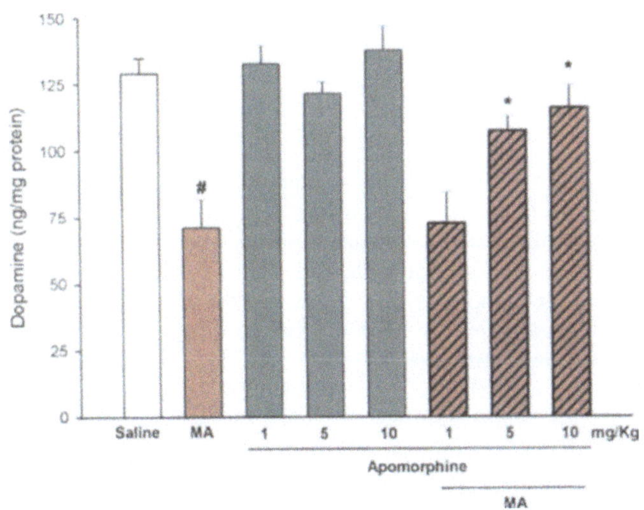

Fig. 3. Dose-dependent effects of a single administration of apomorphine on striatal dopamine loss induced by methamphetamine. Values are means±SEM. Comparisons among groups were performed by using ANOVA and Sheffe's post-hoc analysis. * $p < 0.05$ as compared to control or methamphetamine alone

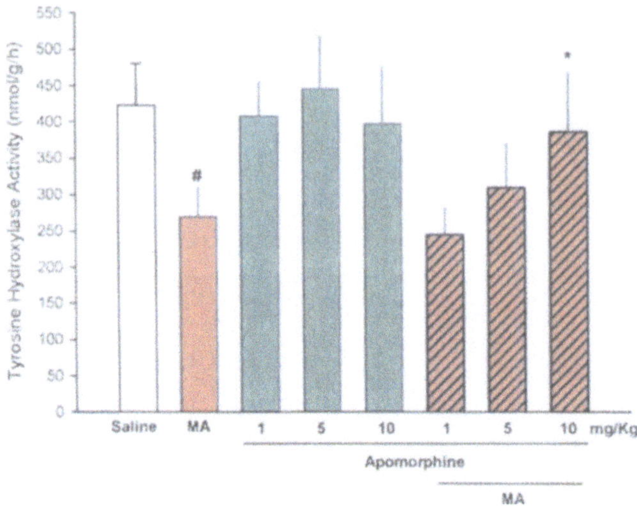

Fig. 4. Dose-dependent effects of a single administration of apomorphine on striatal decrease of TH activity induced by methamphetamine. Values are means ± SEM. Compaprisons among groups were performed by using ANOVA and Sheffe's post-hoc analysis. * p < 0.05 as compared to control or methamphetamine alone

The pre-treatment with intraperitoneal administration of apomorphine produced a dose-dependent protection on the methamphetamine toxicity, as assessed both by striatal dopamine levels (Fig. 3) and analysis of TH (Fig. 4.).

Discussion

The data show that apomorphine *in vivo* enhances spontaneous recovery of nigro-striatal neurons in mice administered MPTP, or protects against acute toxicity by methamphetamine. The two toxins produce similar neuronal damage by different mechanisms. MPP+ enters the dopaminergic cells and inhibits oxidative metabolism causing ATP depletion. Methamphetamine promotes the formation of free radicals, a process in which dopamine may have a prominent role. In both cases oxidative stress seems to play a relevant role.

The protective effect of apomorphine is not associated with changes in the pharmacokinetics of striatal methamphetamine, nor with changes in the primary mechanism of action of MPTP. Intraperitoneal administration of the drug did not interfere with the time course of the striatal methamphetamine content, hours following the toxin injection (data not shown). We also rule out a role for temperature in the neuroprotective effect of apomorphine, as the drug does not prevent the increase of body temperature induced by methamphetamine, and the protective effect of apomorphine are dose dependent, while the effect of apomorphine *per se* on body temperature it is not. The direct involvement of dopamine receptors in the neuroprotective effects also appears unlikely, on the basis of a number of considerations: a) protection occurs at doses higher than those required for dopamine agonist effects [11]; b) protective effects are not common to all the dopamine agonists [6]; c) protective effects are not opposed by dopamine antagonists, as shown in our study and in previous ones [41]; d) both enantiomers of apomorphine (which have similar antioxidant properties, but different dopamine agonist effects) are neuroprotective when tested in cell cultures [11]. Apomorphine inhibits MAO-B [15], the enzyme responsible for the conversion of MPTP to the toxic MPP+, but this property is not relevant to the neuroprotection, as the inhibition of the enzyme should worsen the striatal damage induced by methamphetamine [43]. Instead, apomorphine had a dose-dependent protective effect on the toxicity induced by methamphetamine. The highest dose of apomorphine completely opposed the dopamine loss caused by methamphetamine, an effect which was not modified by haloperidol. It is, therefore, likely that the neuroprotective effect of apomorphine relies mostly on the antioxidant properties of the drug [12, 40].

The study also shows that a continuous, subcutaneous infusion of apomorphine enhances the spontaneous recovery of the striatal damage caused by a single dose of MPTP. We should stress that the infusion employed in this study started 40 h following the MPTP injection, when MPTP and MPP+ are cleared from striatum [13], and when the MPTP damage has already occurred. In this contest, it is worth stressing that apomorphine did not prevent the MPTP-induced loss of neurons within the substantia nigra compacta. The data suggest that apomorphine induces sprouting of survival striatal dopaminergic terminals, or increases the number of intrinsic neuron expressing a catecholamine phenotype. The neurorescuing effects required a continuous infusion. The finding seems consistent with trophic properties of the drug [36].

All together the data indicate that apomorphine has neuroprotective and neurorescuing effects, at doses that can be used in a clinical setting. The finding may be relevant to brain ischemia, as well to chronic neurodegeneration conditions.

References

1. Agid Y, Javoy F, Glowinski J (1973) Hyperactivity of remaining dopaminergic neurones after partial destruction of the nigro-striatal dopaminergic system in the rat. Nat New Biol 245:150–151
2. Blum D, Torch S, Lambeng N, Nissou M, Benabid AL, Sadoul R, Verna JM (2001) Molecular pathways involved in the neurotoxicity of 6-OHDA, dopamine and MPTP: contribution to the apoptotic theory in Parkinson's disease. Prog Neurobiol 65:135–172
3. Caldwell MA, Reymann JM, Bentue-Ferrer D, Allain H, Leonard BE (1996) The dopamine agonists lisuride and piribedil protect against behavioural and histological changes following 4-vessel occlusion in the rat. Neuropsychobiology 34:117–124
4. Caldwell MA, Reymann JM, Allain H, Leonard BE, Bentue-Ferrer D (1997) Lisuride prevents learning and memory impairment and attenuates the increase in extracellular dopamine induced by transient global cerebral ischemia in rats. Brain Res 771:305–318
5. Carvey PM, Pieri S, Ling ZD (1997) Attenuation of levodopa-induced toxicity in mesencephalic cultures by pramipexole. J Neural Transm 104:209–228
6. Carvey PM, McGuire SO, Ling ZD (2001) Neuroprotective effects of D3 dopamine receptor agonists. 7:213–223
7. Coldwell MC, Boyfield I, Brown T, Hagan JJ, Middlemiss DN (1999) Comparison of the functional potencies of ropinirole and other dopamine receptor agonists at human D2(long), D3 and D4.4 receptors expressed in Chinese hamster ovary cells. Br J Pharmacol 127:1696–1702
8. De Keyser J, Sulter G, Luiten PG (1999) Clinical trials with neuroprotective drugs in acute ischaemic stroke: are we doing the right thing? Trends Neurosci 22:535–540

9. Delbarre B, Delbarre G, Rochat C, Calinon F (1995) Effect of piribedil, a D-2 dopaminergic agonist, on dopamine, amino acids, and free radicals in gerbil brain after cerebral ischemia. Mol Chem Neuropathol 26:43–52
10. Felten DL, Felten SY, Fuller RW, Romano TD, Smalstig EB, Wong DT, Clemens JA (1992) Chronic dietary pergolide preserves nigrostriatal neuronal integrity in aged-Fischer-344 rats. Neurobiol Aging 13:339–351
11. Gassen M, Gross A, Youdim MB (1998) Apomorphine enantiomers protect cultured pheochromocytoma (PC12) cells from oxidative stress induced by H_2O_2 and 6-hydroxydopamine. Mov Disord 13:661–667
12. Gassen M, Gross A, Youdim MB (1999) Apomorphine, a dopamine receptor agonist with remarkable antioxidant and cytoprotective properties. Adv Neurol 80:297–302
13. Giovanni A, Sieber BA, Heikkila RE, Sonsalla PK (1994) Studies on species sensitivity to the dopaminergic neurotoxin 1-methyl-4-phenyl-1,2,3,6-tetrahydropyridine. Part 1: Systemic administration. J Pharmacol Exp Ther 270:1000–1007
14. Globus MY, Busto R, Dietrich WD, Martinez E, Valdes I, Ginsberg MD (1988) Effect of ischemia on the in vivo release of striatal dopamine, glutamate, and gamma-aminobutyric acid studied by intracerebral microdialysis. J Neurochem 51:1455–1464
15. Grunblatt E, Mandel S, Berkuzki T, Youdim MB (1999a) Apomorphine protects against MPTP-induced neurotoxicity in mice. Mov Disord 14:612–618
16. Grunblatt E, Mandel S, Gassen M, Youdim MB (1999b) Potent neuroprotective and antioxidant activity of apomorphine in MPTP and 6-hydroxydopamine induced neurotoxicity. J Neural Transm Suppl 55:57–70
17. Grunblatt E, Mandel S, Maor G, Youdim MB (2001) Effects of R- and S-apomorphine on MPTP-induced nigro-striatal dopamine neuronal loss. J Neurochem 77:146–156
18. Hall ED, Andrus PK, Oostveen JA, Althaus JS, Von Voigtlander PF (1996) Neuroprotective effects of the dopamine D2/D3 agonist pramipexole against postischemic or methamphetamine-induced degeneration of nigrostriatal neurons. Brain Res 742:80–88
19. Hantraye P, Varastet M, Peschanski M, Riche D, Cesaro P, Willer JC, Maziere M (1993) Stable parkinsonian syndrome and uneven loss of striatal dopamine fibres following chronic MPTP administration in baboons. Neuroscience 53:169–178
20. Heikkila RE, Sieber BA, Manzino L, Sonsalla PK (1989) Some features of the nigrostriatal dopaminergic neurotoxin 1-methyl-4-phenyl-1,2,3,6-tetrahydropyridine (MPTP) in the mouse. Mol Chem Neuropathol 10:171–183
21. Hunter AJ, Mackay KB, Rogers DC (1998) To what extent have functional studies of ischaemia in animals been useful in the assessment of potential neuroprotective agents? Trends Pharmacol Sci 19:59–66
22. Iida M, Miyazaki I, Tanaka K, Kabuto H, Iwata-Ichikawa E, Ogawa N (1999) Dopamine D2 receptor-mediated antioxidant and neuroprotective effects of ropinirole, a dopamine agonist. Brain Res 838:51–59
23. Kohutnicka M, Lewandowska E, Kurkowska-Jastrzebska I, Czlonkowski A, Czlonkowska A (1998) Microglial and astrocytic involvement in a murine model of Parkinson's disease induced by 1-methyl-4-phenyl-1,2,3,6-tetrahydropyridine (MPTP). Immunopharmacology 39:167–180
24. Kondo T, Ito T, Sugita Y (1994) Bromocriptine scavenges methamphetamine-induced hydroxyl radicals and attenuates dopamine depletion in mouse striatum. Ann N Y Acad Sci 738:222–229
25. Kurkowska-Jastrzebska I, Wronska A, Kohutnicka M, Czlonkowski A, Czlonkowska A (1999a) MHC class II positive microglia and lymphocytic infiltration are present in the substantia nigra and striatum in mouse model of Parkinson's disease. Acta Neurobiol Exp (Warsz) 59:1–8
26. Kurkowska-Jastrzebska I, Wronska A, Kohutnicka M, Czlonkowski A, Czlonkowska A (1999b) The inflammatory reaction following 1-methyl-4-phenyl-1,2,3,6-tetrahydropyridine intoxication in mouse. Exp Neurol 156:50–61
27. Langston JW, Forno LS, Tetrud J, Reeves AG, Kaplan JA, Karluk D (1999) Evidence of active nerve cell degeneration in the substantia nigra of humans years after 1-methyl-4-phenyl-1,2,3,6-tetrahydropyridine exposure. Ann Neurol 46:598–605
28. Lin MT, Kao TY, Chio CC, Jin YT (1995) Dopamine depletion protects striatal neurons from heatstroke-induced ischemia and cell death in rats. Am J Physiol 269:H487–490
29. Ling ZD, Tong CW, Carvey PM (1998) Partial purification of a pramipexole-induced trophic activity directed at dopamine neurons in ventral mesencephalic cultures. Brain Res 791:137–145
30. Ling ZD, Robie HC, Tong CW, Carvey PM (1999) Both the antioxidant and D3 agonist actions of pramipexole mediate its neuroprotective actions in mesencephalic cultures. J Pharmacol Exp Ther 289:202–210

31. Liu XH, Kato H, Chen T, Kato K, Itoyama Y (1995) Bromocriptine protects against delayed neuronal death of hippocampal neurons following cerebral ischemia in the gerbil. J Neurol Sci 129:9–14
32. McGeer PL, Itagaki S, Boyes BE, McGeer EG (1988) Reactive microglia are positive for HLA-DR in the substantia nigra of Parkinson's and Alzheimer's disease brains. Neurology 38:1285–1291
33. Melamed E, Hefti F, Wurtman RJ (1982) Compensatory mechanisms in the nigrostriatal dopaminergic system in Parkinson's disease: studies in an animal model. Isr J Med Sci 18:159–163
34. Muralikrishnan D, Mohanakumar KP (1998) Neuroprotection by bromocriptine against 1-methyl-4-phenyl-1,2,3,6-tetrahydropyridine-induced neurotoxicity in mice. Faseb J 12:905–912
35. Ogawa N, Tanaka K, Asanuma M, Kawai M, Masumizu T, Kohno M, Mori A (1994) Bromocriptine protects mice against 6-hydroxydopamine and scavenges hydroxyl free radicals in vitro. Brain Res 657:207–213
36. Ohta M, Mizuta I, Ohta K, Nishimura M, Mizuta E, Hayashi K, Kuno S (2000) Apomorphine up-regulates NGF and GDNF synthesis in cultured mouse astrocytes. Biochem Biophys Res Commun 272:18–22
37. Olanow CW, Jenner P, Brooks D (1998) Dopamine agonists and neuroprotection in Parkinson's disease. Ann Neurol 44:S167–174
38. O'Neill MJ, Hicks CA, Ward MA, Cardwell GP, Reymann JM, Allain H, Bentue-Ferrer D (1998) Dopamine D2 receptor agonists protect against ischaemia-induced hippocampal neurodegeneration in global cerebral ischaemia. Eur J Pharmacol 352:37–46
39. Opacka-Juffry J, Wilson AW, Blunt SB (1998) Effects of pergolide treatment on in vivo hydroxyl free radical formation during infusion of 6-hydroxydopamine in rat striatum. Brain Res 810:27–33
40. Sam EE, Verbeke N (1995) Free radical scavenging properties of apomorphine enantiomers and dopamine: possible implication in their mechanism of action in parkinsonism. J Neural Transm Park Dis Dement Sect 10:115–127
41. Sonsalla PK, Gibb JW, Hanson GR (1986) Roles of D1 and D2 dopamine receptor subtypes in mediating the methamphetamine-induced changes in monoamine systems. J Pharmacol Exp Ther 238:932–937
42. Varastet M, Riche D, Maziere M, Hantraye P (1994) Chronic MPTP treatment reproduces in baboons the differential vulnerability of mesencephalic dopaminergic neurons observed in Parkinson's disease. Neuroscience 63:47–56
43. Wagner GC, Walsh SL (1991) Evaluation of the effects of inhibition of monoamine oxidase and senescence on methamphetamine-induced neuronal damage. Int J Dev Neurosci 9:171–174
44. Wenning GK, Bosch S, Luginger E, Wagner M, Poewe W (1999) Effects of long-term, continuous subcutaneous apomorphine infusions on motor complications in advanced Parkinson's disease. Adv Neurol 80:545–548

V Factors Modulating Neuronal Plasticity and the Course of Maturation Phenomenon in Cerebral Ischemia (Metabolic and Inflammatory Factors)

Marrow Stromal Cells as Restorative Treatment of Neural Injury

M. Chopp and Y. Li

Summary. Cellular treatment of brain after stroke, traumatic brain injury or neurodegenerative diseases has primarily focused on the use of cell populations, whether embryonic stem cells or others, to replace injured cerebral tissue [1, 2, 15]. Thus, the logic of these approaches resides in the ability of certain stem and progenitor-like cells to differentiate into neural tissue [15]. This approach, even if it can replace tissue, may be appropriate and could provide therapeutic benefit for certain neurodegenerative diseases, like Parkinson's disease in which specific populations of neurons are lost and products of these cells are thereby diminished [14, 34]. However, when loss of cerebral tissue occurs under conditions of stroke or traumatic brain injury, the use of cell replacement therapy to confer restoration of neurological function is less likely.

Key words. Cellular treatment – cell replacement therapy – endogenous remodeling – neural injury – bone marrow stromal cells – functional recovery

Introduction

In this chapter, we describe our work with cells deployed to activate endogenous remodeling of brain. We demonstrate that cell therapy for stroke [6, 7, 19, 20, 23] and traumatic brain injury [25–28] provides significant functional benefit and restores neurological function. This restoration of function is mediated by compensatory responses of brain, which are amplified by the cellular therapy [7].

The cells to be discussed are primarily bone marrow stromal cells (MSCs) [6, 7, 25, 26]. We will demonstrate that this cell population provides a highly effective treatment for neural injury. Ischemic lesion volume is not altered by treatment with these cells, yet significant functional benefit is engendered when therapy is instituted 1 or 7 or more days after treatment [7].

Michael Chopp[1,2], Yi Li[1]
[1]Department of Neurology, Henry Ford Health Sciences Center, Detroit, MI 48202, USA
[2]Department of Physics, Oakland University, Rochester, MI 48309, USA.

Correspondence to: Michael Chopp, PhD, Professor & Vice Chairman, Department of Neurology (E&R3056), Henry Ford Health Sciences Center, 2799 West Grand Boulevard, Detroit, MI 48202, Tel.: 313-916-3936, Fax: 313-916-1318, E-Mail: chopp@neuro.hfh.edu

Maturation Phenomenon in Cerebral Ischemia V
A.M. Buchan et al. (Eds.)
© Springer-Verlag Berlin Heidelberg 2004

Bone marrow stromal cells, also referred to as bone marrow mesenchymal cells [36], are a population of cells extracted from bone marrow placed in a plastic dish. The subpopulations of cells that adhere to the plastic are the MSCs. They are a highly heterogeneous population of cells. These cells also contain stem-like cells that can differentiate into cells of multiple origins [3, 5, 13, 33]. However, we focus, not on the stem cells, but on the role these cells have as sources of trophic factors and cytokines. MSCs serve to support the production of hematopoietic cells and provide the resources for the highly efficient production of these cells [29, 31].

We initially began our studies with the treatment of stroke and traumatic brain injury by injection of populations of bone marrow cells directly into the brain 1 day after stroke [23] or injury [27]. The cells were placed in the penumbral regions of injured brain, with cells in subcortical and cortical tissue. Functional improvement in animals subjected to stroke [19, 23], traumatic brain injury [27], spinal cord injury [11], and 1-methyl-4-phenyl-1,2,3,6-tetrahydropyridine (MPTP) Parkinson's [21] models was evident. Outcome measurements included neurological examination consisting of an 18 point neurological scale composed of motor function, reflex and balance tests. In addition, somatosensory and motor function tests were evaluated using an adhesive-tab removal test and an accelerating rotor rod test, respectively [7]. These studies were followed by treatment of stroke with cells injected into the carotid artery ipsilateral to the ischemic [21] and traumatic injury [27]. Again, functional benefit was evident. Since MSCs are nurtured within an environment in bone marrow which is a source of inflammatory cells, we injected MSCs into the tail vein in rodents after stroke [7] and traumatic brain injury [26] and found that these cells selectively migrate to the site of injury and co-localize primarily to the boundary zone of the ischemic lesion, with approximately 70–80% of MSCs residing in the ipsilateral hemisphere compared to the contralateral non injured or ischemic hemisphere. These cells were then shown to target injured tissue. They respond to chemotactic signals and adhesion molecules, similarly to inflammatory cells [44, 45]. Thus, MSCs can pass throughout the blood-brain barrier (BBB) and home their way selectively to compromised tissue. But more importantly, the MSCs were found to confer highly significant functional benefit, even when treating animals one week after stroke [7]. These data do not exclude the possibility that MSCs may be an effective therapeutic intervention when administered more than one week post injury, simply these cells have not yet been tested at these time points. Functional benefit was evident within days of treatment. By 7 days after onset of treatment, significant restoration of function was found compared to control treated animals [7]. The control animals are treated with dead cell MSCs, fibroblast or with vehicle phosphate-buffered saline (PBS). There was evidence of a dose-response effect, with 1×10^6 cells injected intravenously showing no significant functional benefit, yet 3×10^6 cells showing highly significant functional benefit on an array of functional outcome measures [7]. This was true for both young and old ischemic animals (unpublished data).

The main question that arises is, what are the underlying mechanisms that promote this remarkable improvement in functional outcome? It is highly unlikely that functional benefit is derived from the cell replacement of injured and dead tissue. Less than 5% of the cells injected intravenously were found in the ischemic brain [7]. Thus, tissue replacement even if present would be miniscule. The functional benefit was evident rapidly after treatment within a few days, a time point far to early for cells to make functional connections and to integrate and find their role in the brain circuitry. Only a small percentage of the cells expressed protein phenotypic characteristics of the neural parenchymal cells. This percentage ranged for 1 to 8%, depending on how the cells were cultured prior to injection. The protein phenotypic behavior also does not confer neural functional identity resemblance to these MSCs. Morphologically, few if any cells resembled parenchymal cells. And preliminary electrophysiological recording of membrane potential from brain slices failed to detect membrane potential or action potential resembling those of neurons. Thus, functional benefit derives not from the differentiation of these cells into brain cells, but from the action of these cells on the endogenous brain tissue.

What are the primary mechanisms that underlie the functional benefit derived from MSC therapy? Functional recovery likely derives from a network of interacting changes in brain. Therefore, the most parsimonious hypothesis is that MSCs produce or activate the production brain of factors that remodel the tissue. MSCs have the capacity to produce many cytokines and trophic factors [29, 31]. They essentially exist to support the production of hematopoietic cells within the bone marrow and they likely, in bone marrow secrete factors that facilitate the production of blood cells. Thus, we hypothesized that MSCs produce trophic factors or evoke the production of trophic factors within the brain. In-vitro experiments were performed that demonstrated production of an array of growth and tropic factors in cultured medium, including brain-derived neurotrophic factor (BDNF), nerve growth factor (NGF), basic fibroblast growth factor (bFGF), vascular endothelial growth factor (VEGF) and hepatocyte growth factor (HGF) [10]. Production of these factors appeared to be sensitive to the microenvironment in which these cells were placed. We also do not exclude the possibility, and it is eminently reasonable that this cell population secretes many more trophic factor cytokines.

The next question to be asked is whether there is an effect of MSC treatment on the levels of trophic factors measured in ischemic brain. Rats were subjected to stroke and were treated, as in many of our studies, at 1 day after stroke onset. Brain tissue obtained at 14 days after treatment with MSCs was analyzed for growth factor levels using the quantitative enzyme-linked immunosorbent assay (ELISA) for BDNF, bFGF and VEGF [9]. Significant increases in trophic factor levels were found compared to rats treated with control fibroblasts, or PBS. These data demonstrate that MSCs cause an increase in levels of important trophic factors that can contribute to brain remodeling. Preliminary studies, using immunohistochemistry for analysis of cellular expression of trophic factors- suggest that endogenous cells such as astrocytes [8] likely express increased levels of trophic factors and it is not only the MSCs that produce and secrete these factors (unpublished data). More likely, it is the paracrine response of endogenous and altered cells to the MSC presence that respond and produce factors that alter brain.

If trophic factor production is the key to reviving function of brain, why is it necessary to employ cells? Why not simply administer trophic factors exogenously? The answer to this question provides insight into the future of pharmacological therapy for the treatment of neurological diseases is going. We should be entering an era of "smart" therapy, of tailored delivery of agents. Cells, like MSCs, act as living distributed sources of therapeutic cocktails. These cells target injured tissue. They produce not one, but an array of factors. These factors, based on in-vitro studies, are expressed in a continuous and time-dependent way. The levels and temporal profiles of interaction of MSCs with the resident cells and the production of trophic factors appear sensitive to the needs of the tissue and the microenvironment in which the cells reside. This "smart" production of therapeutic compounds stands in sharp contrast to pharmacological administration of trophic factor agents, which suffer from inability to enter the brain, are single compounds, and tissue insensitive. Thus, the future for cell therapy and hopefully for drug therapy, resides in the ability to deliver compounds truly needed by the injured tissue to compensate for the injury.

The brain responds to the production of trophic factors. Many interactive events are ignited in response to the MSC therapy. There is significant production of new cells within the generative regions of brain, including the subventricular zone, subgranular layer in the hippocampus, and olfactory bulb [12, 32, 37]. Many of these cells are neuroblasts, and they therefore may contribute to functional recovery, particularly in the long term after stroke. Although, at this point in time there is no evidence that cells within the subventricular zone contribute to functional recovery, there are convincing data that increased numbers of cells within the dentate gyrus evoke improved function [39, 43]. We also have preliminary evidence that brain in response to injury and treatment produces new cells [19] even within the boundary zone of the ischemic lesion. Along with the production of new cells, we find a robust increase in angiogenesis in the injured brain [9]. This angiogenesis is primarily present within the ipsilateral hemisphere. Angiogenesis, fosters increased tissue perfusion and likely promotes the viability and enhances electrical and molecular activity within the compromised tissue [42]. There is also evidence that the angiogenesis is associated with the production of new brain cells, possibly, both populations deriving from the same source of progenitor-like cells [24]. For functional benefit to be present, it is essential that the neuronal structure be modified. Brain plasticity and improved function likely require enhanced dendritic arborization, synaptogenesis, and changes in neuronal substructure, including numbers, sizes and distribution of dendrite spines.

Clinical applications of this therapy for the treatment of stroke and neural injury should be considered, although additional preclinical data are required. MSCs have been employed to treat patients with osteogenesis imperfecta [16], breast cancer [18]; and bone marrow cells have been employed to treat patients with multiple sclerosis [17, 30, 40]. To date, there have been no reports of adverse effects or toxicity suffered by these patients. There is a long history of treatment of patients with bone marrow cells and the toxicity and biology of these cells have been very well characterized [38]. There are two options for the use of bone marrow in the treatment of stroke. One is autologous transplantation. Thus, after stroke, possibly within one or two days post ictus, bone marrow can be extracted from the patient. These cells would be cultured and amplified to treat the patient. At this point in time, we have no knowledge of the dose required to promote effective neurological recovery in the patient. The animal studies, if scaled to unity, suggest that 5–6 million cells per kilogram may be needed to benefit in the human. Preliminary studies in our laboratory suggest that at least 20 times this dose has no toxicity in rodents. However, it is essential that a Phase I study be performed in the human to test this hypothesis. Although autologous transplantation is very appealing, there may be some downside; the cells have to be amplified. Although human MSCs have been expanded by 10^{13} [35], the time it takes for expansion may vary and possibly cause a delay in treatment out to a month or more. Although our studies in the animals indicate that therapeutic benefit is robustly present to at one week, this therapeutic window may extend beyond this time point. We can also consider allogenic administration of MSCs, and possibly a universal MSC donor cell line can be established. The animal data suggest that donor, allogeneic MSCs, even across species, are not rejected. Extensive experiments have been performed in our laboratory using human MSCs to treat the rat [22]. Functional benefit, at least equal to that of using rodent MSCs are present. In addition, there is no evidence for immunorejection. MSCs have been reported not to have major histocompatibility complex receptors [4]. Therefore, they may not be rejected. The use of human MSCs for the treatment of the rodent has been validated in other laboratories [41, 46]. Further studies on the immunology of these cells are needed to fully address the question of rejection. Allogeneic transplantation would be an ideal approach, allowing cells to be stored and pulled off the shelf and be given to patients within days after stroke. The beauty of this therapeutic approach is that the treatment can be delayed for a few days, to stabilize and to evaluate the recovery of the patient.

Conclusion

MSC therapy provides a way of activating endogenous restorative mechanisms within brain, to remodel brain and thereby to significantly improve function. Given the urgent need for a therapy to treat stroke, further studies of the application of this promising therapy are imperative.

Acknowledgement. This work has been funded by NINDS grant PO1 NS23393.

References

1. Abe K (2000) Therapeutic potential of neurotrophic factors and neural stem cells against ischemic brain injury. J Cereb Blood Flow Metab 20:1393–1408
2. Akamatsu W, Okano H (2001) [Neural stem cell, as a source of graft material for transplantation in neuronal disease]. No To Hattatsu 33:114–120
3. Ashton BA, Allen TD, Howlett CR, Eaglesom CC, Hattori A, Owen M (1980) Formation of bone and cartilage by marrow stromal cells in diffusion chambers in vivo. Clin Orthop 294–307
4. Bartholomew A, Sturgeon C, Siatskas M, Ferrer K, McIntosh K, Patil S, Hardy W, Devine S, Ucker D, Deans R, Moseley A, Hoffman R (2002) Mesenchymal stem cells suppress lymphocyte proliferation in vitro and prolong skin graft survival in vivo. Exp Hematol 30:42–48
5. Bennett JH, Joyner CJ, Triffitt JT, Owen ME (1991) Adipocytic cells cultured from marrow have osteogenic potential. J Cell Sci 99:131–139
6. Chen J, Li Y, Wang L, Lu M, Zhang X, Chopp M (2001) Therapeutic benefit of intracerebral transplantation of bone marrow stromal cells after cerebral ischemia in rats. J Neurol Sci 189:49–57

7. Chen J, Li Y, Wang L, Zhang Z, Lu D, Lu M, Chopp M (2001) Therapeutic benefit of intravenous administration of bone marrow stromal cells after cerebral ischemia in rats. Stroke 32:1005–1011

8. Chen J, Zhang ZG, Li Y, Wang L, Xu YX, Lu M, Chopp M (2002) Intravenous administration of human bone marrow stromal cells induces angiogenesis in the ischemic boundary zone after stroke in rats (Submitted)

9. Chen X, Li Y, Wang L, Katakowski M, Zhang L, Chen J, Xu Y, Gautam S, Chopp M (2002) Ischemic rat brain extracts induce human marrow stromal cell growth factor production. Neuropathology (submitted)

10. Chen XG, Katakowski M, Li Y, Lu D, Wang L, Zhang L, Chen J, Xu YX, Gautam S, Mahmood A, Chopp M (2002) Human bone marrow stromal cell cultures conditioned by traumatic brain tissue extracts: Growth factor production Journal of Neuroscience Research, in press

11. Chopp M, Zhang XH, Li Y, Wang L, Chen J, Lu D, Lu M, Rosenblum M (2000) Spinal cord injury in rat: treatment with bone marrow stromal cell transplantation. Neuroreport 11:3001–3005

12. Doetsch F, Garcia-Verdugo JM, Alvarez-Buylla A (1997) Cellular composition and three-dimensional organization of the subventricular germinal zone in the adult mammalian brain. J Neurosci 17:5046–5061

13. Ferrari G, Cusella-De Angelis G, Coletta M, Paolucci E, Stornaiuolo A, Cossu G, Mavilio F (1998) Muscle regeneration by bone marrow-derived myogenic progenitors. Science 279:1528–1530

14. Freed CR, Greene PE, Breeze RE, Tsai WY, DuMouchel W, Kao R, Dillon S, Winfield H, Culver S, Trojanowski JQ, Eidelberg D, Fahn S (2001) Transplantation of embryonic dopamine neurons for severe Parkinson's disease. N Engl J Med 344:710–719

15. Gage FH (2000) Mammalian neural stem cells. Science 287:1433–1438

16. Horwitz EM, Prockop DJ, Gordon PL, Koo WW, Fitzpatrick LA, Neel MD, McCarville ME, Orchard PJ, Pyeritz RE, Brenner MK (2001) Clinical responses to bone marrow transplantation in children with severe osteogenesis imperfecta. Blood 971227–1231

17. Ikehara S (1998) Bone marrow transplantation for autoimmune diseases. Acta Haematol 99:116–132

18. Koc ON, Gerson SL, Cooper BW, Dyhouse SM, Haynesworth SE, Caplan AI, Lazarus HM (2000) Rapid hematopoietic recovery after coinfusion of autologous-blood stem cells and culture-expanded marrow mesenchymal stem cells in advanced breast cancer patients receiving high-dose chemotherapy. J Clin Oncol 18:307–316

19. Li Y, Chen J, Chopp M (2001) Adult bone marrow transplantation after stroke in adult rats. Cell Transplant 10:31–40

20. Li Y, Chen J, Wang L, Lu M, Chopp M (2001) Treatment of stroke in rat with intracarotid administration of marrow stromal cells. Neurology 56:1666–1672

21. Li Y, Chen J, Wang L, Zhang L, Lu M, Chopp M (2001) Intracerebral transplantation of bone marrow stromal cells in a 1-methyl-4-phenyl-1,2,3,6-tetrahydropyridine mouse model of Parkinson's disease. Neurosci Lett 316:67–70

22. Li Y, Chen J, Chen XG, Wang L, Gautam SC, Xu YX, Katakowski M, Zhang LJ, Lu M, Janakiraman N, Chopp M (2002) Human marrow stromal cell therapy for stroke in rats: neurotrophins and functional recovery. Neurology, in press

23. Li Y, Chopp M, Chen J, Wang L, Gautam SC, Xu YX, Zhang Z (2000) Intrastriatal transplantation of bone marrow nonhematopoietic cells improves functional recovery after stroke in adult mice. J Cereb Blood Flow Metab 20:1311–1319

24. Louissaint A, Rao S, Leventhal C, Goldman SA (2002) Coordinated interaction of neurogenesis and angiogenesis in the adult songbird brain. Neuron 34:945–960

25. Lu D, Li Y, Wang L, Chen J, Mahmood A, Chopp M (2001) Intraarterial administration of marrow stromal cells in a rat model of traumatic brain injury,. J Neurotrauma 18:813–819

26. Lu D, Mahmood A, Wang L, Li Y, Lu M, Chopp M (2001) Adult bone marrow stromal cells administered intravenously to rats after traumatic brain injury migrate into brain and improve neurological outcome. Neuroreport 12:559–563

27. Mahmood A, Lu D, Yi L, Chen JL, Chopp M (2001) Intracranial bone marrow transplantation after traumatic brain injury improving functional outcome in adult rats. J Neurosurg 94:589–595

28. Mahmood A, Lu D, Wang L, Li Y, Chen J, Chopp M (2001) Adult male bone marrow stromal cells administered intravenously to female rats after traumatic brain injury migrate into brain and improve neurological outcome, Neurosurgery

29. Majumdar MK, Thiede MA, Haynesworth SE, Bruder SP, Gerson SL (2000) Human marrow-derived mesenchymal stem cells (mscs) express hematopoietic cytokines and support long-term hematopoiesis when differentiated toward stromal and osteogenic lineages. J Hematother Stem Cell Res 9:841–848

30. Mandalfino P, Rice G, Smith A, Klein JL, Rystedt L, Ebers GC (2000) Bone marrow transplantation in multiple sclerosis. J Neurol 247:691–695
31. Mayani H, Guilbert LJ, Janowska-Wieczorek A (1992) Biology of the hematopoietic microenvironment. Eur J Haematol 49:225–233
32. Palmer TD, Willhoite AR, Gage FH (2000) Vascular niche for adult hippocampal neurogenesis. J Comp Neurol 425:479–494
33. Pereira RF, Halford KW, O'Hara MD, Leeper DB, Sokolov BP, Pollard MD, Bagasra O, Prockop DJ (1995) Cultured adherent cells from marrow can serve as long-lasting precursor cells for bone, cartilage, and lung in irradiated mice., Proc Natl Acad Sci USA 92:4857–4861
34. Piccini P, Brooks DJ, Bjorklund A, Gunn RN, Grasby PM, Rimoldi O, Brundin P, Hagell P, Rehncrona S, Widner H, Lindvall O (1999) Dopamine release from nigral transplants visualized in vivo in a Parkinson's patient [see comments]. Nat Neurosci 2:1137–1140
35. Prockop DJ, Azizi SA, Colter D, Digirolamo C, Kopen G, Phinney DG (2000) Potential use of stem cells from bone marrow to repair the extracellular matrix and the central nervous system. Biochem Soc Trans 28:341–345
36. Prockop DJ, Sekiya I, Colter DC (2001) Isolation and characterization of rapidly self-renewing stem cells from cultures of human marrow stromal cells. Cytotherapy 3:393–396
37. Rousselot P, Lois C, Alvarez-Buylla A (1995) Embryonic (PSA) N-CAM reveals chains of migrating neuroblasts between the lateral ventricle and the olfactory bulb of adult mice. J Comp Neurol 351:51–61
38. Santos GW (1983) History of bone marrow transplantation, Clin Haematol, 12:611–639
39. Song HJ, Stevens CF, Gage FH (2002) Neural stem cells from adult hippocampus develop essential properties of functional CNS neurons. Nat Neurosci 5:438–445
40. Sullivan KM, Parkman R, Walters MC (2000) Bone Marrow Transplantation for Non-Malignant Disease, Hematology (Am Soc Hematol Educ Program) 319–338
41. Toma C, Pittenger MF, Cahill KS, Byrne BJ, Kessler PD (2002) Human mesenchymal stem cells differentiate to a cardiomyocyte phenotype in the adult murine heart. Circulation 105:93–98
42. Vale PR, Losordo DW, Milliken CE, Maysky M, Esakof DD, Symes JF, Isner JM (2000) Left ventricular electromechanical mapping to assess efficacy of phVEGF(165) gene transfer for therapeutic angiogenesis in chronic myocardial ischemia. Circulation 102:965–974
43. van Praag H, Schinder AF, Christie BR, Toni N, Palmer TD, Gage FH (2002) Functional neurogenesis in the adult hippocampus. Nature, 415:1030–1034
44. Wang L, Li Y, Chen J, Gautam SC, Zhang ZG, Lu M, Chopp M (2002) Ischemic cerebral tissue and MCP-1 enhance bone marrow stromal cell migration in interface culture. Experimental Hematology, in press
45. Wang L, Li Y, Chen XG, Chen J, Gautam SC, Xu YX, Chopp M (2002) MCP-1, MIP-1, IL-8 and ischemic cerebral tissue enhance human bone marrow stromal cell migration in interface cultures. Hematology, in press
46. Zhao LR, Duan WM, Reyes M, Keene CD, Verfaillie CM, Low WC (2002) Human bone marrow stem cells exhibit neural phenotypes and ameliorate neurological deficits after grafting into the ischemic brain of rats. Exp Neurol 174:11–20

Protein Aggregation, Unfolded Protein Response and Delayed Neuronal Death after Brain Ischemia

B. R. Hu, M. E. Martone, and C. L. Liu

Key words. Brain ischemia – secondary neuronal death – protein aggregation – ubiquitin – molecular chaperone – unfolded protein response – endoplasmic reticulum – electron microscopy

Abbreviations. EPTA = ethanolic phosphotungstic acid, ubi-proteins = ubiquitin-conjugated proteins, DG = dentate gyrus

Introduction

It was first found in the 1970's that neuronal death does not occur immediately but takes place over several days following an initial period of transient cerebral ischemia [31, 33]. Such secondary neuronal damage following initial insult defines the maturation phenomenon, or delayed neuronal death [1, 33, 51]. In forebrain rat ischemia models, transient cerebral ischemia followed by reperfusion causes neuronal death selectively in hippocampal CA1 pyramidal neurons after 48 to 72 h of reperfusion while leaving dentate gyrus (DG) granule cells largely intact [1, 33]. These models have been utilized to study delayed neuronal death, i.e., the maturation phenomenon in the CA1 region and neuronal survival in the DG area after ischemia. During the 48 to 72 h maturation or delayed period, CA1 neurons destined to die appeared essentially normal under the light microscope [1, 33]. At the ultrastructural level, however, disaggregation of polyribosomes, deposition of dark substances, abnormalities in the Golgi apparatus and modification of synapses have been reported [6, 30, 33, 38, 47. Secondary neuronal death after an initial insult also occurs after other brain injuries as for instance in the penumbra region after focal ischemia and in some brain regions after hypoglycemia [1]. Many hypotheses have been postulated to account for delayed neuronal death [51]. In this article, we will present evidence that ischemia induces a massive accumulation of abnormal proteins which is followed by their aggregation in neurons destined to

Bingren R. Hu[1], Maryann E. Martone[2], Chunli Liu[1]
[1]Dept of Neurology, University of Miami School of Medicine, Miami, FL 33136, USA
[2]Department of Neurosciences, University of California, San Diego, CA 92093, USA

Corresponding: Bingren Hu, PhD, Assistant Professor, Cerebral Vascular Disease Research Center, Department of Neurology D4–5, University of Miami School of Medicine, 1501 NW 9th Ave, Miami, FL 33136, USA, Tel.: (305) 243-4857, Fax: (305) 243-5330, E-Mail: bhu@med.miami.edu

Maturation Phenomenon in Cerebral Ischemia V
A M.. Buchan et al. (Eds.)
© Springer-Verlag Berlin Heidelberg 2004

die. We hypothesize that accumulation of abnormal proteins and their aggregates contributes to delayed neuronal degeneration after brain ischemia.

Protein Aggregation and the Unfolded Protein Response

Accumulation of ubiquitinated abnormal proteins and their aggregates has commonly been observed in almost all neurodegenerative diseases including Alzheimer's, Parkinson's and Huntington's diseases, as well as prion diseases and amyotrophic lateral sclerosis [3]. It has been proposed that inherent neurotoxicities of abnormal proteins and their aggregates may be commonly responsible for neurodegeneration in these chronic neurodegenerative diseases [10, 54]. Polypeptide chains of proteins must to be folded into their native conformations to avoid aggregation and to retain functions. When newly synthesized polypeptides fail to fold completely or correctly, or mature proteins are denatured [unfolded], damaged by proteases or abnormally modified as for example by reactive oxygen species [ROS], their sticky hydrophobic segments are exposed on the surface. Without protection by cell defense systems, the newly synthesized, improperly folded, denatured, damaged or abnormally modified proteins remain abnormal and tend to aggregate. Protein aggregation is widely considered to be a nonspecific coalescence of abnormal proteins, driven by interactions between solvent-exposed hydrophobic surfaces that are normally buried within a protein's interior or inserted into lipid membranes [23].

There are several defense systems for protecting abnormal proteins and preventing abnormal protein aggregation in cells. A group of proteins or molecules known as molecular chaperones can shield hydrophobic sequences and thereby foster the proper folding of unfolded proteins and prevent their aggregation. Most stress proteins or heat shock proteins (HSPs) are molecular chaperones [18]. Abnormal proteins can be scavenged by the ubiquitin-proteasome system. A signal sequence in the hydrophobic segment can be recognized by the ubiquitin system. When the signal sequence is exposed, it will be ubiquitinated through a series of ATP-dependent reactions to form an isopeptidyl bond ligating ubiquitin to the abnormal protein (ubi-protein). Ubiquitination tags proteins for degradation rather than chaperone-like protection. It has been estimated that up to 90% of proteolysis is blocked in cultured cells by inhibitors of proteasomes [14].

The protein maturation pathway for newly synthesized membranous and secretory proteins begins with the endoplasmic reticulum (ER), and links the Golgi apparatus and a series of vesicular compartments together to provide a framework through which new proteins are matured and transported. The ER orchestrates the synthesis and processing of nearly all proteins that pass through the framework system, and contains high concentrations of molecular chaperones and folding enzymes. If proteins fail to fold or are misfolded in the ER, a mechanism called ER-associated degradation (ERAD) can recognize and transport the abnormal proteins to the cytosol for degradation [46].

Many stress conditions will increase the level of abnormal proteins in the ER lumen and cytosol [45]. These conditions include depletion of cellular ATP, production of reactive oxygen species (ROS), altered redox status, changes in protein glycosylation and depletion of the ER calcium [5, 20], all of which can be seen during and after ischemia [43, 51]. The common result of these ischemia-induced changes is a state of stress in which the burden of abnormal proteins exceeds the capacity of the processing machinery in the ER and cytosol. Very recently, several studies have demonstrated that the ER responds to stress by finely-tuned signaling pathways that regulate protein translation, gene transcription and protein degradation. These regulatory processes are commonly referred to as the 'unfolded protein response' (UPR). The UPR is an adaptive or negative feedback mechanism for attenuating lethal accumulation of abnormal proteins during ER stress [21]. Three main pathways of the UPR have been recently elucidated: (i) The UPR quickly shuts off overall mRNA translation, thus decreasing overload of the protein-folding machinery. A new gene named PERK encoding a type I transmembrane ER-resident protein (PERK; for PKR-like Endoplasmic Reticulum Kinase) has recently been cloned [21]. PERK can phosphorylate or inactivate eukaryotic initiation factor-2a (eIF2a) on serine residue 51 [22], thus blocking protein synthesis at its initiation. Activation of PERK and inactivation or phosphorylation of eIF-2a has been observed after brain ischemia [2, 11, 15, 26, 34, 37]. (ii) The UPR also increases transcription of genes encoding for the ER molecular chaperones and folding enzymes, thus increasing the folding and processing capacity [7, 32, 57]. The ER sensor for the transcriptional control is recently discovered IRE1 that transduces the UPR signal from the ER to the nucleus through activation of ATF6 during ER stress [55]. ATF6 is a basic leucine-zipper transcription factor that translocates into the nucleus and binds to the ER stress response element (ERSE) to induce genes encoding molecular chaperones and folding enzymes under ER stress [35]. (iii) The UPR enhances degradation of abnormal proteins. The abnormal proteins are retained in the ER, but must be degraded by the ubiquitin-proteosomal system that resides in the cytosol. This requires transport of abnormal proteins from the ER back to the cytosol through SEC61, the central component of the ER protein import channel. This ER-associated degradation or ERAD attenuates a potentially lethal accumulation and aggregation of abnormal proteins in the ER [41, 52]. The UPR induction increases ERAD capacity. Conversely, loss of ERAD leads to constitutive UPR induction [53].

In addition to the ER defense system, cytosolic molecular chaperones such as the HSP70, HSP90, HSP60 and HSP40 families can further prevent abnormal protein aggregation and toxicities in the cytosol [23]. Expression of inducible cytosolic molecular chaperones is regulated by the heat shock transcription factors (HSFs). HSFs are a family of transcription factors consisting of HSF1–HSF4, and they normally exist as inactive monomers. HSFs are bound, thus negatively regulated by their own C-terminal hydrophobic repeat domain, as well as by HSP90, and HSP70 or HSC70 repressor complexes. In response to stress, abnormal proteins bind to the negative regulators to dissociate HSF monomers. Dissociated HSFs are then trimerized, translocated into nuclei, and bound to the DNA heat shock elements (HSEs) such that mRNA induction of the inducible cytosolic molecular chaperone and folding enzyme ensues [9, 40]. Thus, abnormal proteins can serve as eukaryotic stress signals and trigger the expression of stress genes [4].

Regular or sublethal environmental stress produces small amounts of abnormal proteins in the ER and cytosol that can be protected or removed by the above-mentioned negative feedback UPR mechanisms and the cytosolic defense systems. However, under more severe pathological conditions, when abnormal proteins accumulate over a limit that surpasses the defense machinery, they will aggregate and/or associate nonspecifically with their nearby lipid membranes through their hydrophobic segments [28, 29]. These protein aggregates are difficult to process and the aggregation process is virtually irreversible [9]. Genetic mutations of UPR defense molecules such as PERK, IRE1 or SEC61 in cells impair their ability to survive even under moderate stress conditions [7, 22, 55, 57]. Abnormal protein aggregates are highly toxic to cells and may exert their deleterious effects over extended periods of time to long-lived neurons after initial injuries [10, 54].

Protein Aggregation and Secondary Neuronal Death after Brain Ischemia: A New Hypothesis

Many hypotheses about the molecular mechanisms underlying secondary neuronal death after brain ischemia have been proposed [51]. Among others, persistent depression of protein synthesis [25], impairment of protein ubiquitination [17, 24, 36, 39] and expression of stress proteins [1, 42, 49] have been postulated to be important for neuronal death/survival after ischemia. Are these events independent or are they associated with a common underlying event? Very recent progress in the field of cell biology and our recent findings support a notion that these intracellular events are commonly related to overproduction of abnormal proteins and their aggregates after brain ischemia. We recently used an ethanolic phosphotungstic acid (EPTA) staining method to study synaptic modification after ischemia [30, 38]. At the same time, we unexpectedly found that EPTA strongly stained abnormal aggregates associated with membranous structures in the somas and dendrites of CA1 neurons destined to die (Fig. 1, also see [28, 29]). The EPTA staining method has frequently been used as an EM-selective staining method for synaptic structures and nuclei [8]. Because EPTA preferentially reacts with proteins [12], the aggregates stained with EPTA are likely composed of proteins. This conclusion is supported by the fact that the aggregates contain strong ubiquitin immunoreactivity as demonstrated by immunogold EM (Fig. 2). Through a series of studies, we have concluded that proteins are severely and progressively unfolded, abnormally modified or damaged, and thus aggregated in postischemic neurons destined to die but not in the neurons that survived the same ischemic insult during the postischemic phase. Abnormal protein aggregates were mainly observed to be associated with the cytoplasmic faces of membranous structures including intracellular vesicles, endoplasmic reticulum, Golgi and mitochondria, as well as the dendritic plasmalemma [28, 29].

Consistent with the data obtained from above EPTA EM and the immunogold EM studies, we have also found that 15 min transient cerebral ischemia induces a massive and persistent increase in aggregation of ubiquitin-conjugated proteins (ubi-proteins) during the postischemic phase. By using high-resolution confocal microscopy and an antibody that recognizes both free ubiquitin and ubi-proteins, we have demonstrated that the pattern of ubiquitin immunostaining in CA1 neurons is markedly changed from a relatively even distribution in the controls and

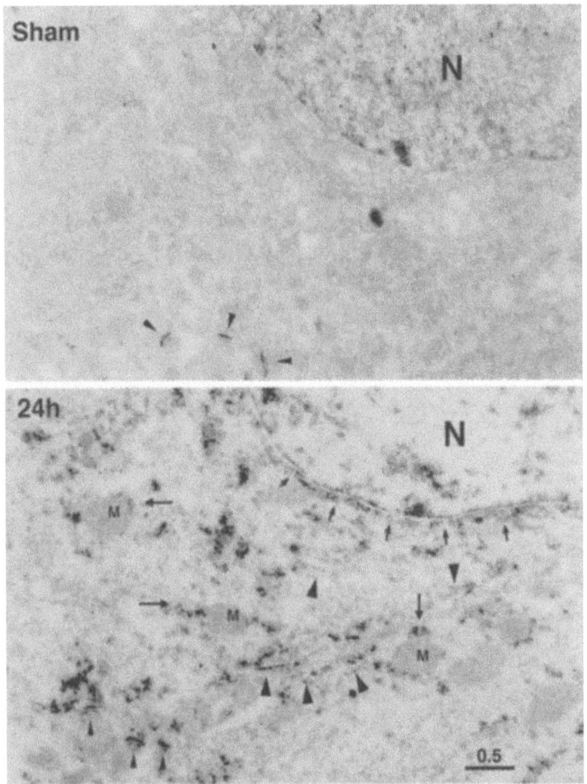

Fig. 1. Electron micrographs of EPTA staining in the cell soma of CA1 pyramidal neurons in sham-operated controls (sham) and rats subjected to 15 min cerebral ischemia followed by 24 h of reperfusion. EPTA strongly stained synaptic structures (arrowheads) and nucleus (N), but very weakly reacted with other subcellular structures in sham-control neurons. In the post-ischemic brain, EPTA not only stained synaptic structures (small arrowheads) and nucleus (N), but also stained intracellular abnormal protein aggregates. EPTA-stained abnormal proteins aggregates were extensively attached to the cytoplasmic faces of the nuclear membrane (small arrows), intracellular vesicles, endoplasmic reticulum (large arrowheads) and mitochondria (M) at 24 h post-ischemia. N=nucleus; Scale bar=0.5 m

before 30 min of reperfusion, to an uneven pattern after 2 h of reperfusion. The aggregates increase in size from 2 h to 4 h of reperfusion. By 24 h of reperfusion, the ubi-proteins form large, patchy aggregates surrounding the nuclei and attached to the dendritic membranes (Fig. 3). This pattern at 24 h of reperfusion is exactly the same as the distribution of the protein aggregates observed by EPTA EM (see Fig. 1). The aggregated pattern is severe and sustained starting from 2 h of reperfusion until cell death in CA1 neurons, but soon recovers to the control distribution in DG neurons during reperfusion [28, 29].

The results of the morphological and immunocytochemical studies were supported by biochemical evidence. On Western blots, ubi-proteins were dramatically increased as early as 30 min of reperfusion and persisted until cell death in CA1 neurons, but protein ubiquitination was transient in surviving DG neurons after

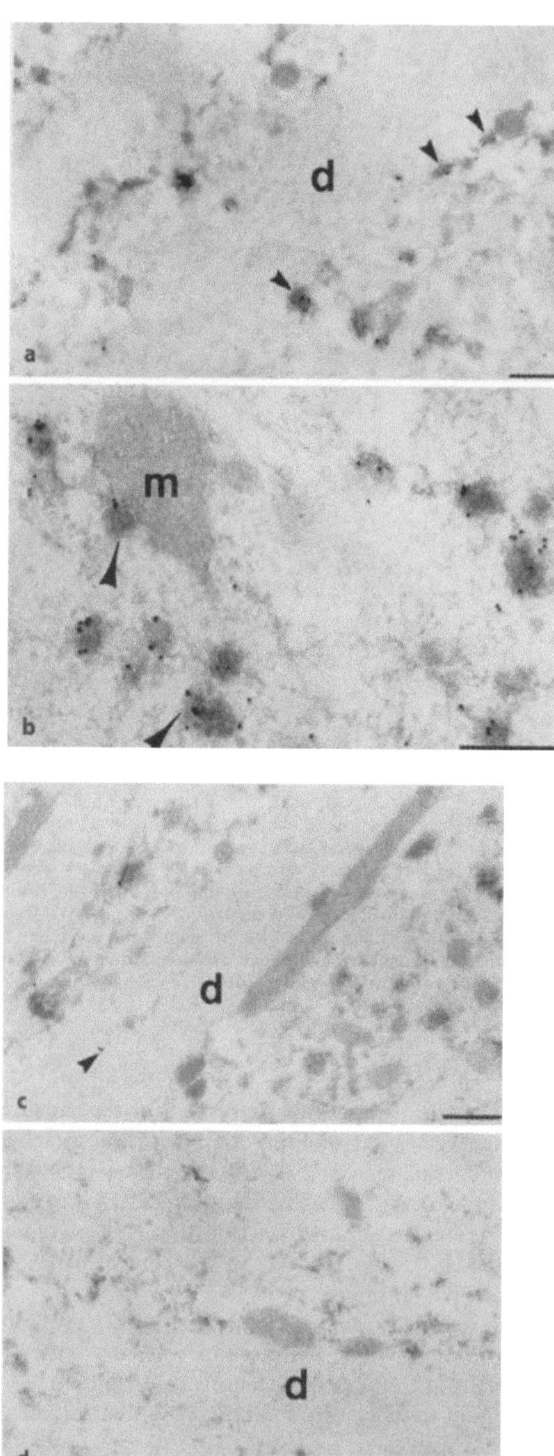

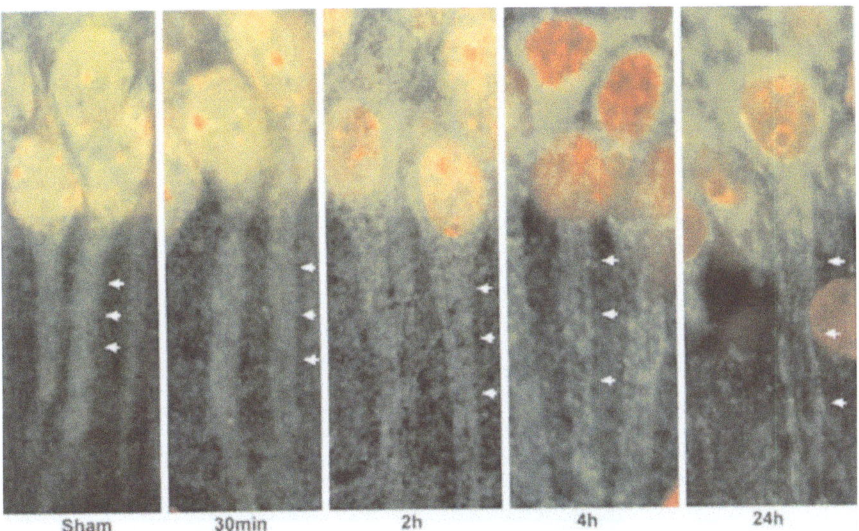

Fig. 3. Higher magnification laser-scanning confocal microscopic images of CA1 region double labeled with anti-ubiquitin (green) and propidium iodide (red). Sections are shown from rats subjected to 15 min of ischemia followed by 30 min, 2, 4 and 24 h of reperfusion. Magnification was increased using the zoom function in the confocal software (Lasersharp). Ubiquitin-labeled protein aggregates first appear as small dots at 2 h of reperfusion, and progressively increase in size over time. By 24 h of reperfusion, the ubi-protein aggregates form a patchy pattern surrounding the nuclei and attached to the dendritic membrane (arrows). Ubiquitin immunostaining in nuclei disappears after 4 h of reperfusion after 15 min of ischemia

ischemia (Fig. 4 A). As discussed above, an increase in protein ubiquitination reflects accumulation of abnormal proteins. Most ubi-proteins have not irreversibly aggregated at 30 min of reperfusion because they are still soluble in 2% Triton X100 buffer (Fig. 4 B). However, abnormal protein aggregates gradually become Triton-insoluble, suggesting that they are progressively and irreversibly aggregated in CA1 neurons after ischemia (Fig. 4 B) [28]. These results are consistent with data from several earlier studies demonstrating depletion of intracellular free ubiquitin and increase in ubi-proteins after ischemia [17, 24, 36, 39]. Evidence suggests that depletion of free ubiquitin immunostaining reflects an overproduction of abnormal proteins that are ubiquitinated and aggregated after ischemia [28, 29].

To investigate further whether accumulation of ubi-protein aggregates contributes to delayed neuronal death after ischemia, we produced 3, 7 and 20 min of ce-

Fig. 2. Ubiquitin immunogold labeling in the apical dendrites (**d**) of CA1 pyramidal neurons in the post-ischemic brain at 24 h reperfusion (**a, b**) and in sham-operated control (**c**). Ubiquitin immunoelectron microscopy was performed on sections of brain that were not osmicated prior to embedding and were then counterstained with uranyl acetate and lead citrate. Heavy immunolabeling for ubiquitin was observed over the dark aggregates distributed along the dendritic plasmalemma (arrowheads in **a**) and associated with mitochondria (m in **b**) and vesicles (arrowhead in **b**). Immunolabeling in the control brain was usually present in the cytoplasm (arrowhead in **c**) and not with electron dense structures. Immunostaining controls in which the primary antibody was omitted showed very little non-specific labeling (**d**). Scale bar = 0.5 µm

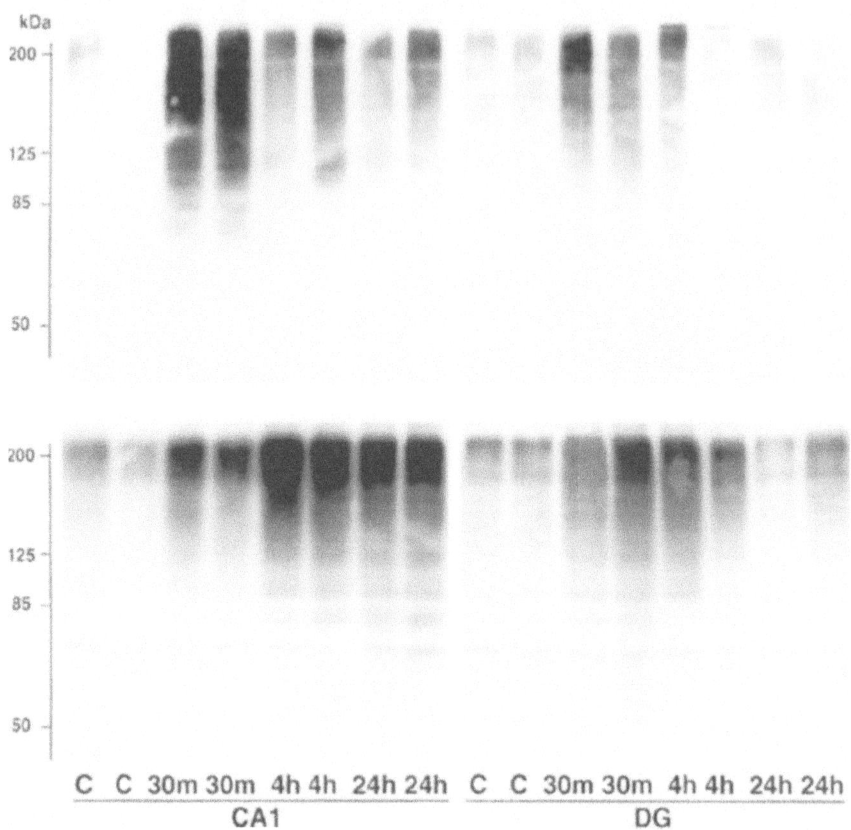

Fig. 4. Uniquitin immunoblots in the membrane fractions before (upper) and after (lower) wash with 2% Triton X100 after brain ischemia. Samples of the CA1 region and DG area from sham-control rats (C) and rats subjected to 15 min of cerebral ischemia followed by 30 min, 4 and 24 h of reperfusion. Each sample derived from one rat. Two separate samples in each experimental group were run in the SDS-PAGE. The blots were labeled with an anti-ubiquitin antibody and visualized with the ECL system. Molecular size markers are indicated on the left

rebral ischemia in rat and examined different brain regions with differing suscepti-bilities to delayed neuronal death using light and electron microscopy. At 3 days of reperfusion, about 60% of CA1 neurons were dead after 7 min of ischemia. Damaged neurons were rarely observed at this time point after 3 min of ischemia, while nearly 100% of CA1 neurons, about 50% of DG granule cells and about 20% cortical neurons were dead after 20 min of ischemia (data not shown). Rat brains subjected to 3, 7 and 20 min of ischemia followed by 24 h of reperfusion were examined using confocal microscopy after ubiquitin immunostaining, and electron microscopy after EPTA staining. As shown above, both methods stain abnormal protein aggregates after ischemia. No protein aggregates were found in CA1 neurons after 3 min of ischemia. After 7 min of ischemia, protein aggregates were found only in a proportion of CA1 neurons, but not in DG neurons. After 20 min of ischemia, however, protein aggregates were present in almost all CA1 neurons, a proportion of both DG granule cells and neocortical neurons. The percentages of neurons in these brain regions containing protein aggregates are similar to the number exhibiting delayed neuronal death (data not shown). Electron microscopic tomography further demonstrated that intra-neuronal protein aggregates were mainly associated with the cytoplasmic face of the plasmalemma (data not shown). These results suggest that visible protein aggregates are only seen in neurons destined to die after transient cerebral ischemia.

What are mechanisms underlying accumulation of abnormal proteins after ischemia? Recent emerging studies have shown that ER stress leads to accumulation of unfolded proteins and triggers the UPR in the ER. Many signs of UPR in the ER after brain ischemia have been reported. (i) PERK is activated [34], eIF-2α is phosphorylated [2, 11, 15, 37], and protein synthesis initiation is depressed [26]. This indicates that the UPR pathway of PERK–eIF-2α is activated in neurons after ischemia. (ii) Several leucine zipper transcription factors such as ATF2 and c-Jun are preferentially activated in CA1 neurons prior to their death [27]. In addition, more severe ER stress turns on transcription of a CHOP/GADD153 gene by activation of ATF6 [55, 57]. Transcription of CHOP/GADD153 is increased in neurons after brain ischemia [44]. The evidence suggests that the IRE1-ATF6 pathway might be turned on in neurons after ischemia. (iii) Free ubiquitin is severely decreased and ubi-proteins are highly increased after ischemia. Protein aggregates in postischemic neurons are composed of ubi-proteins that are associated with membranes on the cytosolic face [28, 29]. All of these signs indicate that abnormal proteins are accumulated in the ER and are retrogradely transported to the cytosol. Together with abnormal proteins in the cytosol, abnormal proteins produced during and after ischemia may surpass the capacity of the cytosolic defense and degradation systems, thereby becoming aggregated with each other and associated nonspecifically with lipid membranes [28, 29].

Accumulation of abnormal proteins and protein aggregation found in postischemic neurons may represent a new mechanism underlying delayed neuronal death (Fig. 5). We have proposed that dramatic changes in cellular homeostasis including depletion of ATP, changes in cellular metabolism and ionic environment, and production of ROS during and after ischemia will lead to the overproduction of abnormal proteins and formation of protein aggregates that are associated with lipid membranous structures. The abnormal proteins and their aggregates are major targets for cellular scavenger systems including lysosomal and non-lysosomal proteases, and microglia [3]. Thus, accumulation of abnormal proteins and their aggregates on cellular membranes may damage or destroy the membranes by the action of cellular scavenger systems. When over a certain limit, the membrane damage may ultimately lead to neuronal death. This hypothesis seems to fit nicely with several existing hypotheses: (i) Depression of protein synthesis causes neuronal death [25]. Accumulation of abnormal proteins in the ER shuts off overall protein synthesis by inactivation or phosphorylation of eIF-2α [2, 11, 15, 34, 37]. Protein synthesis initiation is persistently depressed in CA1 neurons after ischemia [26]. (ii) Depletion of intracellular free ubiquitin may cause neuronal death [17, 24, 36, 39]. The abnormal proteins consume intracellular free ubiquitin causing a persistent decrease of free ubiquitin in CA1 neurons after ischemia [29]. (iii) Ex-

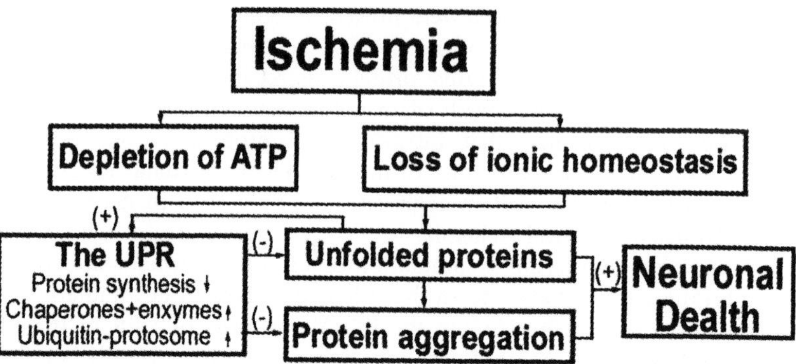

Fig. 5. Our hypothesis can be summarized in this flow chart: Ischemia induces a cascade of events including energy failure, loss of ionic homeostasis, change in cellular metabolism and production of reactive oxygen species, which cause accumulation of abnormal proteins and their aggregates in neurons during and after ischemia. Accumulation of abnormal proteins switches on (+) the UPR-defensive mechanisms to attenuate (-) the overload of abnormal proteins by decreasing synthesis of new immature proteins, by promoting transcription of chaperones and folding enzymes, and by enhancing degradation of abnormal proteins. However, when abnormal proteins and their aggregates accumulate to surpass the capacity of cellular processing and degradation systems, they will in the course of time exert deleterious effects on neurons and eventually cause neuronal death after ischemia

pression of stress proteins or heat shock proteins protects neurons against ischemia [1, 42, 48, 49, 56]. Stress proteins shield abnormal proteins in order to prevent their aggregation after brain ischemia. (iv) Induction of stress proteins by either preconditioning, or viral infection or transgenic overexpression prior to ischemia protects the neurons against ischemia [13, 49, 56]. (v) Activation of microglia after cerebral ischemia is an early indicator of tissue damage. Attachment of microglia to the cell body and dendrites of CA1 neurons prior to their death has been observed after transient cerebral ischemia [19]. Abnormal proteins can serve as activators signaling for cellular scavengers like microglia [3].

As mentioned, intracellular accumulation of ubi-protein aggregates is a common pathological feature of neurodegenerative disorders. However, it is still a subject of debate as to whether protein aggregates themselves are pathogenic. As discussed, overexpression of molecular chaperones in a cell protects the cell from death in virtually all pathological conditions, and the protection should be through shielding abnormal proteins to prevent their aggregation [50]. Several recent studies have provided direct evidence that protein aggregates at the intermediate stage are highly toxic to living cells [6, 10, 54]. These studies support the idea that protein aggregation in living neurons contributes to neuronal death in all pathological conditions. The following results obtained from in-vivo animal ischemia models further suggest that accumulation of abnormal proteins and their aggregates contribute to delayed neuronal death after ischemia. (i) Abnormal proteins and their aggregates are accumulated progressively from the early stage of reperfusion, starting as early as 30 min of reperfusion, about 2–3 days prior to ischemic neuronal death. This suggests that accumulation of abnormal proteins may play a causative role in neuronal death after ischemia [28, 29]. (ii) We have recently examined several brain regions in rats after different severities of ischemia, and found that the visible protein aggregates seen by EM are only present in the brain regions where neurons are destined to die. These visible protein aggregates were rarely found in brain regions where neurons survive the same periods of ischemia. (iii) Ischemic preconditioning prevents protein aggregation as well as neuronal death after ischemia [Hu et al., unpublished data]. Ischemic preconditioning is known to induce expression of molecular chaperones that would protect abnormal proteins and prevent their aggregation. We should also point out that although the question as to whether abnormal proteins induced by ischemia have to be aggregated to become toxic to cells remains to be answered conclusively, our results strongly suggest that accumulation of abnormal proteins and their aggregates after ischemia are detrimental to neurons. We believe that both abnormal proteins and their aggregates are highly toxic to neurons after ischemia (Fig. 5). Accordingly, we have suggested that association of abnormal proteins with cell lipid membranes underlies the neurotoxicities of protein aggregates and contributes to delayed neuronal death after brain ischemia [28, 29].

Acknowledgements. This work was supported by National Institute of Health grant NS47470 to B.R.H. We would like to thank Dr. B. Watson who read the manuscript.

References

1. RR (2001) Impairment of the ubiquitin-proteasome system by protein. In: Abe K, Kawagoe J, Aoki M, Kogure K, Itoyama Y (1998) Stress protein inductions after brain ischemia. Cell Mol Neurobiol 18:709–719
2. Althausen S, Mengesdorf T, Mies G, Olah L, Nairn AC, Proud CG, Paschen W (2001) Changes in the phosphorylation of initiation factor eIF-2alpha, elongation factor eEF-2 and p70 S6 kinase after transient focal cerebral ischaemia in mice. J Neurochem 78:779–787
3. Alves-Rodrigues A, Gregori L, Figueiredo-Pereira ME (1998) Ubiquitin, cellular inclusions and their role in neurodegeneration. Trends Neurosci 21:516–520
4. Ananthan J, Goldberg AL, Voellmy R (1986) Abnormal proteins serve as eukaryotic stress signals and trigger the activation of heat shock genes. Science 232:522–524
5. Angelidis CE, Lazaridis I, Pagoulatos GN (1999) Aggregation of hsp70 and hsc70 in vivo is distinct and temperature-dependent and their chaperone function is directly related to non-aggregated forms. Eur J Biochem 259:505–512
6. Bence NF, Sampat RM, Kopito aggregation. Science 292:1552–1555
7. Bertolotti A, Zhang Y, Hendershot LM, Harding HP, Ron D (2000) Dynamic interaction of BiP and ER stress transducers in the unfolded-protein response. Nat Cell Biol 2:326–332
8. Bloom FE, Aghajanian GK (1966) Cytochemistry of synapses: a selective staining method for elecytron microscopy. Science 154:1575–1577
9. Brostrom CO, Brostrom MA (1998) Regulation of translational initiation during cellular response to stress. Prog Nucleic Acid Res 58:79–125
10. Bucciantini M, Giannoni E, Chiti F, Baroni F, Formigli L, Zurdo J, Taddei N, Ramponi G, Dobson CM, Stefani M (2002) Inherent toxicity of aggregates implies a common mechanism for protein misfolding diseases. Nature 416:507–511
11. Burda J, Martin ME, Garcia A, Alcazar A, Fando JL, Salinas M (1994) Phosphorylation of the alpha subunit of initiation factor 2 correlates with the inhibition of translation following transient cerebral ischaemia in the rat. Biochem J 302:335–338
12. Burry RW, Lasher RS (1978) A quantitative electron microscopic study of synapse formation in dispersed cell cultures of rat cerebellum stained either by Os-UL or by EPTA. Brain Res 147:1–15
13. Chen J, Simon R (1997) Ischemic tolerance in the brain. Neurology 48:306–311

14. Coux O, Tanaka K, Goldberg AL (1996) Structure and functions of the 20S and 26S proteasomes. Annu Rev Biochem 65:801–847
15. DeGracia DJ, Neumar RW, White BC, Krause GS (1996) Global brain ischemia and reperfusion: modifications in eukaryotic initiation factors associated with inhibition of translation initiation. J Neurochem. 67:2005–2012
16. Deshpande J, Bergstedt K, Linden T, Kalimo H, Wieloch T (1992) Ultrastructural changes in the hippocampal CA1 region following transient cerebral ischemia: evidence against programmed cell death. Exp Brain Res 88:91–105
17. Dewar D, Graham DI, Teasdale GM, McCulloch J (1994) Cerebral ischemia induces alterations in tau and ubiquitin proteins. Dementia 5:168–173
18. Fink AL (1999) Chaperone-mediated protein folding. Physiol Rev 79:425–449
19. Gehrmann J, Banati RB, Wiessner C, Hossmann KA, Kreutzberg GW (1995) Reactive microglia in cerebral ischaemia: an early mediator of tissue damage? Neuropathol Appl Neurobiol 21:277–289
20. Hampton RY (2000) ER stress response: getting the UPR hand on misfolded proteins. Curr Biol 10:R518–521
21. Harding HP, Zhang Y, Bertolotti A, Zeng H, Ron D (2000) Perk is essential for translational regulation and cell survival during the unfolded protein response. Mol Cell 5:897–904
22. Harding HP, Zhang Y, Ron D (1999) Protein translation and folding are coupled by an endoplasmic-reticulum-resident kinase. Nature 397:271–274
23. Hartl FU, Hayer-Hartl M (2002) Molecular chaperones in the cytosol: from nascent chain to folded protein. Science 295:1852–1858
24. Hayashi T, Tanaka J, Kamikubo T, Takada K, Matsuda M (1993) Increase in ubiquitin conjugates dependent on ischemic damage. Brain Res 620:171–173
25. Hossmann K-A (1993) Disturbances of cerebral protein synthesis and ischemic cell death. Prog Brain Res 96:167–177
26. Hu BR, Wieloch T (1993) Stress-induced inhibition of protein synthesis initiation: Modulation of initiation factor 2 and guanine nucleotide exchange factor activity following transient cerebral ischemia in the rat. J Neurosci 13:1830–1838
27. Hu BR, Fux CM, Martone ME, Zivin JA, Ellisman MH (1999) Persistent phosphorylation of cyclic AMP responsive element-binding protein and activating transcription factor-2 transcription factors following transient cerebral ischemia in rat brain. Neuroscience 89:437–452
28. Hu BR, Janelidze S, Ginsberg MD, Busto R, Perez-Pinzon M, Sick TJ, Siesjo BK, Liu CL (2001) Protein aggregation after focal brain ischemia and reperfusion. J Cereb Blood Flow Metab 21:865–875
29. Hu BR, Martone ME, Jones YZ, Liu CL (2000a) Protein aggregation after transient cerebral ischemia. J Neurosci 20(9):3191–3199
30. Hu BR, Park M, Martone ME, Fischer WH, Ellisman MH, Zivin JA (1998) Assembly of proteins to postsynaptic densities after transient cerebral ischemia. Neurosci 18:625–633
31. Ito U, Spatz M, Walker JT Jr, Klatzo I (1975) Experimental cerebral ischemia in mongolian gerbils. I. Light microscopic observations. Acta Neuropathol (Berl) 32:209–223
32. Kaufman RJ (1999) Stress signaling from the lumen of the endoplasmic reticulum: coordination of gene transcriptional and translational controls. Genes Dev 13:1211–1233
33. Kirino T (1982) Delayed neuronal death in the gerbil hippocampus following ischemia. Brain Res 239:57–69
34. Kumar R, Azam S, Sullivan JM, Owen C, Cavener DR, Zhang P, Ron D, Harding HP, Chen JJ, Han A, White BC, Krause GS, DeGracia DJ (2001) Brain ischemia and reperfusion activates the eukaryotic initiation factor 2alpha kinase, PERK. J Neurochem 77:1418–1421
35. Li M, Baumeister P, Roy B, Phan T, Foti D, Luo S, Lee AS (2000) ATF6 as a transcription activator of the endoplasmic reticulum stress element: thapsigargin stress-induced changes and synergistic interactions with NF-Y and YY1. Mol Cell Biol 20:5096–5106
36. Magnusson K, Wieloch T (1989) Impairment of protein ubiquitination may cause delayed neuronal death. Neurosci Lett 96:264–270
37. Martin de la Vega C, Burda J, Nemethova M, Quevedo C, Alcazar A, Martin ME, Danielisova V, Fando JL, Salinas M (2001) Possible mechanisms involved in the down-regulation of translation during transient global ischaemia in the rat brain. Biochem J 357:819–826
38. Martone ME, Jones YZ, Young SJ, Ellisman MH, Zivin JA, Hu BR (1999) Modification of postsynaptic densities after transient cerebral ischemia: A quantitative and three-dimensional ultrastructural study. Neurosci 19:1988–1997
39. Morimoto T, Ide T, Ihara Y, Tamura A, Kirino T (1996) Transient ischemia depletes free ubiquitin in the gerbil hippocampal CA1 neurons. Am J Pathol 148:249–257

40. Morimoto RI (1998) Regulation of the heat shock transcriptional response: Cross talk between a family of heat shock factors, molecular chaperones, and negative regulators. Genes Dev 12:3788–3796

41. Nehls S, Snapp EL, Cole NB, Zaal KJ, Kenworthy AK, Roberts TH, Ellenberg J, Presley JF, Siggia E, Lippincott-Schwartz J (2000) Dynamics and retention of misfolded proteins in native ER membranes. Nat Cell Biol 2:288–295

42. Nowak TS Jr (1985) Synthesis of a stress protein following transient ischemia in the gerbil. J Neurochem 45:1635–1641

43. Paschen W (2000) Role of calcium in neuronal cell injury: which subcellular compartment is involved? Brain Res Bull 2000 Nov 1, 53(4):409–413

44. Paschen W, Gissel C, Linden T, Althausen S, Doutheil J (1998) Activation of gadd153 expression through transient cerebral ischemia: evidence that ischemia causes endoplasmic reticulum dysfunction. Brain Res Mol Brain Res 60:115–122

45. Patil C, Walter P (2001) Intracellular signaling from the endoplasmic reticulum to the nucleus: the unfolded protein response in yeast and mammals. Curr Opin Cell Biol 13:349–355

46. Plemper RK, Wolf DH (1999) Retrograde protein translocation: ERADication of secretory proteins in health and disease. Trends Biochem Sci 24:266–270

47. Rafols JA, Daya AM, O'Neil BJ, Krause GC, Neumar RW, White BC (1995) Global brain ischemia and reperfusion: Golgi apparatus ultrastructure in neurons selectively vulnerable to death. Acta Neuropathol (Berl) 90:17–30

48. Rajdev S, Hara K, Kokubo Y, Mestril R, Dillmann W, Weinstein PR, Sharp FR (2000) Mice overexpressing rat heat shock protein 70 are protected against cerebral infarction. Ann Neurol 2000 Jun 47(6):782–791

49. Sharp FR, Massa SM, Swanson RA (1999) Heat-shock protein protection. Trends Neurosci 22:97–99

50. Sherman MY, Goldberg AL (2001) Cellular defenses against unfolded proteins: a cell biologist thinks about neurodegenerative diseases. Neuron 29:15–32

51. Siesjö BK, Siesjö P (1996) Mechanisms of secondary brain injury. Eur J Anaesthesiol 13:247–268

52. Sommer T, Wolf DH (1997) Endoplasmic reticulum degradation: reverse protein flow of no return. FASEB J 11:1227–1233

53. Travers KJ, Patil CK, Wodicka L, Lockhart DJ, Weissman JS, Walter P (2000) Functional and genomic analyses reveal an essential coordination between the unfolded protein response and ER-associated degradation. Cell 101:249–258

54. Walsh DM, Klyubin I, Fadeeva JV, Cullen WK, Anwyl R, Wolfe MS, Rowan MJ, Selkoe DJ (2001) Naturally secreted oligomers of amyloid beta protein potently inhibit hippocampal long-term potentiation in vivo. Nature 416:535–539

55. Wang Y, Shen J, Arenzana N, Tirasophon W, Kaufman RJ, Prywes R (2000) Activation of ATF6 and an ATF6 DNA binding site by the endoplasmic reticulum stress response. J Biol Chem 275(35):27013–27020

56. Yenari MA, Dumas TC, Sapolsky RM, Steinberg GK (2001) Gene therapy for treatment of cerebral ischemia using defective herpes simplex viral vectors. Ann NY Acad Sci 939:340–357

57. Yoshida H, Okada T, Haze K, Yanagi H, Yura T, Negishi M, Mori K (2000) ATF6 activated by proteolysis binds in the presence of NF-Y (CBF) directly to the cis-acting element responsible for the mammalian unfolded protein response. Mol Cell Biol 20:6755–6767

Slow Progression of Neurologic Impairment after Mild Ischemic Insult in Rodents: Relationship to Metabolic and Histologic Changes

T. Kuroiwa, G. Mies, K. Ohno, I. Yamada, S. Endo, R. Okeda, and U. Ito

Key words. Cerebral ischemia – maturation phenomenon – neurological deficit

Index terms. Slow neurological impairment, metabolic and histological change, cerebral ischemia, cerebral infarction, selective neuronal death, maturation phenomenon, Mongolian gerbil, C57BL6 mouse, ATP depletion, lactate accumulation, TUNEL staining, energy metabolism, stroke index, neurological deficit score, spontaneous locomotion, faulty foot placement

Summary. We assessed the relationship between slowly progressing neurologic impairment and metabolic and morphologic changes in two rodent models of mild cerebral ischemia. Mild ischemia was induced by occluding the common carotid artery of Mongolian gerbils twice for 10 min each or by permanently occluding the left internal carotid artery of C57BL6 mice. In the gerbils (n = 13), the stroke index (Ohno, 1984) gradually increased during the first 7 days after the ischemic insult (maturation phenomenon). Their neurologic impairment paralleled the slow progression of ATP depletion in the injured hemisphere. Histologic examination revealed cerebral infarction involving the ipsilateral cortex, caudate nucleus, dorsolateral thalamus, and hippocampus. In the mice (n = 12), blood flow at the left parietotemporal skull (measured by using laser Doppler flowmeter) decreased to 40.2 ± 16.4% of the pre-ischemic level, and their neurological deficit score gradually increased during 8 days after the induction of ischemia. Tests of foot placement also revealed gradual impairment. Brains frozen at day 2 post-ischemia revealed normal levels of energy metabolism and cerebral protein synthesis in the ipsilateral hemisphere, except for two mice with a localized area of mild lactate accumu-

Toshihiko Kuroiwa[1,2], Guenter Mies[1], Kikuo Ohno[3], Ichiro Yamada[4], Shu Endo[5], Riki Okeda[2] and Umeo Ito[6]
[1] Department of Experimental Neurology, Max Planck Institute for Neurological Research, Cologne, Germany
[2] Departments of Neuropathology,
[3] Neurosurgery, and
[4] Radiology and the
[5] Animal Research Center, Tokyo Medical and Dental University, Tokyo, Japan
[6] Tokyo Metropolitan Institute for Neuroscience, Japan

Correspondence to: Toshihiko Kuroiwa, MD, Department of Neuropathology, Medical Research Institute, Tokyo Medical and Dental University, Yushima 1-5-45, Bunkyo-ku. Tokyo 113-8510, Japan, Tel.: 81-3-5803-5848, Fax: 81-3-5803-5848, E-Mail: t.kuroiwa.npat@mri.tmd.ac.jp

Maturation Phenomenon in Cerebral Ischemia V
A. M. Buchan et al. (Eds.)
© Springer-Verlag Berlin Heidelberg 2004

lation at the parietotemporal cortex. TUNEL staining of these two samples revealed scattered neuronal death in this area. Therefore, despite the marked differences between the metabolic and histologic outcomes, the slow neurologic impairment was remarkably similar between the two ischemia models. Our results suggest that various processes in the tissue from animals suffering from ischemia, not its irreversible injury, are important for the slow progression.

Introduction

Cerebral infarction and selective neuronal death develop slowly when the ischemic insult is mild [2, 6, 7]. These are typical examples of the maturation phenomenon first observed in the pyramidal cell layer of the hippocampus [6, 7]. In this slowly developing infarction, energy metabolism also is impaired gradually [11]. Recently, we observed that neurologic impairment develops slowly after mild ischemic insult as well [10], suggesting a causal relationship between the metabolic impairment and histologic changes. However, discrepancies between the histologic outcome and neurologic symptoms have occurred in various experimental models of cerebral ischemia. Here, we examined the relationship between neurologic impairment and metabolic and morphologic changes in two rodent models of cerebral ischemia.

Materials and Methods

We used two models of cerebral ischemia in this study. For the first, Mongolian gerbils were subjected to repetitive occlusion of the left common carotid artery. Each animal was anesthetized with 1.5% isoflurane, and the left common carotid artery was occluded by using a mini vascular clip. After placement of the clip, anesthesia was discontinued, the stroke index (SI) [13] was recorded, and the clip was released at 10 min after occlusion. After a 5-h interval, a second 10-min episode of ischemia similarly was induced and monitored. Animals with $SIs = 10$ or >10 $(n = 13)$ were selected for further investigation, during which the SI was measured at 1, 2, and 7 d after the repetitive ischemia. After this assessment period, animals were deeply anesthetized, and each gerbil's brain was removed after transcardiac perfusion with 4% buffered formalin, cut coronally at the levels of optic chiasma and infundibulum, and prepared for light microscopic examination.

In the second model, C57BL6 mice $(n = 12)$ were anesthetized and underwent permanent occlusion of the left internal carotid artery. During surgery, the blood flow at the ipsilateral parietotemporal skull was measured by using laser Doppler flowmeter. Neurologic symptoms before and after the arterial occlusion were assessed by obtaining the 5-point neurologic deficit score [5] and by evaluating foot placement and spontaneous locomotion. For the assessment of locomotion and foot placement, each animal was kept in an observation cage (approximate size, $20 \times 30 \times 20$ cm), in which the floor comprised parallel bars spaced 5 mm apart. The animal was allowed to move freely about the cage for 2 min, during which the distance of spontaneous locomotion (cm/min) and the number of faulty foot placements (no./cm locomotion) were obtained by videotaping. The brains of the mice were frozen in liquid nitrogen at 1, 2, or 3 to 8 d after inducing ischemia, and coronal cryostat sections were prepared for mapping of ATP content [8], pH [1], cerebral protein synthesis (CPS), and cerebral blood flow (CBF) and for TUNEL staining. For CPS and CBF mapping, ^{14}C-leucine and ^{3}H-iodoantipyrin, respectively, were injected intraperitoneally prior to euthanasia [12].

Results

In the Mongolian gerbil model of mild cerebral ischemia, the animals gradually developed neurologic impairment during the first 7 d post-ischemia. The SI was nearly zero before the ischemic insult and gradually increased to 3.4 ± 4.5, 5.0 ± 5.4 (mean $\pm$ S.D.), and 7.4 ± 5.4 at 1, 2, and 7 days after repetitive ischemia, respectively (Fig. 1, upper panel). Histologic examination at day 7 after ischemia revealed infarction at the left caudate nucleus, dorsolateral thalamus, frontoparietal cortex, and hippocampus (Fig. 2, upper panel). Succinic dehydrogenase activity, measured

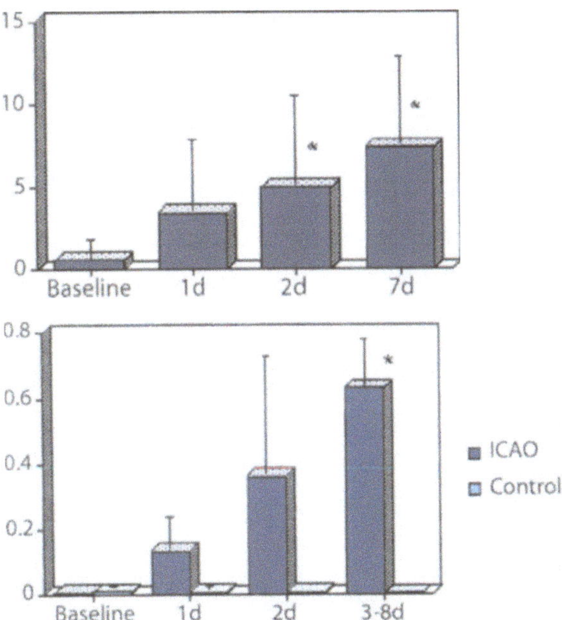

Fig. 1. Time course of stroke index in gerbil after repetitive carotid artery occlusion (upper panel) and faulty foot placement in mice (no./cm locomotion) after internal carotid artery occlusion (lower panel). These two models yielded similar neurologic profiles despite their different histologic outcomes. * p < 0.05

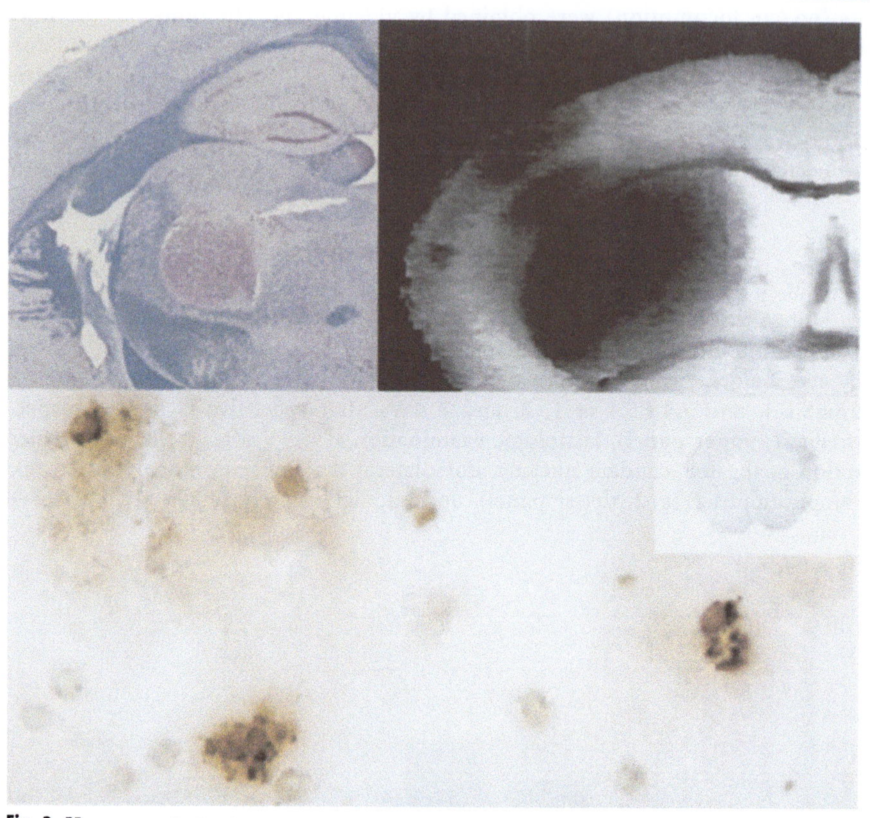

Fig. 2. Upper panel: Cerebral infarction in gerbil brain 7 d after repetitive common carotid artery occlusion. Upper left: Cortical and caudate infarction by Kluever Barrera staining. Upper right: ATP depletion in the infarcted area. Lower panel, Localized area of lactate accumulation (inset) in a mouse brain 2 d after occlusion of the internal carotid artery and scattered TUNEL-positive cells in this area

in a previous study [9], was markedly reduced in the infarcted area (Fig. 2, upper right). The infarct volume was 43 ± 12 mm^3.

In the mice undergoing permanent occlusion of the left internal carotid artery, the blood flow in the cerebral cortex decreased to $40.2 \pm 16.4\%$ of the pre-ischemic level shortly after occlusion, and neurologic impairment developed gradually (Fig. 1). The neurologic deficit score increased from nearly 0 before ischemia to 1.0 ± 1.0 at 1 d after occlusion and further increased to 1.5 ± 1.3 and 2.2 ± 1.3 at days 2 and 3 through 8, respectively. Foot placement also revealed gradual impairment – 0.018 ± 0.007 (no. of faulty placement/cm locomotion) before ischemia compared with 0.13 ± 0.17, 0.36 ± 0.51, and 0.63 ± 0.15 on days 1, 2, and 3 through 8 post-ischemia, respectively. Spontaneous locomotion decreased from 203 ± 90 cm/min before ischemia to 45 ± 17, 39 ± 29, and 16 ± 2 on days 1, 2, and 3 through 8 after induction of ischemia, respectively. Brains frozen at day 2 revealed normal levels of ATP content, pH, and cerebral protein synthesis in the ipsilateral hemisphere, except for two animals that showed only a localized area of mild lactate accumulation in this area (Fig. 2, lower panel, inset). TUNEL staining revealed a localized area of scattered neuronal death in the ipsilateral hemisphere. Therefore, the metabolic and histologic changes that occurred 1 to 2 d after occlusion of the left internal cerebral artery were very mild despite the overt impairment of neurologic function.

Discussion

We compared the temporal profiles of neurologic changes in two rodent models of cerebral ischemia. Despite differences between the histologic outcomes (i.e., cerebral hemispheric infarction and localized neuronal death), the time courses of neurologic impairment were remarkably similar.

In the gerbil model, repetitive occlusion of the common carotid artery (two 10-min occlusions separated by a 5-h interval) typically induces wide areas of cerebral infarction involving both the cerebral cortex and the basal ganglia [3, 4]. Using this ischemia model, we previously documented gradual impairment of energy metabolism during days 2 through 4 post-ischemia. Specifically, the ATP content of the ipsilateral cerebral hemisphere remained within the normal range during the first 5 h after the second episode of occlusion but gradually decreased thereafter and reached the lowest level at 2 to 4 d after ischemia [11]. The activity of succinic dehydrogenase [9] (a mitochondrial respiratory enzyme) and tissue pH also showed similar gradual decrease [11]. In the present study, we observed a gradual impairment of neurologic function during this maturation process (Fig. 1). This association appears to indicate a close causal relationship between the progressing tissue injury and the neurologic symptoms.

In mice subjected to permanent occlusion of the internal carotid artery, we found a similar slow progression of neurologic deficits. However, the decrease in the CBF after arterial occlusion was mild (approximately 40% of the preischemic level) and returned to the pre-ischemia level 1 to 2 d after occlusion probably by the collateral flow (data not shown). TUNEL staining revealed a localized area of mild acidosis with selective neuronal death in only two out of 12 animals evaluated (Fig. 2). Therefore the histologic and metabolic changes in the mice were very mild, although their neurologic impairment was pronounced, and the time-course to the development of this impairment was similar to that of the gerbils.

From these observations, we summarize our results as follows:

1. Two different types of mild ischemia, resulting either in energy impairment and cerebral infarction or a small area of selective neuronal death with slight energy impairment, induced slowly progressing neurologic impairment.
2. Despite the marked differences between the metabolic and histologic outcomes in the two models, the time course and features of the neurological impairment were similar.
3. Our results suggest that processes in the tissue undergoing ischemia without irreversible injury are important for the slow progression of the post-ischemic neurologic impairment.

The mechanism underlying this slowly progressing neurologic impairment is unknown. Synaptic and membrane changes of the neurons in the peri-ischemic tissue might be important for the neurologic changes. Systemic factors including body weight change may not be negligible.

Acknowledgement. We thank Ms Tayoko Tajima and Ms Hiromi Tanizawa for their technical support.

References

1. Csiba L, Paschen W, Hossmann K-A (1985) A topographic quantitative method for measuring tissue pH under physiological and pathophysiological conditions. Brain Res 289:334–337
2. Du C, Hu R, Csernansky CA, Hsu CY, Choi DW (1996) Very delayed infarction after mild focal cerebral ischemia: A role for apoptosis? J Cereb Blood Flow Metabol 16:195–201
3. Hanyu S, Ito U, Hakamata Y, Yoshida M (1993) Repeated unilateral carotid occlusion in Mongolian gerbils: quantitative analysis of cortical neuronal loss. Acta Neuropathol 86:16–20
4. Hanyu S, Ito U, Hakamata Y, Yoshida M (1995) Transition from ischemic neuronal necrosis to infarction in repeated ischemia. Brain Res 686:44–48
5. Huang ZH, Huang PL, Panahian N, Dalkara T, Fishman MC, Moscowitz MA (1994) Effects of cerebral ischemia in mice deficient in neuronal nitric oxide synthase. Science 265:1883–1884
6. Ito U, Spatz M, Walker JT, Klatzo I (1975) Experimental cerebral ischemia in Mongolian gerbils. I. Light microscopical observations. Acta Neuropathol 32:209–223
7. Kirino T (1982) Delayed neuronal death in the gerbil hippocampus following ischemia. Brain Res 239:57–69
8. Kogure K, Alonso OF (1978) A pictorial representation of endogenous brain ATP by a bioluminescent method. Brain Res 154:273–284
9. Kuroiwa T, Terakado M, Yamaguchi T, Endo S, Ueki M, Okeda R (1996) The pyramidal cell layer of sector CA 1 shows the lowest hippocampal succinic dehydrogenase activity in normal and postischemic gerbils. Neurosci Lett 206:1–4
10. Kuroiwa T, Yamada I, Endo S, Hakamata Y, Ito U (2000) 3-Nitropropionic acid preconditioning ameliorates delayed neurological deterioration and infarction after transient focal cerebral ischemia in gerbils. Neurosci Lett 283(2):145–148
11. Kuroiwa T, Mies G, Hermann D, Hakamata Y, Hanyu S, Ito U (2000) Regional differences in the rate of energy impairment and evolving infarction after repeated induction of cerebral ischemia in gerbils. Acta Neuropathol 100:587–594
12. Mies G, Paschen W, Hossmann K-A (1990) Cerebral blood flow, glucose utilization, regional glucose and ATP content during the maturation period of delayed ischemic injury in gerbil brain. J Cereb Blood Flow Metab 10:638–645
13. Ohno K, Ito U, Inaba Y (1984) Regional cerebral blood flow and stroke index after left carotid artery ligation in the conscious gerbil. Brain Res 297:151–157
14. Paschen W, Niebuhr I, Hossmann K-A (1991) A bioluminescence method for the demonstration of regional glucose utilization in brain slices. J Neurochem 36:513–517

Comparison of Light Transparency Changes in the Cerebral Cortex during Focal Ischemia and Death

M. Tomita, Y. Fukuuchi, T. Amano, N. Tanahashi, M. Kobari, Y. Tomita, and M. Ohtomo

Summary. The optical method with computer analysis detected light transparency (LT) increase in the cerebral cortex broadly correlated with capillary flow decrease (CBF) during ischemia and at death in 10 a-chloralose- and urethane-anesthetized cats. The LT increase at the time of death was found to be much greater (-16.6 ± 4.2 in gray scale, $n=5$) than that during ischemia (-11.7 ± 2.9, $n=8$). The LT increase/CBF decrease relationship was broadly linear during the maturation of ischemia, but not so well correlated at the boundary regions of ischemia (penumbra) due to flow recovery tendency through collateral circulation. In about half of the cases, the LT increase in the penumbral zone accompanied relative flow increase, especially during the early phase of ischemia. The LT increase was reversed in two cats during the late phase of ischemia. This reversal was probably due to intracellular protein degeneration of brain cells which results in an increase in opacity of the cerebral cortex and therefore an increase in LT. Changes in LT of the cerebral cortex with ischemia were thus comprised of several factors at least changes in blood volume, RBC aggregation, neuronal depolarization (light scattering by brain cell membrane), cell swelling (edema), and intracellular protein degeneration. Further analysis of light transparency may provide an important key to demarcate a point of no return of brain tissue undergoing infarct.

Key words. Light transparency – cerebral blood volume – microflow – neuronal depolarization – spreading depression – ischemia – penumbra – death

Introduction

Light transparency of the cerebral cortex has seldom been studied adequately because of methodological difficulties. In a previous study employing a new optical method [3], we reported that light transparency (LT) changes spread heterogeneously in a concentric manner with the evolution of microcirculatory derange-

Minoru Tomita, Yasuo Fukuuchi, Takahiro Amano, Norio Tanahashi, Marahiro Kobari[2], Yataka Tomita[1], Manabu Ohtomo
Department of Neurology, Keio University School of Medicine, Tokyo, Japan
[1] Tachikawa Hospital, Tokyo; [2]Urawa Municipal Hospital, Urawa-city, Saitama, Japan
Correspondence to: Minoru Tomita, MD, Department of Neurology, School of Medicine, Keio University, 35 Shinanomachi, Shinjuku-ku, Tokyo 160-8582, Japan, Tel.: 03-3353-1211 (Ext. 62316), Fax: 03-0564-48-4885, E-Mail: mtomita@gol.com

Maturation Phenomenon in Cerebral Ischemia V
A.M. Buchan et al. (Eds.)
© Springer-Verlag Berlin Heidelberg 2004

ment in a small region of ischemic tissue in rats [4]. This study further examined the LT changes occurring during ischemia produced by laser-beam arteriolar occlusion and at death as a representative state of infarct in cats [10].

Methods

Experiments were carried out on 10 cats weighing 2.5 to 4.5 kg under general anesthesia using α-chloralose and urethane. The details of the experimental set-up, surgical procedure, and data analysis were the same as those reported previously for rats [4], based on the photoelectric method [6, 7]. After craniectomy, the dura was removed and a small pial arterial branch of approximately 50 µm in diameter in the region of interest (ROI) over the sensorimotor cortex was selected. Focal

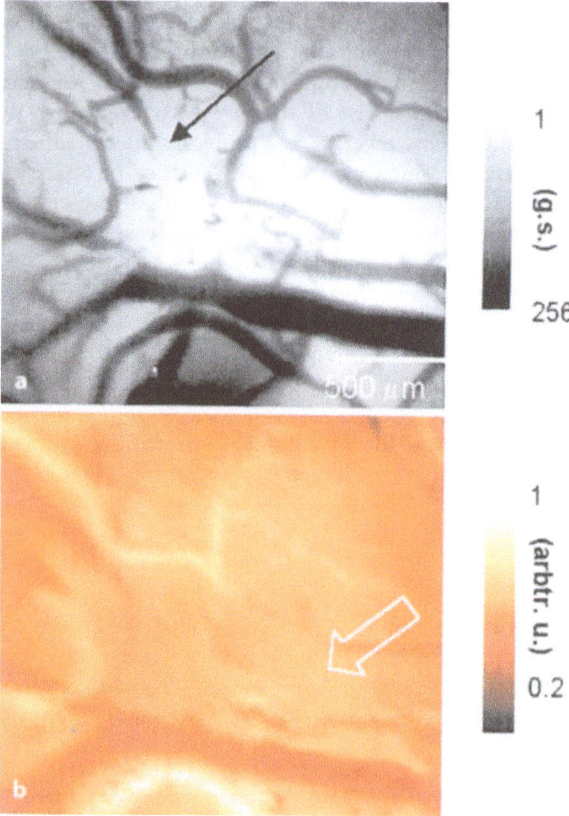

Fig. 1. Light transparency (LT) increase with arteriolar occlusion by laser beam at **a** arrow and **b** a variety of flow changes displayed on 2-D microflow map at 10 min after occlusion. Microflow decrease in the central area broadly co-located at the LT increase area. There was an area of microflow increase as indicated by the empty arrow. The discrepancy appeared to be due to ischemic neuronal depolarization

ischemia was produced by occlusion (coagulation) using an infra-red laser beam transported through an optical fiber (200 µm in diameter) and directed towards the target arteriole [10]. The ROI (ca. 3×3 mm) was videotaped continuously, and the selected portion was fed into a computer via a Scion frame grabber (8 bit) [4]. Each frame was subjected to subtraction of the control frame for LT changes and microvascular changes (vasomotion) during a short interval at a frame rate. The relative intensity changes in LT (ΔLT) were expressed as LT (t) – LT (control), in which differences were scaled to eight-bit images, i.e., 1 (brightest)–256 (darkest), with an intermediate level of 128, and displayed as pictures on the 256 gray scale, for which we employed arbitrary units of g.s. for convenience. To obtain microflow maps, 0.3-ml saline was injected spikely into the internal carotid artery through the lingual artery. A 2500 set of pixelar microflow values was calculated from individual hemodilution curves produced by the saline transit through the pixels; these values were displayed two-dimensionally (2-D) with the aid of Matlab software. The values of ΔLT at death were obtained when the animal was destroyed by an I.V. potassium injection.

Results

The flash of the laser beam caused the small arteriole to occlude and the surrounding vessels to disappear (Fig. 1a). LT increased immediately around the area of coagulation, followed by various changes over time and space (Fig. 2). The maximum increase of LT in the center of ischemia in eight rats reached 11.7 ± 2.9 (mean ± SD) at various times after occlusion. In general, the LT increase expanded amorphously over time during ischemia (Fig. 2), and then became less demarcated. In two cats, LT changes were biphasic, with an initial increase followed by a decrease (probably due to "low perfusion hyperemia," a paradoxical increase of blood content in ischemic tissue [5]) followed by an additional increase. In the other two cats, LT increased progressively, but leveled-off a few hours after occlusion and then decreased to a point just above the preocclusive level. The brain surface looked turbid or opal on video images, suggesting that intracellular protein degeneration or coagulation took place during ischemia. The change was, however, often observed when the brain was left undisturbed for several hours after death. The concomitant LT-microflow changes decreased markedly, but flow changed heterogeneously in the circumferential area (so-called penumbra) as reported previously [10]; some portions even showed an increase in flow with dilation of the inflow arterioles (Fig. 3). The expanded LT area generally correlated with the microflow-decreased area displayed on the 2-D microflow map. However, the correlation, especially in the penumbra, was variable in all cases. At the time of death, the brain looked pale with narrowed vessels and strong RBC aggregation. The LT changes at death were 16.6 ± 4.2 (n = 5).

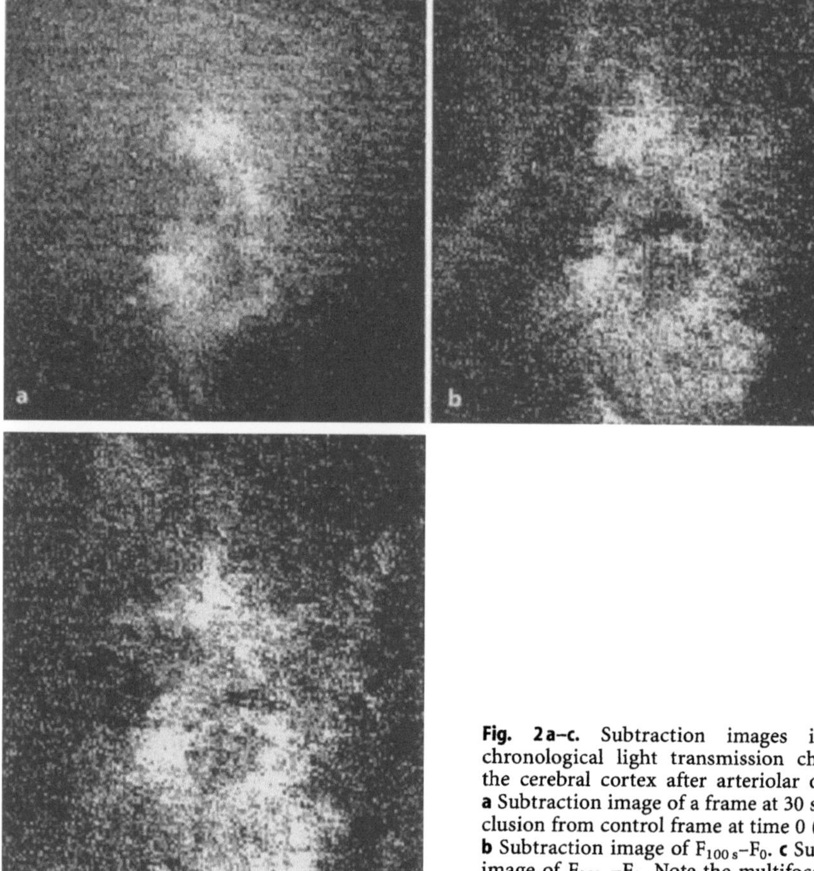

Fig. 2a–c. Subtraction images indicating chronological light transmission changes in the cerebral cortex after arteriolar occlusion. **a** Subtraction image of a frame at 30 s after occlusion from control frame at time 0 ($F_{30\,s}-F_0$). **b** Subtraction image of $F_{100\,s}-F_0$. **c** Subtraction image of $F_{200\,s}-F_0$. Note the multifocal irregular expansion of light transparency

Discussion and Conclusion

Blood is the major chromophore in the tissue [7]. With respect to ischemia, the LT changes were due to a CBV decrease (assuming that CBV/CBF = constant), since the shape and size of the LT changes were generally similar to those in the corresponding 2-D microflow map. However, on closer examination, the co-location was not exact, since an increase in microflow occasionally occurred in the LT-increased spots. The discrepancy could be explained by neuronal depolarization which has also been reported to cause LT changes [1, 2]. Another possibility was the separation of flow layers in the interfacial zone to the normal area: the intraparenchymal vessels became sluggish, while the leptomeningeal vessel network attempted to supply rescue-flow to the ischemic area (a "sink" effect) (Fig. 3). Unlike the round wave-ring spread of light transparency during spreading depression reported previously [8–10], the shape of the light transparency during ischemia was irregular

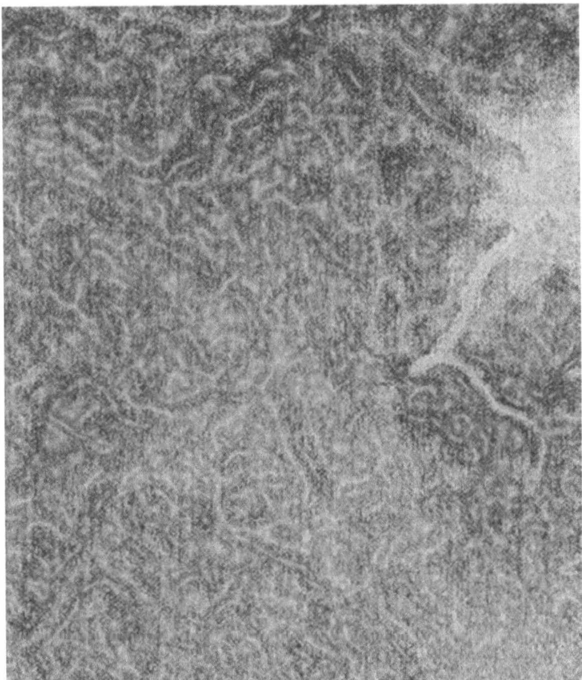

Fig. 3. Subtraction image of sequential frames of video at 2-s intervals at about 10 min after occlusion ($F_{602\,s}-F_{600\,s}$). The microvascular network most likely changed as a result of vasomotion during the 2-s interval. Images were processed by subtraction, contrast-enhancement, and wrap of the resultant image. Note the light transparency increase at upper left-hand corner and a feeding arteriole, which dilated during the 2-s intervals

(Fig. 2). This irregularity could be due to multi-ischemic spots along the affected arterioles, which could trigger ischemic neuronal depolarization and vascular changes independently. The large LT changes at death (final state of ischemia) were probably due to a CBV decrease in association with cell swelling. Again, the LT changes could have been caused by protein degeneration. In conclusion, LT increases in the cerebral cortex with respect to ischemia were not simple, consisting of several components, including changes in blood volume, RBC aggregation, neuronal depolarization (light scattering by the brain cell membrane), cell swelling (edema), and intracellular protein degeneration. However, further analysis of these factors may provide a clue to determining reversibility of the tissue (point of no return) or transition time from ischemic tissue to infarct.

References

1. Andrew RD, Jarvis CR, Obeidat AS (1999) Potential sources of intrinsic optical signals imaged in live brain slices. Methods 18:185–196
2. Joshi I, Andrew RD (2001) Imaging anoxic depolarization during ischemia-like conditions in the mouse hemi-brain slice. J Neurophysiol 85:414–424

3. Schiszler I, Tomita M, Fukuuchi Y, Tanahashi N, Inoue I (2000) New optical method for analyzing cortical blood flow heterogeneity in small animals – validation of the method. Am J Physiol 279:H1291–1298

4. Tomita M, Fukuuchi Y, Tanahashi N, Tanaka K, Kobari M, Takao M, Tomita Y, Ohtomo M, Inoue M, Schiszler I (2001) Evolution of microvascular derangement in a small area of the rat cerebral cortex following occlusion of a pial arterial branch as observed by the novel photoelectric method. In: Bazan NG, Ito U, Marcheselli VL, Kuroiwa T, Klatzo I (eds) Maturation Phenomenon in Cerebral Ischemia IV. Springer, pp 165–170

5. Tomita M, Gotoh F, Amano T, Tanahashi N, Tanaka K (1980) Low perfusion hyperemia following middle cerebral arterial occlusion in cats of different age groups. Stroke 11:629–636

6. Tomita M, Gotoh F, Amano T, Tanahashi N, Kobari M, Shinohara T, Mihara B (1983) Transfer function through regional cerebral cortex evaluated by a photoelectric method. Am J Physiol 245:H385–H398

7. Tomita M, Gotoh F, Sato T, Amano T, Tanahashi N, Tanaka K, Yamamoto M (1978) Photoelectric method for estimating hemodynamic changes in regional cerebral tissue. Am J Physiol 235:H56–H63

8. Tomita M, Schiszler I, Fukuuchi Y, Amano T, Tanahashi N, Kobari M, Takeda H, Tomita Y, Ohtomo Y, Inoue K (2002) A time-variable concentric wave-ring increase in light transparency and associated microflow changes during a potassium-induced spreading depression in the rat cerebral cortex. In: Tomita M, Kanno I, Hamel E (eds) Brain Activation and CBF Control. Excerpt Med, Elsevier, ICS 1235, Amsterdam, in press

9. Tomita Y, Tomita M, Schiszler I, Amano T, Tanahashi N, Kobari M, Takeda H, Ohtomo M, Fukuuchi Y (2002) K^+-induced repetitive concentric wave-ring spread of oligemia/hyperemia in the sensorimotor cortex during spreading depression in rats and cats. Neurosci Lett, in press

10. Tomita Y, Tomita M, Schiszler I, Amano T, Tanahashi N, Kobari M, Takeda H, Ohtomo M, Fukuuchi Y (2002) Moment analysis of microflow histogram in focal ischemic lesion to evaluate microvascular derangement following small pial arterial occlusion in rats. J Cereb Blood Flow Metabol, in press

Sphingolipids Metabolism Following Cerebral Ischemia

M. Nakane, M. Kubota, T. Nakagomi, H. Nakayama, A. Tamura, H. Hisaki, H. Shimasaki, and N. Ueta

Key words. Rat – focal ischemia – gerbil – global ischemia – membrane lipids – sphingomyelin – ceramide

Introduction

It has been well known that ischemic insults induced the hydrolytic breakdown of polyphosphoinositides by calcium-dependent phospholipase C, and diacylglycerol (DAG) was hydrolyzed to free fatty acids (FFAs) by DAG lipase [1, 2, 11, 22, 26]. The composition of FFAs consists mainly of stearic (C18:0) and arachidonic acids (C20:4). These processes are thought to act at early phase of the ischemia. Prolonged ischemia induces the degradation of membrane phospholipids, and the level of various phospholipids decrease at late phase of the ischemia. To determine a therapeutic time window of the ischemic penumbra, it is essential to evaluate the time course of the degradation of glycerophospholipids.

In the ischemic state, the degradation products of some lipids may work as mediators or modulators of neuronal death. Sphingolipids are known to play important roles in cell differentiation and death [7, 9, 20]. Furthermore, exogenous application of ceramide, a hydrolyzed product of sphingomyelin, results in apoptotic cell death in cultured neurons [3, 4, 21]. From the results of these studies it is apparent that the sphingomyelin cycle could play a role in ischemic neuronal cell death. In order to confirm this hypothesis, we selected a gerbil transient forebrain ischemia model, in which neuronal death in the hippocampus is relatively reproducible and homogeneous.

Makoto Nakane[1], Masaru Kubota[2], Tadayoshi Nakagomi[2], Hitoshi Nakayama[1], Akira Tamura[2], Harumi Hisaki[3], Hiroyuki Shimasaki[3], Nobuo Ueta[3]
[1] Department of Neurosurgery, University Hospital, Mizonokuchi, Teikyo University School of Medicine
[2] Department of Neurosurgery and
[3] First Department of Biochemistry, Teikyo University School of Medicine
Correspondence to: Makoto Nakane, Department of Neurosurgery, University Hospital, Mizonokuchi, Teikyo University School of Medicine, 3-8-3 Mizonokuchi, Takatsu-ku, Kawasaki, Kanagawa 213-8507, Japan, Tel.: +81-44-844-3333, Fax: +81-44-844-3442, E-Mail: nsnakane@med.teikyo-u.ac.jp

Maturation Phenomenon in Cerebral Ischemia V
A. M. Buchan et al. (Eds.)
© Springer-Verlag Berlin Heidelberg 2004

Focal Ischemia

Adult male Sprague-Dawley rats (300–350 g) were anesthetized with 2% halothane inhalation. The proximal part of the left MCA was permanently occluded by the transretro-orbital approach [24, 25]. Sham-operated animals were prepared in the same manner except that the exposed MCA was not cauterized. Each experimental group consisted of five rats and durations of ischemia were 0.5, 1, 2, 6, 24, 48 and 96 h, respectively. After treatment with microwaves (Microwave Applicator, Toshiba, Japan; 4 kW, 1.5 s) under halothane inhalation, the left cerebral cortex was dissected from the left cerebral hemisphere for lipids analysis as shown in Fig. 1. This area is called 'penumbra' which receive collateral arterial supply. The cells in the penumbra can survive for some hours, so they may be rescued by pharmacological treatments. All experimental protocols were approved by the Animal Research Committee of Teikyo University School of Medicine.

Total lipids were extracted from the cerebral cortex (70–100 mg) with 20 volume of chloroform-methanol (2:1, by volume) according to the method of Folch et al. [5]. The total lipid extract was subjected to a Bond Elut column (NH_2) (1 cc 100 mg^{-1}, Varian, USA) chromatography and the less polar acidic lipids containing FFAs were eluted with 2% acetic acid in diethylether. The FFAs, methylesterified with trimethylsilyldiazomethane (Tokyo Kasei Corp., Japan), were applied to gas-liquid chromatography (GLC). The amount of fatty acids were calculated by comparison with those of an internal standard (5 µg of heneicosanoic acid; Sigma, St. Louis, MO, USA) [15].

For the measurement of PIPs, cortical tissue was transferred to a glass-glass homogenizer tube containing 10 ml chloroform-methanol (1:1, by volume) and

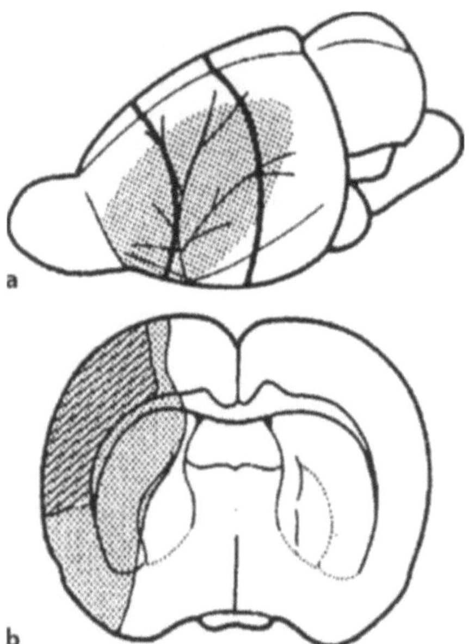

Fig. 1. a The area of ischemic changes is shown. Coronal section of the cerebrum is cut in the thick line division. b Oblique line division on coronal section was used as a tissue sample

100 µg of butylated hydroxytoluene (Sigma, St. Louis, MO, USA), and homogenized for 2 min in crushed ice and then centrifuged at 2000 g for 10 min. After the pellet was washed twice with chloroform-methanol (2:1, by volume), PIP and PIP_2 were extracted from the residual pellet with acidified chloroform-methanol according to the methods of Hauser et al. [10]. The acidified extra was neutralized to pH 7–8 by the addition of 7 M ammonium hydroxide. For separation of PIP and PIP_2, the extract was applied to a silica gel H plate (Whatman, USA), impregnated with 1% potassium oxalate in the solvent system of chloroform-methanol-4 M ammonium hydroxide (9:7:2, by volume). After completion of TLC procedures, the plate was sprayed with 2% 2,7-dichlorofluorescein (Nakarai Chemical, Japan) in methanol and dried. Under an ultraviolet lamp (Vilber Lourmat, France), the bands were scraped off into a tube containing 5 µg of heneicosanoic acid as the internal standard and methanolyzed with 5% anhydrous HCl in methanol (Muto Pure Chemicals Ltd., Japan) at 100 °C for 3 h.

For the separation of phosphatidylcholine (PC), phosphatidylethanolamine (PE) and phosphatidylserine (PS), an aliquot of the above-mentioned total lipid extract was directly subjected to high performance thin-layer chromatography (HPTLC, Merck, Germany) and developed with chloroform-methanol-acetic acid-formic acid-water (35:15:7:2:1, by volume). The spots of these phospholipids were identified from their Rf values on the TLC plate by comparison with an appropriate authentic standard, using the same procedure as that for PIPs. Each phospholipid spot on the plate was scraped directly into a test tube and methanolyzed as described above. For quantitative analysis of PC, PE and PS, 20 µg of heneicosanoic acid was added into the test tube containing silica gel with each lipid before methanolysis.

For the separation of ceramide, the total lipids were subjected to Florisil (1 g) (Kanto Chemical Co. Inc., Japan) column chromatography (activated at 120 °C for 3 h prior to use) and then eluted with chloroform-methanol (2:1, by volume) for removal of glycerophospholipids. The solvent was evaporated under a stream of nitrogen and the lipid moiety was subjected to TLC. The TLC plate was developed with chloroform-methanol-water (65:25:4, by volume) and then with hexane-diethylether-acetic acid (50:50:1, by volume). For the separation of sphingomyelin, the total lipids were directly subjected to TLC and were developed with chloroform-methanol-acetic acid-formic acid-water (35:15:7:2:1, by volume). The spots of ceramide and sphingomyelin were identified from their Rf values on the TLC plates by comparison with a respective authentic standard. Each area of these sphingolipids on the plate was scraped directly into a test tube and methanolyzed with 5% anhydrous HCl in 100 °C for 3 h. For quantitative analysis of ceramide and sphingomyelin, 10 and 50 µg of heneicosanoic acid were added into the test tube before methanolysis, respectively.

The methylesterified fatty acids extracted with hexane were applied to GLC. GLC was carried out with Shimadzu 14A gas liquid chromatograph (Shimadzu Corp., Japan) equipped with a flame ionization detector, using a capillary column of Neutra Bond-1 (30 µm×0.25 mm i.d., GL Sciences Inc., Japan) at 230–250 °C. The amounts of fatty acids were calculated by comparison with each internal standard.

The data were presented as mean ± SD. Groups were compared by 2-way ANOVA followed by post hoc analysis (Fisher's PLSD) to determine significance: $p < 0.05$ was considered to be significant.

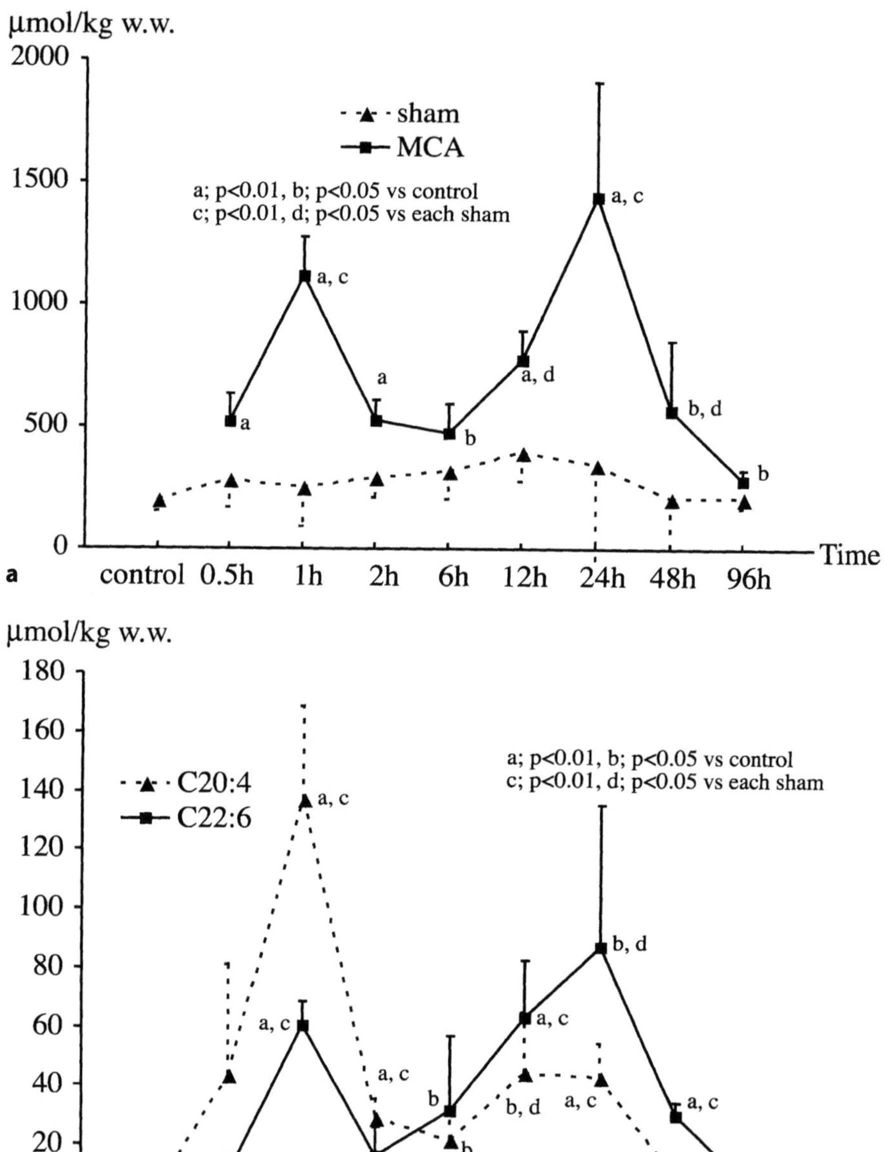

Fig. 2. a Time-dependent changes in the FFA levels in the rat cerebral cortex after MCA occlusion. The FFA levels increased during the early period and returned to the control levels in 2 h. However, they increased again one day after occlusion. **b** The levels of arachidonic and docosahexaenoic acids in the rat cerebral cortex after MCA occlusion. The composition ratio of both fatty acids showed a reversible relationship in the values between the early and late period of ischemia in FFA release. Values are means ± SD.

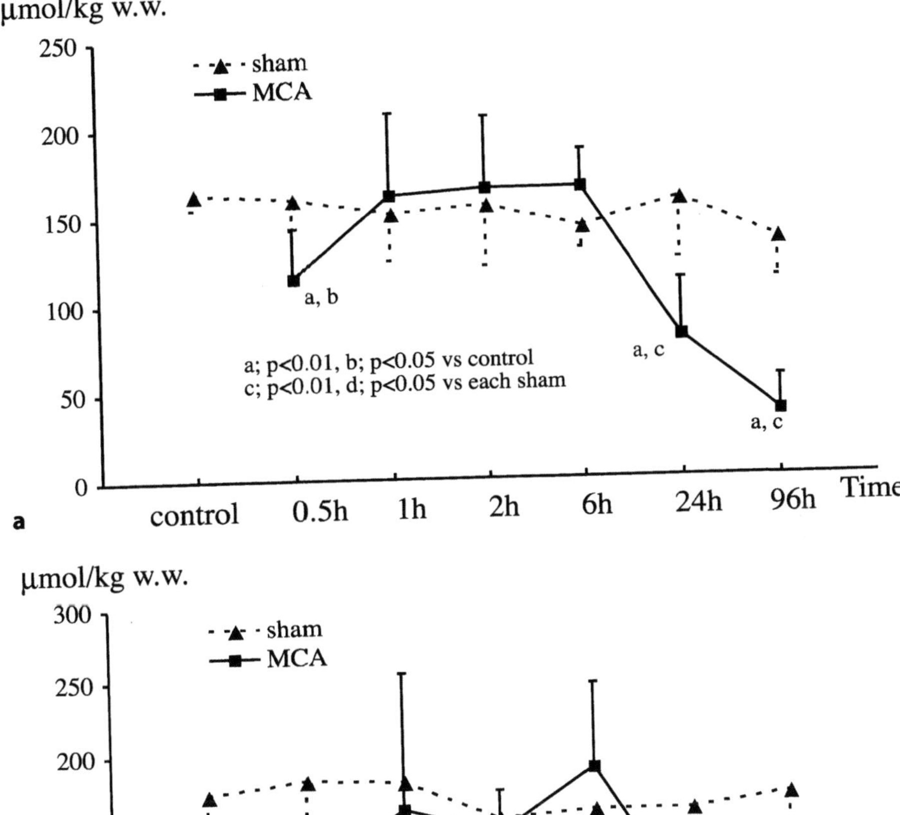

Fig. 3. Time course of changes in the fatty acid composition of the **a** PIP and **b** PIP$_2$ after MCA occlusion. The levels of both polyphosphoinositides significantly decreased at 30 min and returned each sham level in 1, 2 and 6 h after occlusion

Fig. 2a shows time-dependent changes in the FFA levels in penumbra after MCA occlusion. The FFA levels began to increase at 30 min and reached their first peak at 1 h after occlusion. Then, they returned to the control levels in 2 h, and increased again 24 h. These changes are different from those in the ischemic core, in which they might increase immediately after application of the ischemia [11, 27].

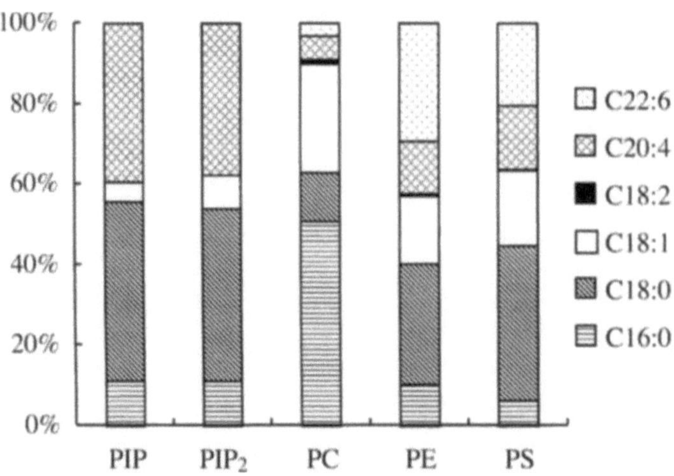

Fig. 4. Differences in levels of fatty acyl-residues of various glycerophospholipids in rat cerebral cortex of pre-ischemic control

In regard to the composition of polyunsaturated fatty acids (PUFAs), the ratio of arachidonic (C20:4) and docosahexaenoic acids (C22:6) showed a reversible relationship in the value between the early and late periods of ischemia (Fig. 2b).

Both the levels of PIP and PIP_2 were significantly decreased at 30 min after occlusion. The levels returned each sham level at 1, 2 and 6 h, then decreased again at 24 and 96 h (Fig. 3). Fig. 4 shows differences in levels of fatty acyl-residues of various glycerophospholipids in rat cerebral cortex of pre-ischemic control. In early phase of cerebral ischemia, the liberated free fatty acids mainly consisted of stearic (C18:0) and arachidonic acids (C20:4). These fatty acids were probably derived from degradation of PIP and PIP_2.

There were no significant differences in PC, PE and PS levels during an early period of ischemia (Fig. 5). However, the levels of PC and PE decreased significantly at 1 day after occlusion. The level of PS did not show any significant difference until 4 days. The levels of FFAs in the penumbra showed a biphasic increase. At second phase with its peak at 24 h, there was a large increase in FFAs, particularly in the levels of C22:6. Since acyl-residues of PE consisted of C18:0 and C22:6, the delayed increase of FFAs were derived from PE, in part PC and PIPs. The degradation of these membrane phospholipids may be associated with irreversible ischemic changes.

The level of sphingomyelin in the penumbra was significantly decreased at 2 h after occlusion, and returned sham level at 6 h, then decreased again at 24 and 96 h (Fig. 6a). On the contrast, the level of ceramide began to increase at 6 h and continued to increase up to 96 h after occlusion (Fig. 6b). The quantity and time course, as well as the species of fatty acid residues involved in these changes, suggest that ceramide was produced by the hydrolysis of sphingomyelin during cerebral ischemia.

The control levels of FFA and phospholipids presented here was higher than those reported by others [2, 26, 28]. It is possible that a high temperature of tissue within 1.5 s after the exposure of microwave may have caused the brain tissue to lose some water.

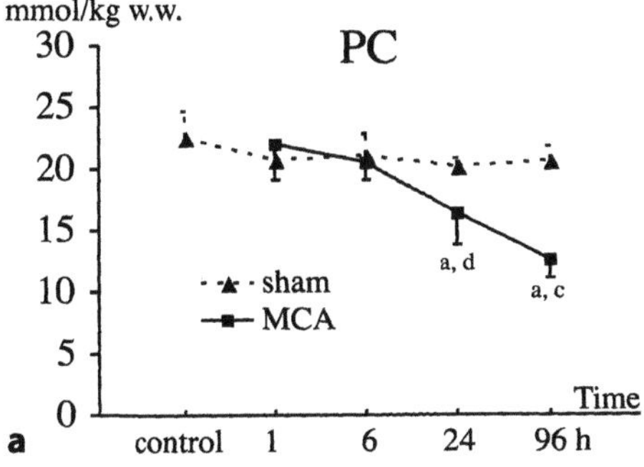

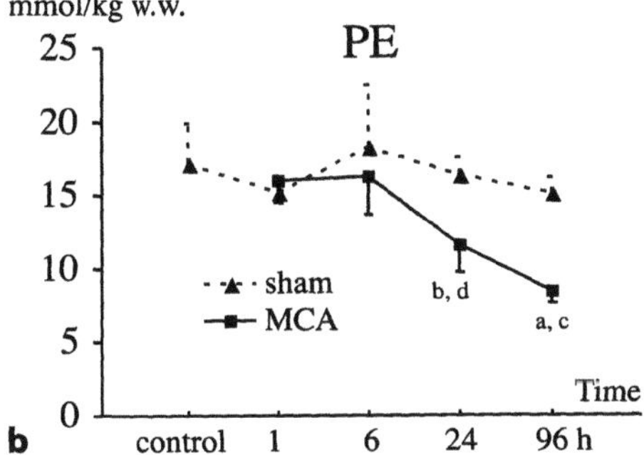

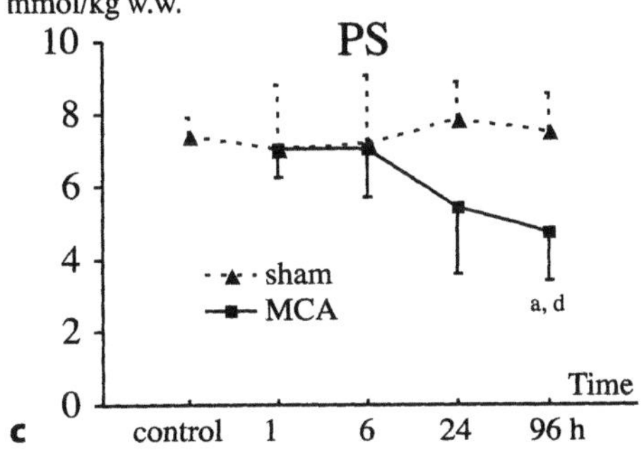

Fig. 5. Changes in **a** PC, **b** PE and **c** PS levels in rat cerebral cortex after MCA occlusion. There were no significant differences in these phospholipids levels during an early period of ischemia. However, the levels of PE and PC decreased significantly at 1 day after occlusion. **a** $p<0.01$, **b** $p<0.05$ vs control, **c** $p<0.01$, **d** $p<0.05$ vs each sham.

Global Ischemia

Male Mongolian gerbils weighing 60–90 g (Nihon Ikagaku, Tokyo) were allowed free access to food and water until the time of the experiments. Under 2% halothane inhalation, the carotid arteries were exposed bilaterally and occluded with Sugita clips for 2 or 5 min. During the operation, the temporal muscle temperature was regulated at $37.5 \pm 0.5\,°C$ with the aid of a servo-controlled heat pad and a heat lamp. Following verification of reflow in the carotid arteries, anesthesia was discontinued and the gerbils were placed in individual boxes maintained at room temperature (20 °C). After various post-ischemic recirculation times (30 or 90 min

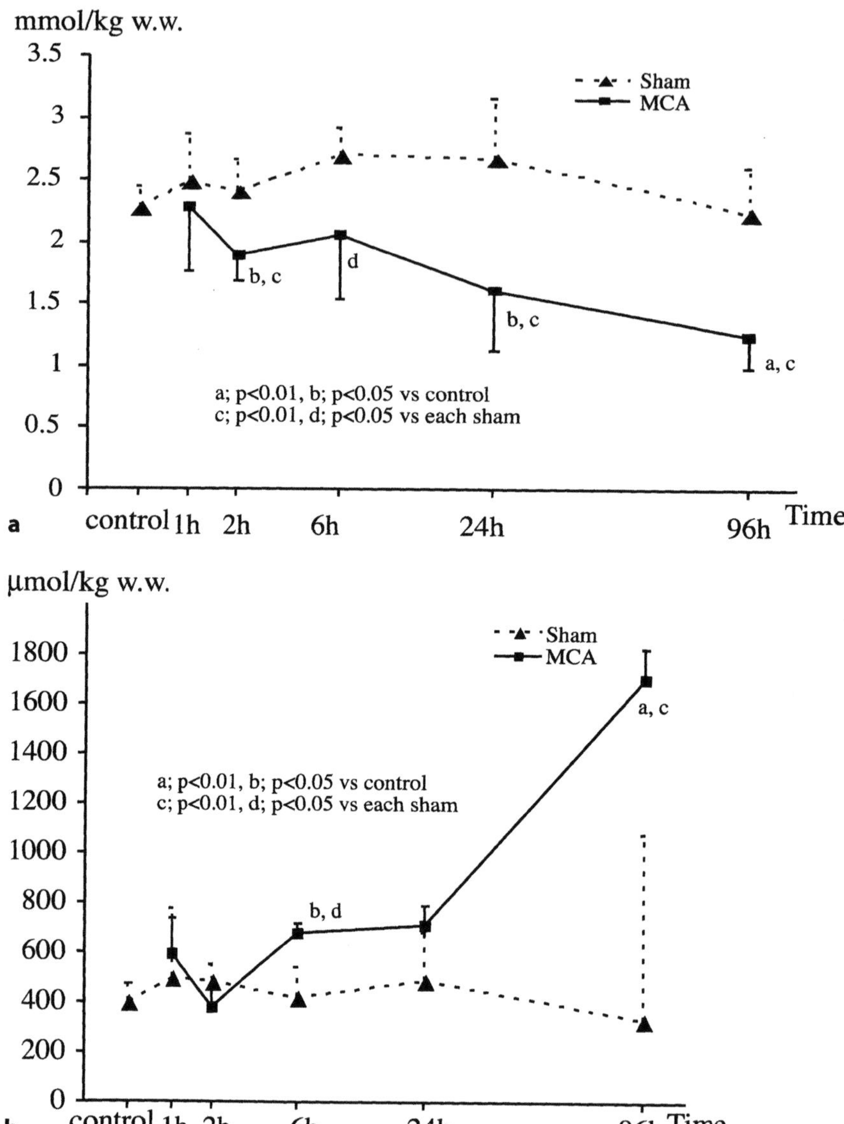

Fig. 6. Changes in **a** sphingomyelin and **b** ceramide levels in rat cerebral cortex after MCA occlusion. Values are means ± SD.

or 24 h; n = 7), a microwave beam (Microwave Applicator, Toshiba, Japan; 4.2 kW, 1.2 s) was directed at the gerbils' heads after the animals had been anesthetized again. Both hippocampi were then dissected under a microscope, and the vessels and fornices were removed. Animals on which no surgical procedure was performed but which were anesthetized in the same manner served as controls (n = 14). Some samples were lost during application of the microwave or due to trouble of the extraction.

About 60 mg of brain tissue were homogenized with 20 volumes of chloroform–methanol (2:1 by volume) [5]. Internal standards, namely 25 µg of N-heptadecanoyl-D-ceramide (C17-sphingomyelin) and 2.5 µg of N-heptadecanoyl-D-sphingosine (C17-ceramide), were added to the homogenized tissue. The C17-sphingomyelin was supplied by Dr. M. Okada from Chugai Pharmaceutical Co., Japan, and the C17-ceramide was synthesized from C17-sphingomyelin hydrolysis with sphingomyelinase (Sigma, USA), as previously described [8, 23]. The extracts were evaporated under a stream of nitrogen. Following evaporation, the lipids were dissolved in a mixture of chloroform and methanol (1:1 by volume).

A fifth of the total extracted lipid was applied to high-performance thin-layer chromatography (HPTLC) plates (Silica gel 60 F254, Merck, Germany) for sphingomyelin separation. All plates were activated at 110 °C for 30–40 min before use. Spotting and subsequent chromatography were performed in a darkroom at constant temperature (18 °C) and humidity (40%). The sphingomyelin was separated on the plates in a solvent system of chloroform-methanol-acetic acid-formic acid (20:5:3:1 by volume). Authentic standard lipids were applied at both ends of each plate. When the HPTLC process was complete the plate was dried with an air gun. The spots of sphingomyelin were identified after primuline had been sprayed on the plate [17], and were then compared with the authentic standards. Each spot was scraped off into a test tube.

Four-fifths of the total extracted lipid was subjected to HPTLC for ceramide separation. The HPTLC plate was developed with chloroform-methanol-acetic acid (95:5:1 by volume), and then with hexane-diethylether-acetic acid (50:50:1 by volume). The spots of ceramide were identified and treated in the same manner as described above.

Two milliliters of 5% anhydrous HCl in methanol were added to the scraped HPTLC, and the tubes were sealed with teflon-lined caps. The acyl residues of sphingomyelin and ceramide were then inter-esterified at 85 °C for 6 h in a water bath. After cooling, 2 ml of *n*-hexane were added to the reaction mixture, and the fatty acid methylesters were extracted with *n*-hexane. This process was repeated twice. The combined extracts were concentrated under nitrogen gas flow and stored at –20 °C until analysis.

The fatty acid methylesters were subjected to GLC for quantification, described above. The levels of these sphingolipids were calculated by comparison with those of the internal standards (C17-sphingomyelin and C17-ceramide).

The results were presented as mean ± SEM values. Data were analyzed by analysis of variance followed by post hoc analysis (Dunnett 2-sided test). The total levels of sphingomyelin and ceramide in the control gerbils were 4.47 ± 0.18 mmol/kg wet weight and 181.8 ± 8.5 µmol/kg, respectively. The sphingomyelin level had decreased and the ceramide level had increased significantly 30 min after 5 min (lethal) ischemia, compared with the control values (3.77 ± 0.21 mmol/kg, $p < 0.05$; 255.6 ± 29.1 µmol/kg, $P < 0.01$, respectively) (Fig. 7). Ninety minutes after lethal ischemia, the levels of sphingomyelin and ceramide had returned to those found in the controls. However, by 24 h after lethal ischemia the sphingomyelin had decreased and the ceramide had increased again (3.60 ± 0.11 mmol/kg, $p < 0.01$;

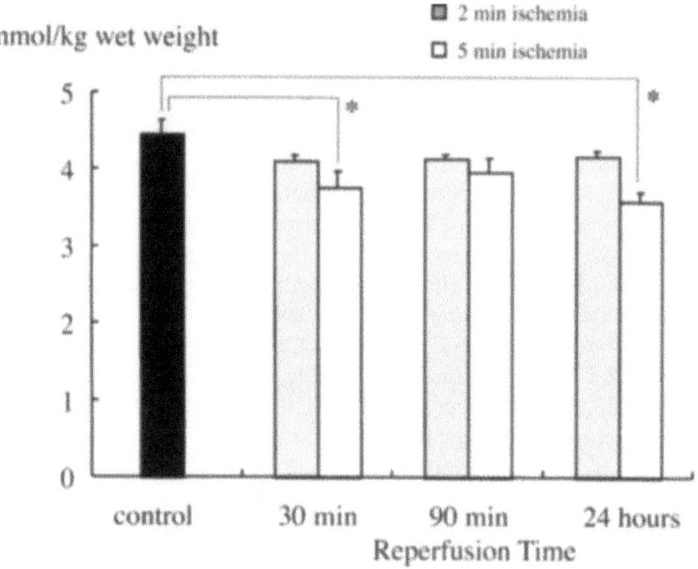

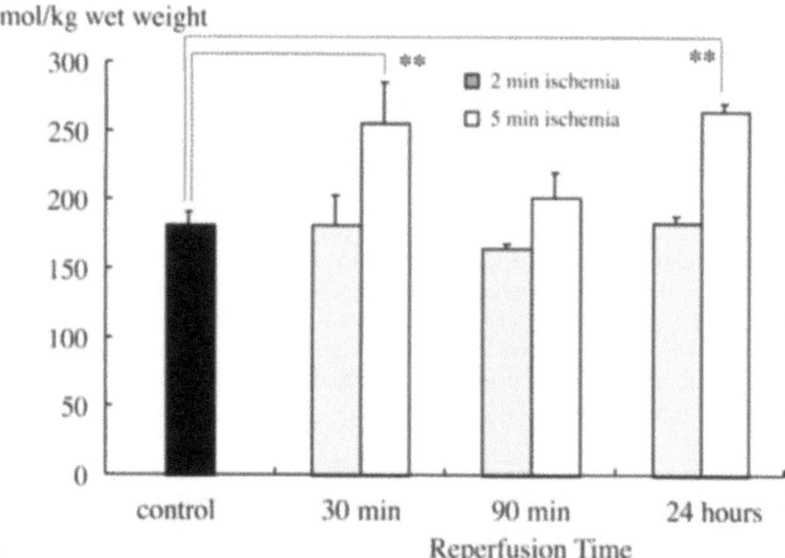

Fig. 7. a Sphingomyelin levels in the gerbil hippocampus after lethal and sublethal forebrain ischemia. Values are means±SEM. $*p<0.05$, $**p<0.01$ (comparison with control group). **b** Ceramide levels in the gerbil hippocampus after lethal and sublethal forebrain ischemia. Values are means±SEM. $**p<0.01$ (comparison with control group)

264.8±4.9 µmol/kg, $P<0.01$, respectively). In contrast, no significant changes in the sphingomyelin and ceramide levels were observed after 2 min (sublethal) ischemia.

CA1 neurons in the hippocampus are vulnerable to cerebral ischemia and show delayed neuronal death (DND) after transient forebrain ischemia of 5 min in gerbils [13]. Brief ischemia, for example for 2 min, however, is in itself not lethal to CA1 neurons [14]. In this study, we observed sphingomyelin hydrolysis and ceramide generation in the hippocampus only after lethal ischemia. Since DND is reported to occur from 2–4 d after ischemia [13], the ceramide elevation is thought to take place before histopathological change in the CA1 neurons. These results suggest that ceramide accumulation in the hippocampus may play an important role in subsequent DND.

We have already reported that ceramide is produced in the ipsilateral cerebral cortex by the hydrolysis of sphingomyelin after middle cerebral artery occlusion in rats [16]. In this rat focal ischemia model, the cortical neurons and glial cells are injured heterogeneously. Moreover, reduction of the sphingomyelin level and elevation of the ceramide level are progressive and irreversible. Therefore, it is possible that these alterations simply reflect the biochemical process of infarction (necrosis).

As mentioned above, morphological changes of the tissue may bring irreversible hydrolysis of sphingomyelin and formation of ceramide. So, in this study, we measured the sphingomyelin and ceramide levels only till 24 h in order that neuronal death might not affect the value of the sphingomyelin and ceramide. Ceramide-induced cell death occurs more quickly in culture experiments than in the CA1 neuron. This is why accumulation of the ceramide may be more gradual *in vivo* than that in culture experiments. It is still unclear in this study.

However, in our study, we observed transiently increased levels of ceramide associated with hydrolysis of sphingomyelin. We speculate that the transient elevation of ceramide seen 30 min after ischemia may trigger neuronal death, and that the delayed elevation seen 24 h after ischemia may represent an irreversible accumulation of ceramide that causes apoptosis of the cell. Hannun suggested in his review that the persistence of ceramide accumulation indicates the reprogramming of cell function through a ceramide-activated pathway [6].

The changes in ceramide levels were of the order of 30%, at most, in this study. However, since the values reflect an average over both hippocampal areas, it is still unclear how much Cer levels increase in the CA1 neurons. Since the levels of ceramide generated were only one tenth those of Cer production calculated by the amount of the decreased sphingomyelin, other sphingolipid derivatives, such as ganglioside and sphingosine, were probably generated too.

Arachidonic acid stimulates the activity of a cytosolic sphingomyelinase [12]. Therefore, arachidonic acid liberated during ischemia [15, 18] may activate the sphingomyelinase, hydrolyzing sphingomyelin into ceramide and choline phosphate.

In conclusion, we have demonstrated a new approach to the study of DND. It has been reported that the DND that occurs in CA1 neurons may be apoptosis [19]. Our results suggest that the sphingomyelin cycle may play a significant role in DND in these cells.

Acknowledgement. This study was supported in part by the Research Grant for Cardiovascular Disease (11C-2) from the Ministry of Health and Welfare and by a grant-in aid for scientific research from the Ministry of Education, Science and Culture of Japan. The authors would like to thank Ms. Noriko Kishino, Ms. Kimi Tatebe, and Ms. Tomomi Iwasawa for their excellent assistance.

Abbreviations

FFA	free fatty acid
DAG	diacylglycerol
MCA	middle cerebral artery
HPTLC	high performance thin-layer chromatography
GLC	gas-liquid chromatography
TLC	thin-layer chromatography
PIP	phosphatidylinositol 4-phosphate
PIP_2	phosphatidylinositol 4,5-bisphosphate
PC	phosphatidylcholine
PE	phosphatidylethanolamine
PS	phosphatidylserine
PUFA	polyunsaturated fatty acid
DND	delayed neuronal death

References

1. Abe K, Kogure K, Yamamoto H, Imazawa M, Miyamoto K (1987) Mechanism of arachidonic acid liberation during ischemia in gerbil cerebral cortex. J Neurochem 48:503–509
2. Bazan NGJ (1970) Effects of ischemia and electroconvulsive shock on free fatty acid pool in the brain. Biochim Biophys Acta 218:1–10
3. Brugg B, Michel PP, Agid Y, Ruberg M (1996) Ceramide induces apoptosis in cultured mesencephalic neurons. J Neurochem 66:733–739
4. Casaccia-Bonnefil P, Aibel L, Chao MV (1996) Central glial and neuronal populations display differential sensitivity to ceramide-dependent cell death. J Neurosci Res 43:382–389
5. Folch J, Lees M, Sloane-Stanley GH (1957) A simple method for the isolation and purification of total lipids from animal tissues. J Biol Chem 226:497–509
6. Hannun YA (1996) Functions of ceramide in coordinating cellular responces to stress. Science 24:1855–1859
7. Hannun YA, Loomis CR, Merrill AHJ, Bell RM (1986) Sphingosine inhibition of protein kinase C activity and phorbol dibutyrate binding in vitro and in human platelets. J Biol Chem 261:12604–12609
8. Hara A, Taketomi T (1983) Detection of D-erythro and L-threo sphingosine bases in preparative sphingosylphosphorylcholine and its N-acylated derivatives and some evidence of their different chemical configurations. J Biochem 94:1715–1718
9. Harel R, Futerman AH (1993) Inhibition of sphingolipid synthesis affects axonal outgrowth in cultured hippocampal neurons. J Biol Chem 268:14476–14481
10. Hauser G, Eichberg J, Gonzalez-Sastre F (1971) Regional distribution of polyphosphoinositides in rat brain. Biochim Biophys Acta 248:87–95
11. Ikeda M, Yoshida S, Busto R, Santiso M, Ginsberg MD (1986) Polyphosphoinositides as a probable source of brain free fatty acids accumulated at the onset of ischemia. J Neurochem 47:123–132
12. Jayadev S, Linardic CM, Hannun YA (1994) Identification of arachidonic acid as a mediator of sphingomyelin hydrolysis in response to tumor necrosis factor. J Biol Chem 269:5757–5763

13. Kirino T (1982) Delayed neuronal death in the gerbil hippocampus following ischemia. Brain Res 239:57–69
14. Kitagawa K, Matsumoto M, Tagaya M, Hata R, Ueda H, Niinobe M, Handa N, Fukunaga R, Kimura K, Mikoshiba K, Kamada T (1990) 'Ischemic tolerance' phenomenon found in the brain. Brain Res 528:21–24
15. Kubota M, Nakane M, Narita K, Nakagomi T, Tamura A, Hisaki H, Shimasaki H, Ueta N (1998) Mild hypothermia reduces the rate of metabolism of arachidonic acid following postischemic reperfusion. Brain Res 779:297–300
16. Kubota M, Narita K, Nakagomi T, Tamura A, Shimasaki H, Ueta N, Yoshida S (1996) Sphingomyelin changes in rat cerebral cortex during focal ischemia. Neurol Res 18:337–341
17. Lee C, Hajra AK (1991) Molecular species of diacylglycerols and phosphoglycerides and the postmortem changes in the molecular species of diacylglycerols in rat brains. J Neurochem 56:370–379
18. Nakano S, Kogure K, Abe K, Yae T (1990) Ischemia-induced alterations in lipid metabolism of the gerbil cerebral cortex: I. Changes in free fatty acid liberation. J Neurochem 54:1911–1916
19. Nitatori T, Sato N, Waguri S, Karasawa Y, Araki H, Shibanai K, Kominami E, Uchiyama Y (1995) Delayed neuronal death in the CA1 pyramidal cell layer of the gerbil hippocampus following transient ischemia is apoptosis. J Neurosci 15:1001–1011
20. Obeid LM, Linardic CM, Karolak LA, Hannun YA (1993) Programmed cell death induced by ceramide. Science 259:1769–1771
21. Schwarz A, Futerman AH (1997) Distinct roles for ceramide and glucosylceramide at different stages of neuronal growth. J Neurosci 17:2929–2938
22. Siesjo BK, Katsura K, Zhao Q, Folbergrova J, Pahlmark K, Siesjo P, Smith ML (1995) Mechanisms of secondary brain damage in global and focal ischemia: a speculative synthesis. J Neurotrauma 12:943–956
23. Stoffel W, Melzner I (1980) Studies in vitro on the biosynthesis of ceramide and sphingomyelin. A re-evaluation of proposed pathways. Hoppe-Seyler's Z Physiol Chem 361:755–771
24. Tamura A, Graham DI, McCulloch J, Teasdale GM (1981) Focal cerebral ischaemia in the rat: 1. Description of technique and early neuropathological consequences following middle cerebral artery occlusion. J Cereb Blood Flow Metab 1:53–60
25. Tamura A, Graham DI, McCulloch J, Teasdale GM (1981) Focal cerebral ischaemia in the rat: 2. Regional cerebral blood flow determined by [^{14}C]iodoantipyrine autoradiography following middle cerebral artery occlusion. J Cereb Blood Flow Metab 1:61–69
26. Yoshida S, Inoh S, Asano T, Sano K, Kubota M, Shimazaki H, Ueta N (1980) Effect of transient ischemia on free fatty acids and phospholipids in the gerbil brain: Lipid peroxidation as a possible cause of postischemic injury. J Neurosurg 53:323–331
27. Yoshida S, Inoh S, Asano T, Sano K, Kubota M, Shimazaki H, Ueta N (1980) Effect of transient ischemia on free fatty acids and phospholipids in the gerbil brain. Lipid peroxidation as a possible cause of postischemic injury. J Neurosurg 53:323–331
28. Zhang JP, Sun GY (1995) Free fatty acids, neutral glycerides, and phosphoglycerides in transient focal cerebral ischemia. J Neurochem 64:1688–1695

Preconditioning of Gerbil Brain Reduces Hippocampal Depolarization Time Following Transient Forebrain Ischemia: Relationship to Hippocampal CA1 Neuron Injury

G. MIES and B. MOMENI

Key words. Global cerebral ischemia – gerbil – ischemic tolerance – electrophysiology – DC potential – hippocampal depolarization time – quantitative morphometry – CA1 neuron injury

Introduction

A brief ischemic episode can increase the resistance of the brain against a second otherwise lethal ischemic event, a phenomenon called "ischemic preconditioning". Induction of ischemic tolerance results in the neuroprotection of selective vulnerable brain regions following global cerebral ischemia [23, 26, 28] or in a decrease of infarct volume after transient focal cerebral ischemia [4, 29, 34, 35, 51, 54]. Various mechanisms have been proposed to play a role in the induction of ischemic tolerance in experimental in-vivo studies, e.g., opening of ATP-sensitive K^+ channels [7, 17, 32], nitric oxide activation of p21-kinase [15], induction of immediate early and stress genes [4, 5, 27] and/or transcriptional factors [4, 6, 8, 22], GFAP activation in astrocytes [10, 21, 25], interleukin-1 upregulation [45], induction of neurotrophic factor [17, 25, 36, 56], and/or a faster postischemic recovery of cerebral protein synthesis following preconditioning [13, 38]. The exact mechanisms involved, however, still remain to be elucidated. In the present study, therefore, the depolarization time of hippocampal DC-potential was correlated with hippocampal CA1 neuron injury in untreated and preconditioned gerbils subjected to various episodes of transient global cerebral ischemia. The intention was to obtain more clues about membrane function possibly involved in selective vulnerability and/or neuroprotection of the hippocampal formation when subjected to an ischemic preconditioning paradigm prior to a second lethal ischemic insult.

Materials and Methods

Experimental procedures were carried out with governmental approval and according to the NIH guidelines for the care and use of laboratory animals. Adult Mon-

Günter Mies, Behrouz Momeni, Max-Planck-Institute for Neurological Research, Department of Experimental Neurology, Cologne, Germany

Correspondence to: Prof. Dr. med. Günter Mies, Max-Planck-Institut für neurologische Forschung, Abteilung für experimentelle Neurologie, Gleueler Str. 50, 50931 Köln, Germany, Tel.: +49 (0)221/4726-216, Fax: +49 (0)221/4726-298, E-Mail: gmies@mpin-koeln.mpg.de

Maturation Phenomenon in Cerebral Ischemia V
A.M. Buchan et al. (Eds.)
© Springer-Verlag Berlin Heidelberg 2004

golian gerbils (Meriones unguiculatus) weighing 50–110 g were divided into the following experimental groups: sham-operated animals (n = 5); animals subjected to global forebrain ischemia for 2, 5, 7.5, and 10 min (n = 5 animals/group); animals pretreated with a 2-min ischemic insult followed by global forebrain ischemia for 5, 7.5, and 10 min (n = 5 animals/group) 24 h later.

Gerbils were anesthetized with 1% halothane (30% O_2, remainder N_2O). Both common carotid arteries (CCA) were exposed and a suture was looped around both arteries. Animals were fixated in a stereotactic holder and a trephine opening was carefully drilled over the left skull (1.5 mm posterior and 3 mm lateral to bregma). A glas microelectrode (tip: 5–10 µm) filled with Ringer's solution was lowered to the hippocampal formation (2.5 mm ventral to bregma) to record the hippocampal DC potential via a miniature calomel electrode and a reference calomel electrode placed over the nose bone using a differential amplifier. Then a LDF probe was positioned next to the microelectrode to verify the onset of ischemia and recirculation. Both signals were registered with a multichannel recorder for later evaluation. Rectal temperature was maintained close to 37.0 °C using a feedback-controlled heating system during surgical preparation of the animals, during the ischemic period, and until 90 min after postischemic recirculation.

Forebrain ischemia was induced for 2–10 min by pulling the sutures placed around the common carotid arteries and recirculation was re-established by release of the sutures, the duration of the ischemic period being verified by the cortical LDF recording. Thirty minutes after the ischemic insult, animals were allowed to recover under normothermic temperature control for further 60 min. Seven days later, gerbils were sacrificed under deep halothane anesthesia and brains were immersion-fixed in 4% paraformaldehyde, embedded in paraffin and cut into 5 µm thick sections at the hippocampal level and stained with cresyl-violet. Quantitative morphometry of hippocampal CA1 neuron injury was achieved by counting the number of intact neurons per mm length of the CA1 sector [16].

Evaluation of LDF, hippocampal DC potential, and quantitative morphometry of hippocampal CA1 neurons were performed in the following way: the onset of critical ischemia was determined as the instantaneous decline of the LDF signal below 30% of control and recirculation latency was defined as the time elapsed between re-opening of the CCAs and return of the LDF signal to 70% of control. The latency time of hippocampal depolarization was defined as the time difference between the onset of CCA occlusion and the decline to 50% of the pre-ischemic DC potential level. The hippocampal DC depolarization time was calculated from the time interval between onset of depolarization and its return to 90% of the control recording.

Statistical evaluations between untreated and preconditioned experimental groups were performed using the Mann-U-Whitney test. A p-value below 0.05 was considered as statistically different. The relationship between hippocampal DC depolarization time and CA1 injury was evaluated using the following equation:

$$y = b \cdot \left\{1 - e^{[-a \cdot (x-c)]}\right\}$$

with the coefficient b equal to the estimated number of intact CA1 neurons, c representing the estimated threshold value of the hippocampal DC-depolarization time for irreversible CA1 neuron damage, and a reflecting the coefficient of the exponential curve.

Results

In all animals, bilateral CCA occlusion resulted in forebrain ischemia as indicated by the decline in the cortical LDF signal to ~ 20% of the control value. Following release of CCA occlusion after ischemia times from 2–10 min, the LDF signal returned to 70% of control within 1.2–3.4 min, revealing no statistical difference between the untreated and preconditioned animal groups.

As shown in Table 1, ischemic preconditioning did not affect the latency time of the hippocampal DC potential which suggests a similar depletion rate of energy-rich metabolites in untreated and preconditioned animals. Ischemic preconditioning of gerbils, however, resulted in a significant reduction of the hippocampal DC-depolarization time, indicating a faster recovery of the cell membrane potential after the second ischemic insult (Fig. 1). The shortening of cell membrane depolarization time, in turn, was associated with a significant increase in surviving hippocampal CA1 neurons (Table 2).

Figure 2 compares the relationship between hippocampal DC-depolarization time and the number of surviving CA1 neurons in preconditioned and untreated animals. Regression analysis of these data revealed similar threshold values for injury in the untreated (16.6±0.2 min, mean±standard error of the estimate, not shown) and in the preconditioned group (16.9±0.6 min, not shown). Thus, the mean DC-depolarization time for induction of CA1 damage is 16.7±0.3 min (Fig. 2). This suggests that induction of ischemic tolerance in hippocampal CA1 neurons by ischemic preconditioning is due to the faster depolarization time but not to an improved tolerance to depolarization per se.

Discussion

In the present study, an established experimental model of transient global forebrain ischemia was used, i.e. the bilateral common carotid artery occlusion in the Mongolian gerbil [26]. To account for the exact duration of the ischemia-induced pathophysiological impact, the hippocampal DC-depolarization time, i.e., the duration of disturbed ion homeostasis, was measured [2, 3]. As is evident from this study, increased hippocampal CA1 neuron survival in preconditioned animals was

Table 1. Hippocampal DC potential latency time in untreated and preconditioned gerbils after various periods of transient forebrain ischemia

Time of CCA occlusion	Untreated gerbils (min)	Preconditioned gerbils (min)
2 min	1.1±0.2	–
5 min	0.8±0.3	1.1±0.5
7.5 min	1.2±0.5	1.0±0.4
10 min	1.0±0.6	1.3±0.5

Values are provided as means±SD. CCA, common carotid artery. Note that ischemic preconditioning (2-min CCA occlusion 24 h prior to second ischemia) did not affect the duration of hippocampal DC-potential latency during the second ischemic episode.

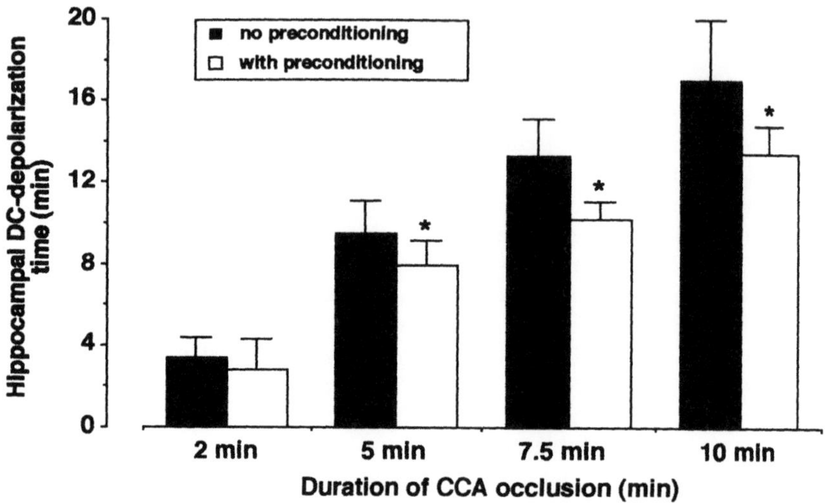

Fig. 1. Hippocampal DC-depolarization time measured in gerbils subjected to 2, 5, 7.5, and 10 min bilateral common carotid artery (CCA) occlusion. Black bars represent experimental groups without preconditioning, white bars with preconditioning (2-min CCA occlusion 24 h prior to second ischemia). Preconditioning induced a significant reduction in hippocampal DC-depolarization time (*p <0.05; values are provided as means±SD)

Table 2. Hippocampal CA1-neuron counting in untreated and preconditioned gerbils after various periods of transient forebrain ischemia

Time of CCA occlusion	Untreated gerbils (CA1 neurons/mm)	Preconditioned gerbils (CA1 neurons/mm)
5 min	193±30	230±25
7.5 min	41±45	200±34 **
10 min	62±48	162±37 *

Values are provided as means±SD. CCA, common carotid artery. Note that ischemic preconditioning (2-min CCA occlusion 24 h prior to second ischemia) increased significantly the number of surviving hippocampal CA1 neurons. **p < 0.01, *p < 0.05

associated with a significantly faster repolarization. Induction of ischemic tolerance, however, did not elevate the resistance of hippocampal CA1 neurons to depolarization per se, i.e., the duration of depolarization required to induce injury was the same in non- and preconditioned animals.

Several paradigms have been shown to increase ischemic tolerance of the brain and to provide subsequent neuroprotection against a secondary otherwise lethal ischemic episode such as hypothermia [9, 42, 43] or spreading depression [24, 35], a brief sublethal period of hypoxia/ischemia [23, 28], or a variety of pharmacological compounds [29, 33, 39, 41, 44, 48, 52]. The common trigger of these preconditioning manoeuvres to induce neuroprotection is a brief nonlethal exposure of the brain to oxidative stress interfering with mitochondrial [6, 44, 54] and/or with en-

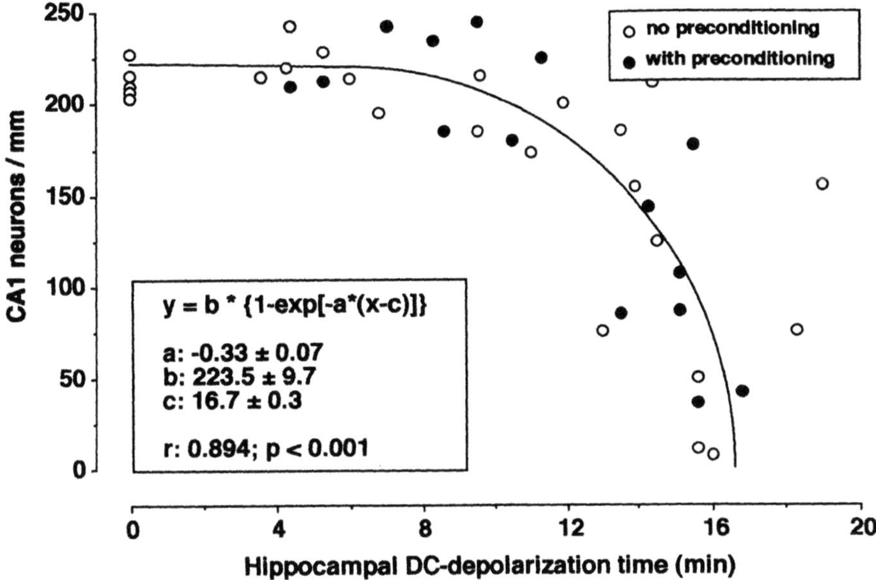

Fig. 2. Relationship between hippocampal DC-depolarization time and the number of intact hippocampal CA1 neurons. A combined threshold value for the hippocampal DC-depolarization time of 16.7±0.3 min was estimated for irreversible CA1 neuron injury. Since no significant difference of the threshold value was found between that measured in animals with and without preconditioning, ischemic preconditioning does not seem to increase the overall resistance of hippocampal CA1 neurons against cerebral ischemia

doplasmatic reticulum (ER) functions [46]. Most importantly, the onset of the neuroprotective effect after cerebral preconditioning occurs after some delay and then lasts for several days. The activation of NMDA receptors and calcium influx-induced free radical generation appear to be necessary for the induction and the manifests of ischemic tolerance [14, 18]. Preconditioning, however, manifestates most robustly after 24 h and later, suggesting that transcriptional [4, 5, 27] and sustained translational responses of brain tissue [4, 6, 8, 10, 21, 22, 25, 45] most likely account for this kind of prolonged neuroprotective state. Following the second ischemic insult, ischemic tolerance of the brain is associated with a faster postischemic restoration of the translational capacity [13, 38, 47] leading to a more rapid cellular availability of survival-associated proteins [20, 40, 49, 51].

The results of the present study provide a new explanation for preconditioning-induced neuroprotection of the hippocampal formation. The improved morphological outcome was due to the faster postischemic repolarization, i.e. a more efficient restoration of ion homeostasis. In this context, mechanisms of tolerance should be addressed that accelerate this process: – postischemic hemodynamics to optimize oxygen delivery; – the generation of energy equivalents by improving mitochondrial function; – optimization of postischemic ion pump efficacy by preventing membrane currents and futile ion cycling; – reduction of energy expenditure for repair of ischemic injury.

As regards postischemic hemodynamics, our measurements did not reveal a difference. However, it cannot be excluded that maintenance of microcirculation, as observed in tolerance to focal ischemia after LPS-pretreatment [11], may have contributed to a faster recovery of oxidative metabolism.

Another important issue could be an improved efficacy in energy utilization. As indicated by our findings on similar latencies of the hippocampal depolarization, the rate of energy metabolite depletion was not affected by ischemic preconditioning. This is in line with earlier observations that ATP-PCr depletion rate and ATP-PCr regeneration did not differ between non- and preconditioned animals. In addition, no alteration in the activity of pyruvate dehydrogenase (PDH), an important intramitochondrial enzyme in the regulation of oxidative metabolism, was observed between experimental groups [19]. It has been speculated that preconditioning may reduce free radical production within the mitochondrial compartment by enhanced coupling to antioxidant defense mechanisms [37] which may then improve oxygen metabolism as suggested by a delayed depletion of intracerebral oxygen [31]. Recently, the activation of mitochondrial ATP-sensitive potassium channels has been discussed to play an important role in the manifestation of ischemic preconditioning by preventing cytochrome c release and thus failure of mitochondrial function via depolarization rather than hyperpolarization of the mitochondrial membrane [17, 32].

Since ischemic preconditioning was shown to induce GFAP activation in astrocytes [10, 21, 25], neuron-glial and/or glial-glial interactions have been suggested to promote ischemic tolerance of the brain. The major function of astrocytes is to maintain the local microenvironment within physiologic limits, i.e., re-uptake and inactivation of neurotransmitters, involvement in blood-brain barrier (BBB) control, and maintenence of water and ion homeostasis [1, 53]. Of particular interest are changes in cellular Na^+/K^+-ATPase activity, because ischemic injury triggers an oxidative modification of Na^+/K^+-ATPase oligomeric structure formation resulting in a subsequent reduction in ion pump activity [12]. Following ischemic preconditioning, however, no disturbance of Na^+/K^+-ATPase activity is observed after brain ischemia suggesting a faster restoration of ion pump function during the postischemic phase [55]. Another mechanism for improved astrocytic Na^+/K^+-ATPase activity could be associated with voltage-dependent Na^+-channels activation known to fuel Na^+/K^+-ATPase for glial homeostasis [50]. An increase in astrocytic Na^+/K^+-ATPase activity may then facilitate K^+ uptake from the extracellular space and thereby reduced the duration of hippocampal depolarization after induction of ischemic tolerance.

An important finding of our study is the observation that preconditioning accelerates postischemic repolarization, but does not modulate the threshold relationship between the duration of depolarization and the manifestation of neuronal injury. This explains that the time window of the protective effect is rather small: with preconditioning, the critical depolarization time of 16.7 min was reached following an ischemia time of about 12 min, and without preconditioning after about 10 min. The expansion of the tolerable ischemia time after preconditioning is therefore just 2 min, i.e. a gain which is equally well achieved by mild hyperthermia or other metabolic inhibitors which delay the onset of depolarization. This suggests that cumulative damage of hippocampal CA1 neurons occurred with increasing length of membrane depolarization by deleterious mechanisms which prevail over cellular defense mechanisms in a threshold-dependent manner. Possible pathomechanisms are the well-documented increase in fatty acids and free radicals known to inhibit Na^+/K^+-ATPase following ischemia [30] which, in concert with prolonged disturbance of intracellular and ER-calcium homeostasis [46] may result in the initiation of neuronal cell injury.

Conclusions

Ischemic preconditioning of the hippocampal formation resulted in a faster recovery of the hippocampal DC-potential during the second otherwise lethal ischemic insult shifting the threshold value of critical hippocampal DC-depolarization time to a non-critical one. This suggest that the exposure time of brain tissue during membrane depolarization to ion derangements determines the severity of the ischemic insult within the hippocampal formation in a threshold-dependent manner and that ischemic preconditioning does not increase the resistance of hippocampal CA1 neurons against an established ischemic insult. The observed reduction in hippocampal DC depolarization time suggest an improved ion clearance in preconditioned hippocampal formation probably related to a concerted action of improved ATP-availability at ionic pumps, opening of ATP-sensitive K^+ channels, and improved Na^+/K^+-ATPase activity.

References

1. Abbott NJ, Revest PA, Romero IA (1992) Astrocyte-endothelial interaction: physiology and pathology [Review] [49 refs]. Neuropathology & Applied Neurobiology 18:424–433
2. Abe H, Nowak TS (1996) Gene expression and induced ischemic tolerance following brief insults. Acta Neurobiologiae Experimentalis 56:3–8
3. Adachi N, Chen JF, Liu KY, Tsubota S, Arai T (1998) Dexamethasone aggravates ischemia-induced neuronal damage by facilitating the onset of anoxic depolarization and the increase in the intracellular CA2+ concentration in gerbil hippocampus. J Cereb Blood Flow Metab 18:274–280
4. Barone FC, White RF, Spera PA, Ellison J, Currie RW, Wang XK, Feuerstein GZ (1998) Ischemic preconditioning and brain tolerance – temporal histological and functional outcomes, protein synthesis requirement, and interleukin-1 receptor antagonist and early gene expression. Stroke 29:1937–1951
5. Belayev L, Ginsberg MD, Alonso OF, Singer JT, Zhao W, Busto R (1996) Bilateral ischemic tolerance of rat hippocampus induced by prior unilateral transient focal ischemia: relationship to c-fos mRNA expression. Neuroreport 8:55–59
6. Bergeron M, Gidday JM, Yu AY, Semenza GL, Ferriero DM, Sharp FR (2000) Role of hypoxia-inducible factor-1 in hypoxia-induced ischemic tolerance in neonatal rat brain. Ann Neurol 48:285–296
7. Blondeau N, Plamondon H, Richelme C, Heurteaux C, Lazdunski M (2001) K-ATP channel openers, adenosine agonists and epileptic preconditioning are stress signals inducing hippocampal neuroprotection. Neuroscience 100:465–474
8. Clemens JA, Stephenson DT, Dixon EP, Smalstig EB, Mincy RE, Rash KS, Little SP (1997) Global cerebral ischemia activates nuclear factor-kappa-b prior to evidence of DNA fragmentation. Mol Brain Res 48:187–196
9. Corbett D, Crooks P (1997) Ischemic preconditioning: A long term survival study using behavioural and histological endpoints. Brain Res 760:129–136
10. Currie RW, Ellison JA, White RF, Feuerstein GZ, Wang X, Barone FC (2000) Benign focal ischemic preconditioning induces neuronal Hsp70 and prolonged astrogliosis with expression of Hsp27. Brain Res 863:169–181

11. Dawson DA, Furuya K, Gotoh J, Nakao Y, Hallenbeck JM (1999) Cerebrovascular hemodynamics and ischemic tolerance: lipopolysaccharide-induced resistance to focal cerebral ischemia is not due to changes in severity of the initial ischemic insult, but is associated with preservation of microvascular perfusion. J Cereb Blood Flow Metab 19:616–623

12. Dobrota D, Matejovicova M, Kurella EG, Boldyrev AA (1999) Na/K-ATPase under oxidative stress: molecular mechanisms of injury. Cellular & Molecular Neurobiology 19:141–149

13. Furuta S, Ohta S, Hatakeyama T, Nakamura K, Sakaki S (1993) Recovery of protein synthesis in tolerance-induced hippocampal CA1 neurons after transient forebrain ischemia. Acta Neuropathol 86:329–336

14. Gidday JM, Shah AR, Maceren RG, Wang QO, Pelligrino DA, Holtzman DM, Park TS (1999) Nitric oxide mediates cerebral ischemic tolerance in a neonatal rat model of hypoxic preconditioning. J Cereb Blood Flow Metab 19:331–340

15. Gonzalez-Zulueta M, Feldman AB, Klesse LJ, Kalb RG, Dillman JF, Parada LF, Dawson TM, Dawson VL (2000) Requirement for nitric oxide activation of p21(ras)/extracellular regulated kinase in neuronal ischemic preconditioning. Proc Natl Acad Sci USA 97:436–441

16. Hallmayer J, Hossmann K-A, Mies G (1985) Low dose of barbiturates for prevention of hippocampal lesions after brief ischemic episodes. Acta Neuropathol (Berl) 68:27–31

17. Heurteaux C, Lauritzen I, Widmann C, Lazdunski M (1995) Essential role of adenosine, adenosine A1 receptors, and ATP-sensitive K^+ channels in cerebral ischemic preconditioning. Proc Natl Acad Sci USA 92:4666–4670

18. Kasischke K, Ludolph AC, Riepe MW (1996) NMDA-antagonists reverse increased hypoxic tolerance by preceding chemical hypoxia. Neurosci Lett 214:175–178

19. Katayama Y, Muramatsu H, Kamiya T, McKee A, Terashi A (1997) Ischemic tolerance phenomenon from an approach of energy metabolism and the mitochondrial enzyme activity of pyruvate dehydrogenase in gerbils. Brain Res 746:126–132

20. Kato H, Chen T, Liu XH, Nakata N, Kogure K (1993) Immunohistochemical localization of ubiquitin in gerbil hippocampus with induced tolerance to ischemia. Brain Res 619:339–343

21. Kato H, Kogure K, Araki T, Itoyama Y (1994) Astroglial and microglial reactions in the gerbil hippocampus with induced ischemic tolerance. Brain Res 664:69–76

22. Kato H, Kogure K, Araki T, Itoyama Y (1995) Induction of Jun-like immunoreactivity in astrocytes in gerbil hippocampus with ischemic tolerance. Neurosci Lett 189:13–16

23. Kato H, Liu K, Araki T, Kogure K (1991) Temporal profile of the effect of pretreatment with brief cerebral ischemia on the neuronal damage following secondary ischemic insult in the gerbil: cumulative damage and protective effects. Brain Res 553:238–242

24. Kawahara N, Croll SD, Wiegand SJ, Klatzo I (1997) Cortical spreading depression induces long-term alterations of BDNF levels in cortex and hippocampus distinct from lesion effects: implications for ischemic tolerance. Neurosci Res 29:37–47

25. Kawahara N, Ruetzler CA, Mies G, Klatzo I (1999) Cortical spreading depression increases protein synthesis and upregulates basic fibroblast growth factor. Exp Neurol 158:27–36

26. Kirino T, Tsujita Y, Tamura A (1991) Induced tolerance to ischemia in gerbil hippocampal neurons. J Cereb Blood Flow Metab 11:299–307

27. Kitagawa K, Matsumoto M, Kuwabara K, Tagaya M, Ohtsuki T, Hata R, Ueda H, Handa N, Kimura K, Kamada T (1991) 'Ischemic tolerance' phenomenon detected in various brain regions. Brain Res 561:203–211

28. Kitagawa K, Matsumoto M, Tagaya M, Hata R, Ueda H, Niinobe M, Handa N, Fukunaga R, Kimura K, Mikoshiba K et al (1990) 'Ischemic tolerance' phenomenon found in the brain. Brain Res 528:21–24

29. Kuroiwa T, Yamada I, Endo S, Hakamata Y, Ito U (2000) 3-Nitropropionic acid preconditioning ameliorates delayed neurological deterioration and infarction after transient focal cerebral ischemia in gerbils. Neurosci Lett 283:145–148

30. Lees GJ (1991) Inhibition of sodium-potassium-ATPase: a potentially ubiquitous mechanism contributing to central nervous system neuropathology. Brain Res Rev 16:283–300

31. Li JY, Ueda H, Seiyama A, Nakano M, Matsumoto M, Yanagihara T (1997) A near-infrared spectroscopic study of cerebral ischemia and ischemic tolerance in gerbils. Stroke 28:1451–1456

32. Liu D, Lu C, Wan R, Auyeung WW, Mattson MP (2002) Activation of mitochondrial ATP-dependent potassium channels protects neurons against ischemia-induced death by a mechanism involving suppression of bax translocation and cytochrome c release. J Cereb Blood Flow Metab 22:431–443

33. Masada T, Xi G, Hua Y, Keep RF (2000) The effects of thrombin preconditioning on focal cerebral ischemia in rats. Brain Res 867:173–179

34. Matsushima K, Hakim AM (1995) Transient forebrain ischemia protects against subsequent focal cerebral ischemia without changing cerebral perfusion. Stroke 26:1047–1052

35. Matsushima K, Hogan MJ, Hakim AM (1996) Cortical spreading depression protects against subsequent focal cerebral ischemia in rats. J Cereb Blood Flow Metab 16:221–226

36. Matsushima K, Schmidtkastner R, Hogan MJ, Hakim AM (1998) Cortical spreading depression activates trophic factor expression, in neurons and astrocytes and protects against subsequent focal brain ischemia. Brain Res 807:47–60

37. Murakami K, Kondo T, Kawase M, Li YB, Sato S, Chen SF, Chan PH (1998) Mitochondrial susceptibility to oxidative stress exacerbates cerebral infarction that follows permanent focal cerebral ischemia in mutant mice with manganese superoxide dismutase deficiency. J Neurosci 18:205–213

38. Nakagomi T, Kirino T, Kanemitsu H, Tsujita Y, Tamura A (1993) Early recovery of protein synthesis following ischemia in hippocampal neurons with induced tolerance in the gerbil. Acta Neuropathol 86:10–15

39. Nakase H, Heimann A, Uranishi R, Riepe MW, Kempski O (2000) Early-onset tolerance in rat global cerebral ischemia induced by a mitochondrial inhibitor. Neurosci Lett 290:105–108

40. Nakata N, Kato H, Kogure K (1993) Inhibition of ischaemic tolerance in the gerbil hippocampus by quercetin and anti-heat shock protein-70 antibody. NeuroReport 4:695–698

41. Nawashiro H, Tasaki K, Ruetzler CA, Hallenbeck JM (1997) TNF-alpha pretreatment induces protective effects against focal cerebral ischemia in mice. J Cereb Blood Flow Metab 17:483–490

42. Nishio S, Chen ZF, Yunoki M, Toyoda T, Anzivino M, Lee KS (1999) Hypothermia-induced ischemic tolerance. Ann NY Acad Sci 890:26–41

43. Nishio S, Yunoki M, Chen ZF, Anzivino MJ, Lee KS (2000) Ischemic tolerance in the rat neocortex following hypothermic preconditioning. J Neurosurg 93:845–851

44. Ohtsuki T, Matsumoto M, Kuwabara K, Kitagawa K, Suzuki K, Taniguchi N, Kamada T (1992) Influence of oxidative stress on induced tolerance to ischemia in gerbil hippocampal neurons. Brain Res 599:246–252

45. Ohtsuki T, Ruetzler CA, Tasaki K, Hallenbeck JM (1996) Interleukin-1 mediates induction of tolerance to global ischemia in gerbil hippocampal CA1 neurons. J Cereb Blood Flow Metab 16:1137–1142

46. Paschen W, Mengesdorf T, Althausen S, Hotop S (2001) Peroxidative stress selectively downregulates the neuronal stress response activated under conditions of endoplasmic reticulum dysfunction. J Neurochem 76:1916–1924

47. Paschen W, Mies G (1999) Effect of induced tolerance on biochemical disturbances in hippocampal slices from the gerbil during and after oxygen glucose deprivation. NeuroReport 10:1417–1421

48. Puisieux F, Deplanque D, Pu Q, Souil E, Bastide M, Bordet R (2000) Differential role of nitric oxide pathway and heat shock protein in preconditioning and lipopolysaccharide-induced brain ischemic tolerance. Eur J Pharmacol 389:71–78

49. Shimazaki K, Ishida A, Kawai N (1994) Increase in bcl-2 oncoprotein and the tolerance to ischemia-induced neuronal death in the gerbil hippocampus. Neurosci Res 20:95–99

50. Sontheimer H, Black JA, Waxman SG (1996) Voltage-gated Na+ channels in glia – properties and possible functions. Trends Neurosci 19:325–331

51. Toyoda T, Kassell NF, Lee KS (1997) Induction of ischemic tolerance and antioxidant activity by brief focal ischemia. NeuroReport 8:847–851

52. Toyoda T, Kassell NF, Lee KS (2000) Induction of tolerance against ischemia/reperfusion injury in the rat brain by preconditioning with the endotoxin analog diphosphoryl lipid A. J Neurosurg 92:435–441

53. Walz W (1989) Role of glial cells in the regulation of the brain ion microenvironment. Prog Neurobiol 33:309–333

54. Wiegand F, Liao W, Busch C, Castell S, Knapp F, Lindauer U, Megow D, Meisel A, Redetzky A, Ruscher K, Trendelenburg G, Victorov I, Riepe M, Diener HC, Dirnagl U (1999) Respiratory chain inhibition induces tolerance to focal cerebral ischemia. J Cereb Blood Flow Metab 19:1229–1237

55. Wyse ATD, Streck EL, Worm P, Wajner A, Ritter F, Netto CA (2000) Preconditioning prevents the inhibition of Na+,K+-ATPase activity after brain ischemia. Neurochem Res 25:971–975

56. Yanamoto H, Mizuta I, Nagata I, Xue JH, Zhang Z, Kikuchi H (2000) Infarct tolerance accompanied enhanced BDNF-like immunoreactivity in neuronal nuclei. Brain Res 877:331–344

The Expression of MMPs in CNS Injury has Beneficial Roles

V. WEE YONG and P. LARSEN

Many matrix metalloproteinase (MMP) members are up-regulated following various CNS injuries including stroke. The increased expression of these MMPs is thought to be detrimental. In this regard, MMPs have been found to disrupt the blood-brain barrier, to be directly cytotoxic to neurons, and to promote inflammation and injury. This talk will highlight potential beneficial functions of MMPs following their expression or up-regulation in CNS injuries.

We demonstrate that MMPs are required for re-myelination following a demyelinating event, and that MMPs may have a critical function in degrading inhibitory proteoglycan extracellular matrix molecules, whose increased expression following injury may be inhibitory for various reparative events. The indiscriminate inhibition of MMPs following events such as stroke may counter attempts at regeneration following CNS injury.

Maturation Phenomenon in Cerebral Ischemia V
A. M. Buchan et al. (Eds.)
© Springer-Verlag Berlin Heidelberg 2004

VI Neurogenesis Stem Cell Activation – Repair and Plasticity

Postischemic Hypothermia Fails to Improve Outcome after a Striatal Hemorrhage in Rats

C. L. MacLellan, A. Shuaib, and F. Colbourne

Key words. Hemorrhage – stroke – hypothermia – striatum – behavior

Summary. Prolonged cooling reduces ischemic brain injury in rodents even when hypothermia is delayed for hours. Presently, we assessed whether hypothermia would reduce functional deficits and the volume of brain damage following an intracerebral hemorrhage (ICH) that was produced by an injection of collagenase into the striatum of rats. Mild hypothermia (33 °C for 24 h and then 35 °C for 24 h), measured with implanted core temperature telemetry probes, was induced starting one hour after collagenase injection. Other rats remained normothermic. The normothermic ICH rats had moderate but statistically significant contralateral reaching deficits, measured in the Montoya staircase test, and a 23-mm^3 volume of tissue lost at 30 days. Hypothermia did not significantly affect either the reaching impairment or the volume of injury (22 mm^3). Accordingly, early hypothermia will not likely be useful in salvaging tissue that is rapidly lost after a hemorrhagic stroke.

Introduction

Intracerebral hemorrhage is one of the most devastating types of stroke, and accounts for approximately 10–15% of strokes in Western populations [21]. Neurological deficits result from direct tissue destruction, space-occupying effects of the hematoma, and edema [27]. Clinical therapies tested include aspiration of the hematoma [2, 3, 15] and reduction of elevated intracranial pressure (ICP [12]) among others. Nonetheless, an ICH remains a clinically difficult problem to treat and despite many experimental therapies none have provided much benefit.

Delayed hypothermia reduces ischemic injury and promotes functional recovery in rodents [7, 13]. For instance, prolonged hypothermia (e.g. 48 h) persistently de-

Crystal L. MacLellan[1], Ashfaq Shuaib[2] and Fred Colbourne[1]
[1] Department of Psychology, Faculty of Science, University of Alberta, Edmonton, Alberta, Canada, T6G 2E9
[2] Division of Neurology, Department of Medicine, University of Alberta, Edmonton, Alberta, Canada, T6G 2B7

Correspondence to: Dr. Fred Colbourne, Department of Psychology, Faculty of Science, University of Alberta, Edmonton, Alberta, Canada, T6G 2E9
Tel.: 780-492-5268, Fax: 780-492-1768, E-Mail: fcolbour@ualberta.ca

Maturation Phenomenon in Cerebral Ischemia V
A. M. Buchan et al. (Eds.)
© Springer-Verlag Berlin Heidelberg 2004

creases CA1 cell death and functional deficits (e.g., working memory errors) after global ischemia in both rats [6] and gerbils [4, 8]. In addition, delayed and prolonged hypothermia reduces infarct size and skilled reaching deficits following middle cerebral artery occlusion in rats [5, 9].

In this study, we hypothesized that prolonged mild hypothermia, as previously found to be especially effective for treating global and focal ischemia would reduce the functional deficits and histological damage of a relatively mild hemorrhagic stroke in rat [18]. Hemorrhagic stroke was produced in the striatum by an infusion of bacterial collagenase, a model pioneered by Rosenberg and colleagues [24].

Materials and Methods

Behavioral Training. Wistar rats were obtained from Charles River (Montreal, Quebec, Canada). Rats were food deprived to 90% of their free-feeding weight, which was between 250 and 300 g. They were then trained in the Montoya staircase test, which measures independent forelimb reaching ability [22], over 15 days (two 15-min trials per day separated by about 4 h, five days per week) [5]. Rats were required to reach at least 9 pellets per side (out of a maximum of 21 available) on three consecutive days during training or they were excluded from the study.

Surgery. Core temperature telemetry probes (model TA10TA-F40, Data Sciences, St. Paul, MN, USA) were implanted under isoflurane anesthesia (1.5–2% maintenance in 70% N_2O and 30% O_2) into the peritoneal cavity six days after the end of staircase training. Rats were then housed individually upon receivers (RPC-1, Data Sciences) interfaced to a computer running DataSciences telemetry software (ART version 2.2), which sampled temperature twice per minute. The average temperature on the day prior to ICH or sham surgery served as a baseline.

Days later, rats were anesthetized with isoflurane (1.5–2% maintenance) and placed in a stereotactic frame. Core temperature was maintained near 37 °C (normothermia) with a heating pad during the brief period of anesthesia. After a midline scalp incision, a small burr hole was made 3 mm contralateral to preferred paw (as determined by performance in the staircase test; random if no clear preference) and 0.2 mm posterior to bregma. Over five minutes 0.7 µl of sterile saline containing 0.14 U bacterial collagenase (to produce ICH) or sterile saline alone (in the sham procedure) was infused into the striatum using a 28-gauge needle lowered 6 mm below the surface of the skull. The needle remained in place for an additional five minutes. All rats were shaved (abdomen and back) to facilitate hypothermia treatment and to prevent knowledge of group identity in the normothermic groups.

Post-operative Temperature Control. The ICH (n=8) and SHAM (n=5) animals were allowed to freely regulate their own temperature and did remain normothermic. Starting one hour after ICH induction, ICH+HYPO (n=8) and SHAM+HYPO (n=6) groups were slowly cooled by a rate of 2 °C per hour to a core temperature of 33 °C and kept at this level for 24 h. Rats were then slowly warmed to 35 °C and maintained at this temperature for an additional 24 h. Following this, core temperature was maintained between 36 and 37 °C for one day. Temperature was measured for 7 days following surgery and the telemetry probes were surgically removed the following day under isoflurane anesthesia.

Behavioral Testing. Rats were again food deprived to 90% of their free-feeding weight. They were then tested twice per day on days 21 to 25 after ICH/SHAM surgery. The total number of pellets retrieved on each side was recorded for each trial and the average over trials was analyzed with ANOVA and planned comparisons.

Assessment of Striatal Injury

Rats were killed 30 days after ICH or sham operation with an overdose of sodium pentobarbital and perfusion fixed with 10% formalin. 40-µm sections were taken with a cryostat starting at +1.7 mm to bregma and extending back to –4.8 mm. Sections were taken every 200 µm and stained with Cresyl Violet. The volume of lesion was quantified using Scion Image J 4.0 as follows and analyzed with a one-way ANOVA:

Volume of infarct = volume of remaining contralateral hemisphere – volume of remaining ipsilateral (injured) hemisphere.

Volume of a hemisphere = average (area of entire hemisphere – area of ventricle – area of damage) × distance between sections × number of sections analyzed.

Results

All groups performed similarly on the training phase of the staircase test (Table 1). SHAM and SHAM + HYPO groups were combined for analysis as there was no difference in pellet retrieval with the ipsilateral or contralateral forelimb (data not shown). ICH and ICH + HYPO groups obtained significantly fewer pellets than SHAMs ($p < 0.05$) with the contralateral forelimb (Table 1). Hypothermia did not lessen this deficit ($p = 0.278$). No impairments in the ipsilateral forelimb were detected in the ICH and ICH + HYPO groups ($p \geq 0.365$).

Injection of bacterial collagenase resulted in a mean total tissue loss of 23.1 mm^3 ± 12.4 SD (Fig. 1). Damage was greatest in the striatum, but also in-

Table 1. Average (± SD) performance (number of pellets retrieved) in the staircase test for the last 10 trials of training, when rats were near asymptotic performance, and the 10 test trials (* = p < 0.05 vs. SHAM). Hypothermia did not significantly affect the number of pellets retrieved with the contralateral forelimb during testing

Group	Training (ipsilateral limb)	Training (contralateral limb)	Testing (ipsilateral limb)	Testing (contralateral limb)
SHAM (total)	15.1 ± 2.5	17.2 ± 1.8	15.3 ± 2.5	16.1 ± 2.2
ICH	16.8 ± 1.9	19.1 ± 2.3	13.9 ± 4.3	11.4 ± 5.3 *
ICH+HYPO	15.9 ± 2.2	16.7 ± 2.0	13.9 ± 2.6	9.3 ± 3.8 *

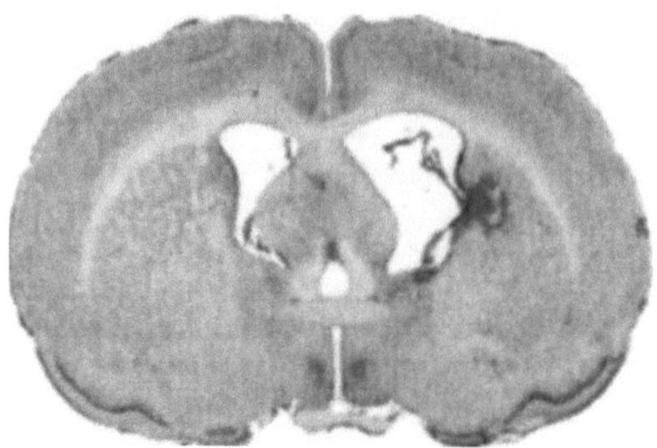

Fig. 1. Photomicrograph showing the extent of subcortical injury in a representative coronal section of a rat injected with collagenase (right hemisphere). Injury was primarily to the striatum but also extended to affect other structures (e.g., thalamus). Ipsilateral ventricular dilation was common (shown). Delayed cooling had no effect on the volume of lesion

cluded the thalamus, globus pallidus and the internal capsule. Mild hypothermia treatment did not significantly lessen the volume of tissue lost (22.2 mm^3 ±9.5 SD; p = 0.982 vs. ICH). Aside from the needle tract, SHAM rats did not sustain any damage.

Discussion

In contrast to other findings of benefit after focal and global ischemia, hypothermia failed to reduce tissue damage and the skilled reaching deficits that accompanied the unilateral striatal hemorrhage. Although unexpected, these results are consistent with lack of protection (i.e. no histological or skilled reaching improvement) after other experimental treatments for ICH [1, 11, 23].

In this study we intentionally used a mild insult (i.e. small volume of tissue damage) as it was expected to maximize the possibility of detecting beneficial effects of hypothermia treatment. Indeed, the skilled reaching impairments were comparable to those seen after a previous focal ischemia study where hypothermia did substantially improve outcome [5]. Therefore, we do not suspect that the hemorrhagic lesion was too severe to treat. Conversely, the injury may not have been severe enough to properly assess hypothermia. For example, cooling may help after more severe insults that produce edema and elevated intracranial pressure as both are known to be amenable to hypothermia [17, 20, 26]. Nonetheless, several studies show that the striatum is difficult to rescue after focal ischemia treated with hypothermia [9, 14, 16, 19]. Therefore, hemorrhagic insults within other brain structures, such as the cortex, might be amenable to hypothermia treatment.

Del Bigio and colleagues [10] have shown that much tissue damage occurs quickly over several hours after ICH. Presumably, therefore, early cooling would be considered essential to optimizing protection. Thus, we induced hypothermia 1 h after the infusion of collagenase. However, bleeding would be occurring at the time these animals became hypothermic. This early hypothermia may have actually aggravated this bleeding. Indeed, hypothermia-induced coagulation abnormalities are well known [25]. Therefore, the deleterious effects on clotting might have countered any beneficial effects. If true, then a greater intervention delay, such as after the bleeding has stopped, might improve outcome since later deleterious events (e.g., edema) would still be treated. Currently, we are investigating the possibility.

In summary, over 48 h of mild hypothermia failed to lessen the skilled reaching impairments after striatal hemorrhagic stroke in rats and failed to significantly affect the volume of tissue lost. Further studies should investigate whether more delayed cooling, such as one that avoids aggravating the initial bleed, is cytoprotective. As well, hypothermia may reduce secondary complications (e.g., edema) and behavioral impairments after more severe insults.

Acknowledgement. The authors gratefully acknowledge research support by the Alberta Heritage Foundation for Medical Research (AHFMR). F. Colbourne is the recipient of an AHFMR scholar award.

References

1. Altumbabic M, Peeling J, Del Bigio MR (1998) Intracerebral hemorrhage in the rat: effects of hematoma aspiration. Stroke 29:1917–1922
2. Auer LM, Deinsberger W, Niederkorn K, Gell G, Kleinert R, Schneider G, Holzer P, Bone G, Mokry M, Korner E et al (1989) Endoscopic surgery versus medical treatment for spontaneous intracerebral hematoma: a randomized study. J Neurosurg 70:530–535
3. Batjer HH, Reisch JS, Allen BC, Plaizier LJ, Su CJ (1990) Failure of surgery to improve outcome in hypertensive putaminal hemorrhage. A prospective randomized trial. Arch Neurol 47:1103–1106
4. Colbourne F, Corbett D (1995) Delayed postischemic hypothermia: a six month survival study using behavioral and histological assessments of neuroprotection. J Neurosci 15:7250–7260
5. Colbourne F, Corbett D, Zhao Z, Yang J, Buchan AM (2000) Prolonged postischemic hypothermia: a long-term outcome study in the rat middle cerebral artery occlusion model. J Cereb Blood Flow Metab 20:1702–1708
6. Colbourne F, Li H, Buchan AM (1999) Indefatigable CA1 sector neuroprotection with mild hypothermia induced 6 h after severe forebrain ischemia in rats. J Cereb Blood Flow Metab 19:742–749
7. Colbourne F, Sutherland G, Corbett D (1997) Postischemic hypothermia: a critical appraisal with implications for clinical treatment. Mol Neurobiol 14:171–201
8. Colbourne F, Sutherland GR, Auer RN (1999) Electron microscopic evidence against apoptosis as the mechanism of neuronal death in global ischemia. J Neurosci 19:4200–4210
9. Corbett D, Hamilton M, Colbourne F (2000) Persistent neuroprotection with prolonged postischemic hypothermia in adult rats subjected to transient middle cerebral artery occlusion. Exp Neurology 163:200–206
10. Del Bigio MR, Yan HJ, Buist R, Peeling J (1996) Experimental intracerebral hemorrhage in rats. Magnetic resonance imaging and histopathological correlates. Stroke 27:2312–2319
11. Del Bigio MR, Yan HJ, Campbell TM, Peeling J (1999) Effect of fucoidan treatment on collagenase-induced intracerebral hemorrhage in rats. Neurol Res 21:415–419
12. Diringer MN (1993) Intracerebral hemorrhage: pathophysiology and management. Crit Care Med 21:1591–1603
13. Gunn AJ, Gunn TR (1998) The 'pharmacology' of neuronal rescue with cerebral hypothermia. Early Human Development 53:19–35
14. Inamasu J, Suga S, Sato S, Horiguchi T, Akaji K, Mayanagi K, Kawase T (2000) Post-ischemic hypothermia delayed neutrophil accumulation and microglial activation following transient focal ischemia in rats. J Neuroimmunol 109:66–74

15. Juvela S, Heiskanen O, Poranen A, Valtonen S, Kuurne T, Kaste M, Troupp H (1989) The treatment of spontaneous intracerebral hemorrhage. A prospective randomized trial of surgical and conservative treatment. J Neurosurg 70:755–758
16. Karibe H, Chen J, Zarow GJ, Graham SH, Weinstein PR (1994) Delayed induction of mild hypothermia to reduce infarct volume after temporary middle cerebral artery occlusion in rats. J Neurosurg 80:112–119
17. Kawai N, Kawanishi M, Okauchi M, Nagao S (2001) Effects of hypothermia on thrombin-induced brain edema formation. Brain Res 895:50–58
18. MacLellan C, Shuaib A, Colbourne F (2002) Failure of delayed and prolonged hypothermia to favorably affect hemorrhagic stroke in rats. Brain Res, in press
19. Maier CM, Ahern K, Cheng ML, Lee JE, Yenari MA, Steinberg GK (1998) Optimal depth and duration of mild hypothermia in a focal model of transient cerebral ischemia: effects on neurologic outcome, infarct size, apoptosis, and inflammation. Stroke 29:2171–2180
20. Marion DW, Obrist WD, Carlier PM, Penrod LE, Darby JM (1993) The use of moderate therapeutic hypothermia for patients with severe head injuries: a preliminary report. J Neurosurg 79:354–362
21. Mayo NE, Neville D, Kirkland S, Ostbye T, Mustard CA, Reeder B, Joffres M, Brauer G, Levy AR (1996) Hospitalization and case-fatality rates for stroke in Canada from 1982 through 1991. The Canadian Collaborative Study Group of Stroke Hospitalizations, Stroke 27:1215–1220
22. Montoya CP, Campbell-Hope LJ, Pemberton KD, Dunnett SB (1991) The "staircase test": a measure of independent forelimb reaching and grasping abilities in rats. J Neurosci Methods 36:219–228
23. Peeling J, Del Bigio MR, Corbett D, Green AR, Jackson DM (2001) Efficacy of disodium 4-[(tert-butylimino)methyl]benzene-1,3-disulfonate N-oxide (NXY-059), a free radical trapping agent, in a rat model of hemorrhagic stroke. Neuropharmacology 40:433–439
24. Rosenberg GA, Mun-Bryce S, Wesley M, Kornfeld M (1990) Collagenase-induced intracerebral hemorrhage in rats. Stroke 21:801–807
25. Schubert A (1995) Side effects of mild hypothermia. J Neurosurg Anesthesiol 7:139–147
26. Shiozaki T, Sugimoto H, Taneda M, Yoshida H, Iwai A, Yoshioka T, Sugimoto T (1993) Effect of mild hypothermia on uncontrollable intracranial hypertension after severe head injury. J Neurosurg 79:363–368
27. Yang GY, Betz AL, Chenevert TL, Brunberg JA, Hoff JT (1994) Experimental intracerebral hemorrhage: relationship between brain edema, blood flow, and blood-brain barrier permeability in rats. J Neurosurg 81:93–102

Activation of NG2-Positive Oligodendrocyte Progenitor Cells after Focal Ischemia in Rat Brain

K. Tanaka, S. Nogawa, D. Ito, S. Suzuki, T. Dembo, A. Kosakai, and Y. Fakuuchi

Summary. The present study examines the alteration of oligodendrocyte progenitor cells which express membrane NG2 chondroitin sulfate proteoglycan after focal ischemia in the rat brain. Adult male Sprague-Dawley rats were subjected to 90-min occlusion of the middle cerebral artery, followed by recirculation time of up to 2 weeks. The distribution and morphological changes in NG2-positive oligodendrocyte progenitor cells were immunohistochemically examined. Stellate-shaped NG2-positive cells with multiple branched processes were abundantly detected in both the gray and white matter of normal brain. After two weeks of recirculation, NG2-positive cells in the area surrounding the infarction site (peri-infarct area) clearly showed enlarged cell bodies with hypertrophied processes. These stained strongly for NG2. Although the number of NG2-positive cells was significantly increased in the peri-infarct area, it was markedly decreased in the infarct core compared to controls. Double immunostaining revealed that these NG2-positive cells were not astrocytes, microglia or mature oligodendrocytes, but NG2-positive oligodendrocyte progenitor cells. Phosphorylation of CREB (cyclic AMP response element-binding protein) was clearly enhanced in these activated NG2-positive cells. These progenitor cells are known to revert to neural stem cells, which can give rise to neurons and astrocytes, as well as to oligodendrocytes. As such, this upregulation of NG2-positive progenitor cells may be an adaptive mechanism attempting to remyelinate or reorganize rat brain tissue after ischemic insult.

Key words. Cerebral infarction – oligodendrocyte – progenitor cell – myelination – chondroitin sulfate proteoglycan

Introduction

In the recent years, antibodies against a chondroitin sulfate proteoglycan, called NG2 (neuron glial antigen 2), have been increasingly used to identify and characterize oligodendrocyte precursor cells in the developing and adult central nervous

Kortaro Tanaka, Shigeru Nogawa, Daisuke Ito, Shigeaki Suzuki, Tomohisa Dembo, Arifumi Kosakai and Yasuo Fukuuchi
Department of Neurology, School of Medicine, Keio University, Tokyo 160-8582, Japan

Correspondence to: Kortaro Tanaka, MD, Department of Neurology, School of Medicine, Keio University, 35 Shinanomachi, Shinjuku-ku, Tokyo 160-8582, Japan, Tel.: #81-3-5363-3788, Fax: #81-3-3353-1272, E-Mail: kortaro@sc.itc.keio.ac.jp

Maturation Phenomenon in Cerebral Ischemia V
A. M. Buchan et al. (Eds.)
© Springer-Verlag Berlin Heidelberg 2004

Oligodendrocyte lineage

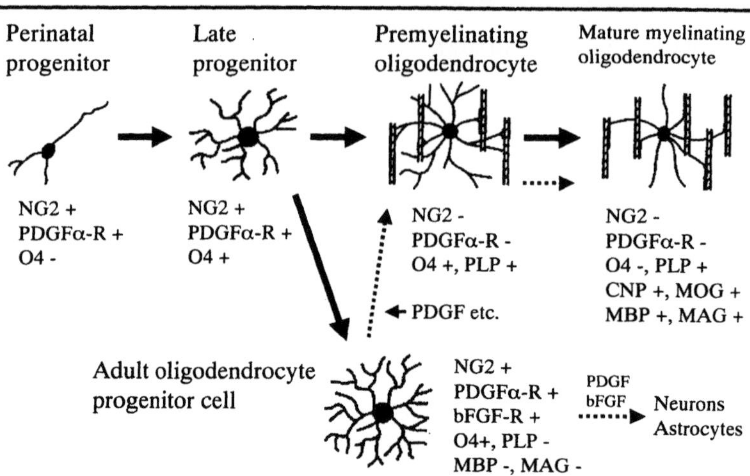

Fig. 1. The development of the oligodendrocyte lineage. Some of the markers that are useful to identify each developmental stage are also shown. Adult oligodendrocyte progenitor cells may differentiate into mature oligodendrocytes, neurons or astrocytes by certain stimulation such as platelet-derived growth factor (PDGF) or basic fibroblast growth factor (bFGF). CNP: 23-cyclic nucleotide 3-phosphohydrolase, MAG: myelin-associated glycoprotein, MBP: myelin basic protein, MOG: myelin oligodendrocyte glycoprotein, O4: O4 antigen, PLP: proteolipid protein, PDGFa-R: a-receptor for PDGF

system [3, 15]. As illustrated in Fig. 1, all NG2-positive perinatal progenitor cells differentiate into late progenitor cells by postnatal day 7. Throughout the second and third postnatal weeks, a portion of late progenitor cells differentiate further into premyelinating oligodendrocytes and lose expression of NG2, followed by maturation to become myelinating oligodendrocytes. On the other hand, a significant portion of late progenitor cells fail to differentiate and remain as NG2-positive progenitor cells in the adult brain [12, 15].

These adult oligodendrocyte progenitor cells (OPCs) are now attracting much attention for many reasons as follows: (1) OPCs have been found to make up 5–8% of the glial population in the central nervous system [12]. (2) OPCs are thought to divide slowly under normal condition [5, 9, 12]. (3) OPCs have been shown to be activated and proliferate following the central nervous system injuries such as multiple sclerosis and viral infection [11, 12]. (4) OPCs produce several of the chondroitin sulfate proteoglycans including NG2, which can be thus utilized as a specific marker for OPCs in the adult brain [3, 15]. (5) Recently, Kondo and Raff [7] have reported that OPCs can revert to multipotential neural stem cells, which can self-renew and give rise to neurons and astrocytes, as well as to oligodendrocytes, in response to certain extracellular signals such as basic fibroblast growth factor (bFGF) and platelet-derived growth factor (PDGF).

In spite of these facts, the behavior of OPCs has never been examined in stroke. The present study was therefore undertaken to clarify the temporal and regional profiles of NG2-positive OPCs after focal ischemia in the rat brain. A portion of the present study has already been published elsewhere [24].

Materials and Methods

The protocol described herein was in accordance with the NIH Guide for Care and Use of Laboratory Animals and received prior approval as being consistent with the Animal Experimentation Guidelines enforced by the Keio University School of Medicine.

Animals and Surgical Procedures

Adult male Sprague-Dawley rats (Japan Laboratory Animals Inc., Tokyo, Japan) weighing 270–330 g (14–16 weeks of age) were randomly divided into two groups: an ischemia group (n = 13) and a control group (n = 4). The animals in the ischemia group were anesthetized with a mixture of 1.0–1.5% halothane and 30% oxygen/70% nitrogen during the operation. A temperature probe (TD-300, Shibaura Electronics, Tokyo, Japan) was inserted into the rectum, and a heat lamp was used to maintain the rectal temperature at 37.0 to 37.5 °C. Next the middle cerebral artery (MCA) was occluded by the intraluminal filament technique, as described previously [21]. In brief, a 3-0 nylon monofilament suture (Matsuda Ikakogyo, Tokyo, Japan) was inserted through the right external carotid artery into the internal carotid artery to obstruct the origin of the right MCA. The suture was advanced to a point 19 mm beyond the origin of the internal carotid artery.

After the intraluminal filament was secured in position, the neck incision was closed with silk suture. After recovery from anesthesia, they were evaluated 1 h post-operation to assess the extent of their neurological deficits. They were stratified based on consistent neurological findings as having severe (grade 3), moderate (grade 2), mild (grade 1), or no deficit (grade 0) by a method devised by Bederson et al. [1]. Briefly, rats found circling towards the left (paretic) side were graded as 3. Those that showed diminished resistance to lateral push towards the left side without circling movements were graded as 2. Rats with any amount of consistent forelimb flexion but no other abnormalities were graded as 1. Rats with no neurological deficits were assigned a grade of 0. Only animals rated grade 2 or 3 were utilized for further study in the ischemia group (n = 12), because they were assumed to have a large infarct area in the territory of the MCA.

After 90 min of MCA occlusion, the rats were reanesthetized with the same anesthetic combination as described above and the intraluminal suture was carefully removed. The animals were allowed to survive for intervals of 24 h (n = 4), 7 days (n = 4), and 14 days (n = 4), with free access to food and water. In the control group, procedures were identical to those employed in the ischemia group except that the operation was stopped after a neck incision exposed the right carotid arteries.

At the end of recirculation period, the animals were given an intraperitoneal injection of pentobarbital at a dose of 100 mg/kg. They were then infused with 300 ml of 50 mM phosphate-buffered saline (PBS) followed by 300 ml of 4% freshly depolymerized paraformaldehyde in 0.1 M phosphate buffer (PB), both at a pH of 7.4. The brains were then removed and postfixed for 3 h in 4% paraformaldehyde in PB at 4 °C before cryoprotection by sequential bathing in 15, 20, and 30% sucrose. The brains were subsequently frozen in liquid Freon 22 (Asahi Glass, Tokyo, Japan) and consecutive 20 μm thick coronal sections were prepared on a cryostat (Cryocut 1800, Leica Instruments GmbH, Nussloch, Germany). The sections were thaw-mounted onto silane-coated slides and then stored at –30 °C.

Immunohistochemistry

Immunohistochemistry was carried out by using an avidin-biotin enzyme complex system (Vectastain ABC Elite Kit, Vector Laboratories, Burlingame, CA, USA), as reported previously [19, 20, 22]. In short, the sections were rehydrated and washed in PBS (pH 7.4) for 5 min. They were then quenched with 0.3% H_2O_2 in methanol for 30 min to block endogenous peroxidase action and washed three times for 5 min each in PBS (pH 7.4). The sections were subsequently preincubated for 60 min in PBS (pH 7.4) containing 2% normal goat serum in order to block non-specific binding of antibody. The specimens were then incubated overnight with the affinity-purified rabbit polyclonal anti-rat NG2 chondroitin sulfate proteoglycan antibody (Chemicon, Temecula, CA, USA) at 1:1000 dilution in PBS with 2% normal goat serum and 0.1% Triton X-100. After treatment, they were thereafter rinsed and incubated with the avidin-biotin peroxidase conjugate. The sections were developed in 0.02% diaminobenzidine with 0.02% H_2O_2. Negative control staining was performed with normal rabbit serum as a substitute for the primary antibody.

Adjacent sections were stained with cresyl violet for conventional histological examination as well as with rabbit polyclonal anti-myelin basic protein (MBP) antibody (Nichirei, Tokyo, Japan) for evaluation of myelination, respectively.

Double Immunohistochemistry for NG2 and GFAP, NG2 and APC, NG2 and Monoclonal NB3C4 Antibody, NG2 and Phosphorylated CREB, or NG2 and Isolectin-B4

Double-immunostaining was carried out to distinguish the NG2-positive cells from other types of glial population and to examine phosphorylation of CREB (cyclic AMP response element-binding protein) in the NG2-positive cells. The sections were rehydrated and washed in PBS (pH 7.4) for 5 min. The samples were subsequently preincubated for 60 min in PBS (pH 7.4) containing 2% normal goat serum and 1% bovine serum albumin to block non-specific binding of antibody. To initiate double-immunostaining for NG2 with glial fibrillary acidic protein (GFAP), adenomatous polyposis coli protein (APC) [2], monoclonal NB3C4 antibody [25], or phosphorylated CREB, the specimens were incubated overnight with the rabbit anti-NG2 chondroitin sulfate proteoglycan antibody (Chemicon) and also with the affinity-purified mouse anti-GFAP monoclonal antibody (Boehringer Mannheim, Philadelphia, PA, USA) at 1:250 dilution, mouse anti-APC antibody (Oncogene, Boston, USA) at

1:20 dilution, mouse monoclonal NB3C4 antibody at 1:100 dilution, or mouse mono-clonal anti-phosphorylated CREB antibody (Upstate Biotechnology, Lake Placid, NY, USA) at 1:100 dilution, respectively. The sections were then washed in PBS, and in-cubated for 2 h with Texas Red-conjugated anti-rabbit IgG antibody (Amersham, Buckinghamshire, UK) at 1:30 dilution and FITC (fluorescein isothiocyanate)-conju-gated anti-mouse IgG antibody (Amersham) at 1:30 dilution.

In order to double-immunostain for NG2 and isolectin-B4, the sections were in-cubated overnight with the anti-NG2 chondroitin sulfate proteoglycan antibody (Chemicon) prepared as above. Subsequently, the specimens were washed in PBS and incubated for 2 h with Texas Red-conjugated anti-rabbit IgG antibody (Amer-sham) at 1:30 dilution and FITC-labeled GSA I-B4 (isolectin-B4 from *Griffonia simplicifolia* seeds) (Sigma, St. Louis, USA). These samples were examined by fluo-rescent microscopy (Eclipse E-800, Nikon, Tokyo, Japan).

Data analysis

Absolute immunoreactive cell counts (per 0.15 mm^2) were made in the coronal sections of the cerebral cortex located 0.20 mm anterior to the bregma with an ocular micrometer (Nikon, Tokyo, Japan) attached to a light microscope. The re-gions of interest were set within the infarct core (A), at the peri-infarct area (B) and in an area remote to the infarct core (C) based on the findings revealed by cresyl violet staining [24]. The counts were performed on three consecutive slices in a blinded manner by one of the co-authors and the mean value was determined.

The data is expressed as the means ± SD. Results were tested by analysis of vari-ance followed by post hoc analysis using the modified t-test with the Bonferroni correction for multiple simultaneous comparisons.

Results

Immunostaining for NG2

In the cerebral cortex in the sham group, many NG2-positive cells were uniformly distributed with small, round or oval cell bodies that give rise to multiple, highly branched processes in all directions, as shown in Fig. 2. In the corpus callosum, many NG2-positive cells were also found with rather longitudinal cell bodies and processes. At 24 h of recirculation, the NG2-positive cells of the peri-infarct area showed only a slight increase in the size of cell bodies and processes (Fig. 2). With 1 week of recirculation, the NG2-positive cells experienced greater enlargement of cell bodies and processes with increased intensity of immunostaining. The same cells were even more prominent in size and stain intensity when sampled at 2 weeks (Fig. 2). They were significantly swollen with very heavy staining and had many short and hypertrophied processes with a beaded, varicose appearance.

The NG2-positive cells were frequently found in pairs like twins not only in the control brain cortex, but also in the peri-infarct area at 2 weeks of recirculation, as shown in Fig. 3.

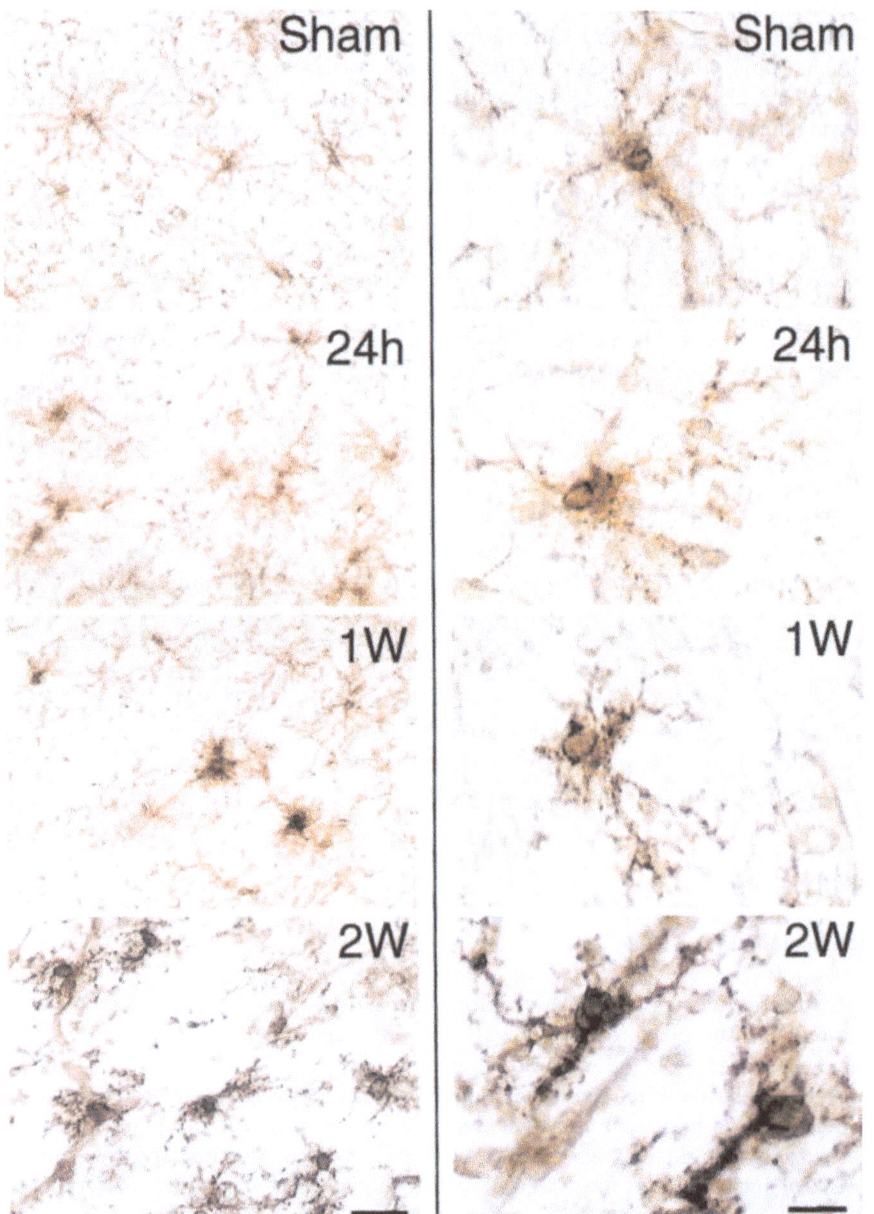

Fig. 2. Representative microscopic pictures in the peri-infarct area, which were stained with anti-NG2 chondroitin sulfate proteoglycan antibody. The left column shows the pictures with low power magnification. Bar = 40 μm. The right column shows the pictures with high power magnification. Bar = 20 μm

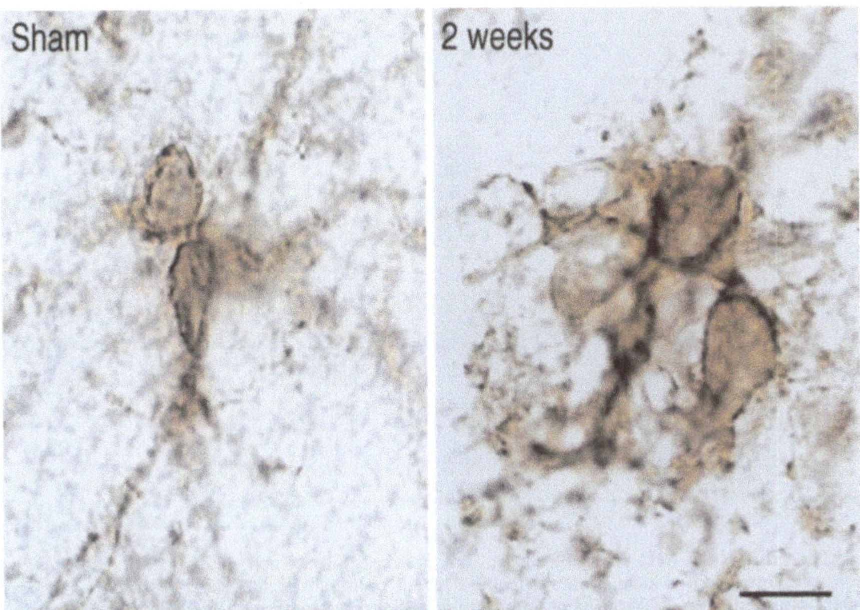

Fig. 3. Representative microscopic pictures in the peri-infarct area, which were stained with anti-NG2 chondroitin sulfate proteoglycan antibody. The left and the right pictures were taken from the sham animal and the animal with 2 weeks of recirculation, respectively. Adult oligodendrocyte progenitor cells were frequently found in pairs. Bar = 10 μm

The number of NG2-positive cells in each region is summarized in Fig. 4. Two way analysis of variance revealed significant differences in the numbers of immunopositive cells between the examined areas (p < 0.001), time points (p < 0.001) and the combination of these two factors (interaction effect) (p < 0.001). In the remote area, no significant changes in cell morphology or staining were noted throughout the experimental period. The peri-infarct area demonstrated a gradual increase in the number of NG2-positive cells after ischemia, resulting in a statistically significant difference at one and two weeks of recirculation as compared to the control comparisons. In contrast, the infarct core showed progressive disappearance of NG2-positive cells with recirculation.

At 2 weeks of recirculation, the peri-infarct area demonstrated a moderate reduction in immunoreactivity for MBP and a few weakly immunoreactive fibers were noted as compared to the controls. Similarly, cresyl violet staining showed a slight reduction in the number of neurons in the peri-infarct area at 2 weeks of recirculation, which was accompanied by a mild increase in glial cells.

Double Immunohistochemistry

In the sham group, NG2-positive cells were not stained at all with anti-GFAP antibody, isolectin-B4, anti-APC antibody and monoclonal NB3C4 antibody, indicating

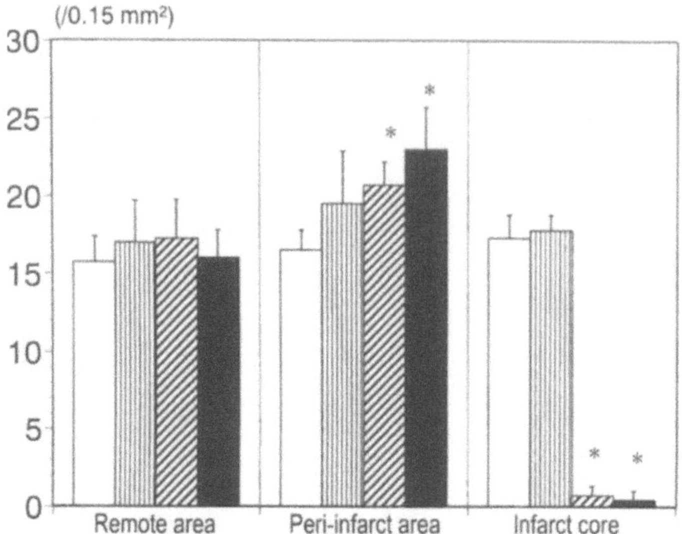

Fig. 4. The numbers of NG2-positive cells in each brain region at each time interval. The open bar represents control group data. The vertically striped, obliquely striped, and solid bars represent the data sampled from subject rats at 24 h, one week, and two weeks of recirculation, respectively. Values are presented as means $\pm$ SD. * $p < 0.05$ compared to the control data

that NG2-positive cells are distinct from astrocytes, microglia and mature oligodendrocytes.

Similarly, during recirculation period, activated NG2-positive cells were not immunoreactive at all against anti-GFAP antibody, isolectin-B4, anti-APC antibody or monoclonal NB3C4 antibody (Fig. 5), indicating that activated NG2-positive cells are not astrocytes, microglia, or mature oligodendrocytes.

Double immunostaining with anti-NG2 antibody and anti-phosphorylated CREB antibody revealed that phosphorylation of CREB was clearly enhanced in the nuclei of activated NG2-positive cells in the peri-infarct area during the recirculation period. In contrast, phosphorylation of CREB was barely observed in the NG2-positive cells in the contralateral cerebral cortex as in the normal brain.

Discussion

This study demonstrates that the size and number of NG2-positive cells are significantly increased during post-ischemic recirculation within the peri-infarct area of rat brains. These cells were not immunoreactive for astrocytic marker GFAP, microglial marker isolectin-B4, and mature oligodendrocyte markers APC and monoclonal NB3C4 antibody. These findings were consistent with qualities typical of oligodendrocyte progenitor cells as previously reported [3, 12, 15]. The NG2-positive cells were evenly distributed in the cerebral cortex of the control rats and possessed small cell bodies from which multiple fine branching processes radiated. These morphological features were compatible with those of oligodendrocyte pro-

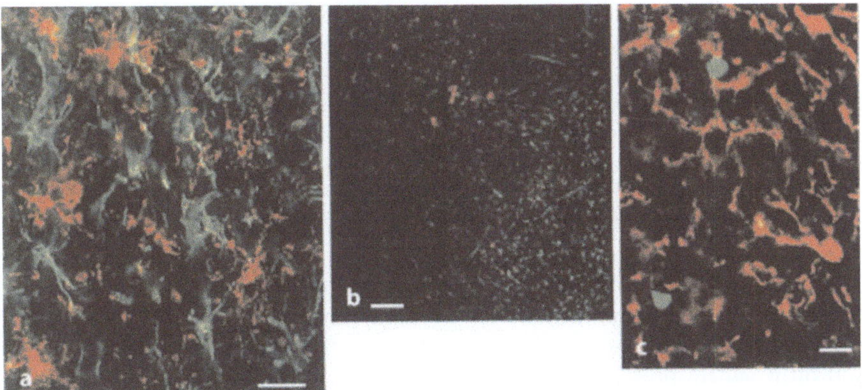

Fig. 5. a Double immunostaining for NG2 with GFAP (bar=20 µm), **b** double immunostaining for NG2 with isolectin-IB4 (bar=50 µm), and **c** double immunostaining for NG2 with monoclonal NB3C4 antibody (bar=10 µm) in the peri-infarct area at 2 weeks of recirculation. NG2 immunoreactivity was visualized with Texas-Red-conjugated anti-rabbit IgG antibody in each picture (red), while immunoreactivity for GFAP, isolectin-IB4 and NB3C4 antibody was visualized with FITC-conjugated anti-mouse IgG antibody (green). These superimposed pictures show that NG2-positive cells were not immunoreactive for GFAP, isolectin-IB4, and monoclonal NB3C4 antibody, which are the markers of astrocytes, microglia and mature oligodendrocytes, respectively

genitor cells reported in the adult rat brain [3, 18]. Although NG2 has been shown to be produced by astrocytes in spinal cord injuries [4], both the lack of the GFAP, isolectin-B4 and APC membrane expression as well as their microscopic appearance provide strong evidence that the NG2-positive cells in the present study are indeed oligodendrocyte progenitor cells.

The heavily stained and enlarged cell bodies and processes of NG2-positive cells, which were noted in the present study, have also been observed after insults such as mechanical brain injury [10], acute experimental autoimmune encephalomyelitis [14], viral infection [11], and kainate injection [16]. These morphological changes could represent the general reaction of NG2-positive cells to brain damage. The NG2-positive cells became more numerous at 1 and 2 weeks of post-ischemic recirculation, suggesting that the NG2-positive cells may have multiplied in response to ischemic stress. Indeed, oligodendrocyte progenitor cells have been shown to have the potential to divide slowly in the adult rat cerebellar cortex [9]. Because of the poor migratory ability of these cells, it is unlikely that the increased number of NG2-positive cells can be attributed to migration of these cells [12]. In fact, the NG2-positive cells were frequently found in pairs in the peri-infarct area as well as in the control brain in the present study, indicating that these cells may be dividing *in situ*. Under pathological conditions such as ischemic stress in the present study, the proliferation rate of oligodendrocyte progenitor cells is expected to increase and some may differentiate into oligodendrocytes [12]. Moreover, cultures of oligodendrocyte progenitor cells, which were purified from postnatal day 6 rat optic nerve, have recently been shown to generate neurons and astrocytes, as well as oligodendrocytes, by certain stimulation such as application of bFGF and PDGF [7]. Oligodendrocyte progenitor cells in the adult brain may also be able to revert in this way, as cultures of adult rat optic nerve, which retains abundant oligodendrocyte progenitor cells, can generate neurons when treated with bFGF [17]. In line with these observations, our preliminary data indicated that the number of mature oligodendrocytes in the peri-infarct area, which was reduced during the early phase of post-ischemic recirculation, was restored during the late phase of recirculation, suggesting a supplement of new mature oligodendrocytes.

In the present study, the peri-infarct area where NG2-positive cells were activated showed a moderate reduction in myelination, suggesting that the activation of NG2-positive cells may contribute to the tissue repair mechanism by supplying myelin-forming cells. However, we have currently no direct evidence that the NG2-positive cells are stimulated by loss of myelin or involved in remyelination. NG2 chondroitin sulfate proteoglycan can itself impede axon development [4]. Therefore, it is difficult to determine if the lack of myelin is an impetus for upregulation of NG2 or if it reflects the inhibition of axonal regeneration caused by increased expression of NG2.

Multiple mitogens of oligodendrocyte progenitor cells have been identified in in-vitro experiments, including bFGF and PDGF. Both of these are released in injured brain tissue as results from infarction [8, 12, 13]. In addition, various cytokines and growth factors such as transforming growth factor β (TGFβ) are known to upregulate production of chondroitin sulfate proteoglycan by oligodendrocyte progenitor cells [12]. Such growth factors are known to regulate the survival and growth of neurons in brain tissue. In the present study, phosphorylation of CREB was clearly enhanced in the activated NG2-positive oligodendrocyte progenitor cells during post-ischemic recirculation. CREB mediates cellular responses to a variety of signals such as mitogenic and differentiative factors including a family of growth factors [23]. Johnson et al. [6] have recently reported that phosphorylation of CREB in the cultured oligodendrocyte progenitor cells is markedly enhanced in response to neurotrophin-3 and plays an important role in cell proliferation.

Taken together, the present study suggests that activation and proliferation of oligodendrocyte progenitor cells take place via CREB-mediated intracellular signal transduction in the peri-infarct area during the post-ischemic recirculation period. These cellular responses may be an adaptive repair reaction of brain tissue that could potentially induce remyelination and neurogenesis. Activation of oligodendrocyte progenitor cells specifically aimed for induction of myelinating mature oligodendrocytes and neurons may be one of the useful strategies for stroke treatment. Future studies should clarify the exact mechanisms underlying the activation of oligodendrocyte progenitor cells and the long-term fates of these activated cells.

Acknowledgements. The authors are grateful to Dr. Kazunori Yoshimura, Dr. Hiroaki Asou, and Jin Nakahara for providing monoclonal NB3C4 antibody and their valuable suggestions. This work was supported by research grants for life sciences and medicine of the Keio University Medical Science Fund, a Grant-in-Aid for Science Research (C) from the Ministry of Education, Science, and Culture of Japan (08670725, 11670639), as well as the Mitsui Life Social Welfare Foundation.

References

1. Bederson JB, Pitts LH, Tsuji M, Nishimura MC, Davis RL, Bartkowski H (1986) Rat middle cerebral artery occlusion: evaluation of the model and development of a neurologic examination. Stroke 17:472–476
2. Bhat RV, Axt KJ, Fosnaugh JS, Smith KJ, Johnson KA, Hill DE, Kinzler KW, Baraban JM (1996) Expression of the APC tumor suppressor protein in oligodendroglia. Glia 17:169–174
3. Dawson MR, Levine JM, Reynolds R (2000) NG2-expressing cells in the central nervous system: are they oligodendroglial progenitors? J Neurosci Res 61:471–479
4. Fawcett JW, Asher RA (1999) The glial scar and central nervous system repair. Brain Res Bull 49:377–391
5. Horner PJ, Power AE, Kempermann G, Kuhn HG, Palmer TD, Winkler J, Thal LJ, Gage FH (2000) Proliferation and differentiation of progenitor cells throughout the intact adult rat spinal cord. J Neurosci 20:2218–2228
6. Johnson JR, Chu AK, Sato-Bigbee C (2000) Possible role of CREB in the stimulation of oligodendrocyte precursor cell proliferation by neurotrophin-3. J Neurochem 74:1409–1417
7. Kondo T, Raff M (2000) Oligodendrocyte precursor cells reprogrammed to become multipotential CNS stem cells. Science 289:1754–1757
8. Krupinski J, Issa R, Bujny T, Slevin M, Kumar P, Kumar S, Kaluza J (1997) A putative role for platelet-derived growth factor in angiogenesis and neuroprotection after ischemic stroke in humans. Stroke 28:564–573
9. Levine JM, Stincone F, Lee YS (1993) Development and differentiation of glial precursor cells in the rat cerebellum. Glia 7:307–321
10. Levine JM (1994) Increased expression of the NG2 chondroitin-sulfate proteoglycan after brain injury. J Neurosci 14:4716–4730
11. Levine JM, Enquist LW, Card JP (1998) Reactions of oligodendrocyte precursor cells to alpha herpesvirus infection of the central nervous system. Glia 23:316–328
12. Levine JM, Reynolds R, Fawcett JW (2001) The oligodendrocyte precursor cell in health and disease. Trends Neurosci 24:39–47
13. Martinez G, Di Giacomo C, Sorrenti V, Carnazza ML, Ragusa N, Barcellona ML, Vanella A (2001) Fibroblast growth factor-2 and transforming growth factor-beta1 immunostaining in rat brain after cerebral postischemic reperfusion. J Neurosci Res 63:136–142
14. Nishiyama A, Yu M, Drazba JA, Tuohy VK (1997) Normal and reactive NG2+ glial cells are distinct from resting and activated microglia. J Neurosci Res 48:299–312
15. Nishiyama A, Chang A, Trapp BD (1999) NG2+ glial cells: A novel glial cell population in the adult brain. J Neuropathol Exp Neurol 58:1113–1124
16. Ong WY, Levine JM (1999) A light and electron microscopic study of NG2 chondroitin sulfate proteoglycan-positive oligodendrocyte precursor cells in the normal and kainate-lesioned rat hippocampus. Neuroscience 92:83–95
17. Palmer TD, Markakis EA, Willhoite AR, Safar F, Gage FH (1999) Fibroblast growth factor-2 activates a latent neurogenic program in neural stem cells from diverse regions of the adult CNS. J Neurosci 19:8487–8497
18. Reynolds R, Hardy R (1997) Oligodendroglial progenitors labeled with the O4 antibody persist in the adult rat cerebral cortex in vivo. J Neurosci Res 47:455–470
19. Suzuki S, Tanaka K, Nogawa S, Ito D, Dembo T, Kosakai A, Fukuuchi Y (2000) Immunohistochemical detection of leukemia inhibitory factor after focal cerebral ischemia in rats. J Cereb Blood Flow Metab 20:661–668
20. Tanaka K, Nagata E, Suzuki S, Dembo T, Nogawa S, Fukuuchi Y (1999) Immunohistochemical analysis of cyclic AMP response element binding protein phosphorylation in focal cerebral ischemia in rats. Brain Res 818:520–526
21. Tanaka K, Nogawa S, Nagata E, Suzuki S, Dembo T, Kosakai A, Fukuuchi Y (1999) Inhibition of cyclic AMP-dependent protein kinase in the acute phase of focal cerebral ischemia in the rat. Neuroscience 94:361–371
22. Tanaka K, Nogawa S, Ito D, Suzuki S, Dembo T, Kosakai A, Fukuuchi Y (2000) Activated phosphorylation of cyclic AMP response element binding protein is associated with preservation of striatal neurons after focal cerebral ischemia in the rat. Neuroscience 100:345–354
23. Tanaka K (2001) Alteration of second messengers during acute cerebral ischemia – adenylate cyclase, cyclic AMP-dependent protein kinase, and cyclic AMP response element binding protein. Prog Neurobiol 65:173–207

24. Tanaka K, Nogawa S, Ito D, Suzuki S, Dembo T, Kosakai A, Fukuuchi Y (2001) Activation of NG2-positive oligodendrocyte progenitor cells during post-ischemic reperfusion in the rat brain. Neuroreport 12:2169–2174
25. Yoshimura K, Kametani F, Shimoda Y, Fujimaki K, Sakurai Y, Kitamura K, Asou H, Nomura M (2001) Antigens of monoclonal antibody NB3C4 are novel markers for oligodendrocytes. Neuroreport 12:417–421

Postischemic Housing in an Enriched Environment Influences Hippocampal Progenitor Cell Differentiation after Focal Cortical Ischemia

B. B. Johansson, M. Komitova, E. Perfilieva, B. Mattsson, and P. Eriksson

Summary. We have tested the hypothesis that environmental factors can influence postischemic progenitor cell survival and differentiation in the dentate gyrus. The proliferation marker bromodeoxyuridine (BrdU) was administered during 7 days starting 24 h after ligation of the right middle cerebral artery. Postoperatively the rats were housed in standard cages or transferred to enriched environment 24 h or 7 days after the ligation. Rats housed in standard cages performed significantly worse than rats housed in an enriched environment in a leg placement and a rotating pole test four weeks after the arterial ligation. Neurogenesis and gliogenesis were determined by triple labeling with antibodies against BrdU, the astrocytic marker glial fibrillary acidic protein (GFAP), and the neuronal markers Calbindin D28k and NeuN. Rats with cortical lesions had a 5- to 6-fold increase in BrdU labeled cells on the ipsilateral side ($p < 0.001$ for the delayed enriched group; $p < 0.01$ for the early enriched and standard groups) and a 2- to 3-fold non-significant increase on the contralateral side with no significant differences between the groups. About 80% of the BrdU-positive cells co-labeled with NeuN and about 70% of the BrdU-positive cells co-labeled with Calbindin D28K. Although housing conditions did not influence neurogenesis it markedly altered gliogenesis. Whereas the standard group did not have more astrocytes than sham-operated rats on the ipsilateral side, the early and delayed enriched group had a 3- to 5-fold increase, respectively, thereby normalizing the severely disturbed neuron to glia ratio in the standard group. We hypothesize that the newly formed neurons in the standard group would have a poor environment in the absence of a concomitant gliogenesis. Astrocytes play an important role in neuronal plasticity, and we propose that more attention should be given to gliogenesis in experimental studies on cell proliferation and differentiation after brain lesions.

Key words: Bromodeoxyuridine (BrdU) – Calbinin – enriched environment – focal brain ischemia – glial fibrillary acidic protein (GFAP) – neurogenesis – gliogenesis – progenitor cells – rat – stem cells

* Barbro B. Johansson, † Mila Komitova, † Ekaterina Perfilieva, * Bengt Mattsson, † Peter Eriksson
* Department of Clinical Neurosciences, Lund University, Sweden and † Department of Clinical Neuroscience, Göteborg University

Correspondence to: Barbro B. Johansson, MD, PhD, Professor of Neurology, Wallenberg Neuroscience Center, Experimental Brain Research, BMC A13, SE 22184 Lund, Sweden, Tel.: +46 46 222 0621, Fax: +46 46 222 0626, E-Mail: barbro.johansson@neurol.lu.se

Maturation Phenomenon in Cerebral Ischemia V
A. M. Buchan et al. (Eds.)
© Springer-Verlag Berlin Heidelberg 2004

Abbreviations. BrdU (bromodeoxyuridine); GFAP (Glial fibrillary acidic protein); NeuN (neuronal nuclei); GCL (granular cell layer); SGZ (subgranular zone); MAP-2 (myelin-associated protein)

Introduction

It is well established that environmental factors can influence brain plasticity in the intact as well as in the lesioned brain [21, 34]. Postischemic housing in an enriched environment can improve outcome [15, 20, 46] alter gene expression for growth factors and other substances in several brain areas [8, 64] and increase dendritic spine density in cortical pyramidal cells contralateral to the infarct [22]. Furthermore, environmental factors interact with therapeutic interventions such as neocortical transplantation [10, 41, 63].

In the present study we have tested the hypothesis that environmental factors can influence postischemic stem cell survival and differentiation. Stem cells or progenitor cells, normally present in the subventricular zone and the dentate gyrus granular cell layer (GCL) of the hippocampus, have the potential for self-renewal and differentiation into neurons, astrocytes, and oligodendrocytes in the adult brain [11, 12]. The proliferation and survival of progenitor cells are dependent on endogenous genetic programs [30, 31, 48] that can be modified by exogenous cues including growth factors, hormones, drugs, stress, food restriction, brain lesions, and environmental factors [14, 37, 38, 49, 56, 60].

This study concerns the possible combined effect of two of the factors mentioned above, lesions and environment. Many brain lesions, in the hippocampus [13, 27, 38, 39, 57] as well as brain lesions not directly affecting the hippocampus [3, 9, 19, 33], have been shown to stimulate hippocampal progenitor cell proliferation and neurogenesis. Enriched environment increases neurogenesis in the dentate gyrus in intact rats and mice [29, 32, 45, 60]. The model we have used is a permanent ligation of the middle cerebral artery distal to the striatal branches in hypertensive rats, a model that induces selective cortical infarcts [15, 20]. The aim of our study was to answer the following questions: To what extent does selective focal cortical infarcts increase cell genesis and survival in the dentate gyrus? Can cell proliferation and differentiation be modified by post-ischemic housing conditions? Does survival or differentiation of hippocampal progenitor cells dividing during the first postischemic week correlate with functional outcome after a neocortical infarct?

Materials and Methods

The design of the study is summarized in Fig. 1 and described in more detail elsewhere [35]. The right middle cerebral artery was ligated distal to the striatal branches in 6 months old male spontaneously hypertensive rats. Bromodeoxyuridine (BrdU), which labels dividing cells, was given during 7 consecutive days starting 24 h after the arterial ligation. Postoperatively the rats were either housed in standard cages (550×350×200 mm, 3 rats in each cage) or transferred to in an enriched environment 24 h or 7 d after the ligation, the latter group after the last BrdU injection. The enriched housing consisted of 815×610×1280 mm cages, equipped with horizontal and vertical boards, chains, swings, wooden blocks and objects of different sizes and materials. The distance between the boards and the objects was changed twice a week.

Preoperatively and 4 weeks after the operation the rats were tested in a limb-placement test and on a rotating horizontal pole [46]. The limb placement test consists of 6 subtests with fore limb grading in all and hind limb grading in two of the subtests. Each subtest was scored as follows: 0 = no placing, 1 = incomplete and/or delayed (> 2 seconds) and 2 = immediate and correct placing. Thus, for each body side, the maximum score i.e. best performance was 16. Coordination and integration of movement was tested on a rotating horizontal pole. The pole, 45 mm in diameter and 1,5 m in length, rotated alternately to the left or to the right with three turns per minute. The score was 0 = the rat falls down; 1 = the rat is unable to traverse the pole but does not fall down; 2 = the rat falls down while attempting to cross the pole; 3 = the rat jumps with both hind limbs together, apparently supporting the weak hind limb with the opposite strong limb; 4 = the affected hind limb is used for less than 50% of the steps; 5 = the rat crosses the pole with a few foot slips; 6 = the rat crosses the pole with no foot slips.

Five weeks after the MCA occlusion the animals were deeply anesthetized with an overdose of sodium pentobarbital and perfused transcardially with saline solution followed by 4.0% paraformaldehyde in 0.1 M phosphate buffer. The brains were postfixed overnight in 4% paraformaldehyde in 0.1 M phosphate buffer, and thereafter transferred to 30% sucrose. Each brain was sectioned coronary (40 μm) through the entire hippocampus on a sliding microtome. Sections were stored individually at –20 °C in a cryoprotecting buffer containing 25% ethylene glycol, 25% glycerol and 0.05 M phosphate buffer until they were processed for immunohistochemistry. For technical details on tissue preparation and antibodies used, see Komitova et al. [35]. Immunolabeling was performed on free-floating sections. Triple immunofluorescence staining for BrdU, a neuronal marker (NeuN or calbindin D28k) and the astrocytic marker glial fibrillary acidic protein (GFAP) was used to detect co-labeling of cell-specific markers with BrdU. NeuN stands for neuronal nuclei and is a transcription factor that is expressed in the nucleus and cytoplasm of mature neurons [40] while calbindin D28k is a neuronal calcium-binding protein and marker for granule cells in the dentate gyrus [54].

BrdU-labeled cells in the GCL and the subgranular zone (SGZ) were counted and phenotyped. The SGZ was defined as a 20 μm band immediately adjacent to the hilar surface of the GCL. The GCL and the SGZ are referred to as the GCL throughout the text, except with respect to volume measurements where only the GCL proper was considered. All cell counts were performed using unbiased stereologic counting techniques [7, 16]. Images were taken with a Nikon Optiphot microscope connected to a video camera and area measurements of the GCL were performed using an Intuos Graphics tablet (Wacom, Japan) and digital image processing software (Nikon, Göteborg, Sweden). BrdU-immunopositive cells were counted on 10 anatomically matched immunoperoxidase-stained sections per animal, 240 μm apart. The GCL sample volume was calculated by multiplying the cross-sectional GCL area with the thickness of the section (40 μm). The number of counted cells was then divided by the GCL sample volume. The total number of BrdU-labeled cells was achieved by multiplying mean cell density with the total GCL volume. Determination of co-labeling of BrdU with NeuN, calbindin D28k and GFAP, respectively, was performed on 5–6 anatomically matched sections 240 μm apart and at least 40 BrdU-labeled cells per animal were analyzed concerning phenotype. Optical Z sectioning of the samples at 0.5 μm intervals was used to verify and quantify co-localization of BrdU with cell-specific markers. A percentage of co-labeling was calculated and the total numbers of newborn neurons and astrocytes were calculated by multiplying that percentage with the total number of BrdU-labeled cells per GCL.

Measurements of ipsi- and contralateral hemisphere cross-sectional area were performed for each animal on 10 Nissl-stained sections 960 μm apart. The hemispheric volume was calculated from the cross-sectional areas and the distance between the sections according to the Cavallieri principle [7]. Measurement of the GCL volume was performed in a similar fashion on 10 Nissl-stained sections, 480 μm apart, encompassing the entire dentate gyrus. Total tissue volume loss was expressed as percentage of the contralateral hemisphere volume.

Statistical analysis. The presented values are mean values ± SD. Comparisons between groups were made with one-way analysis of variance (ANOVA) followed by the Fisher PLSD post-hoc test (Statview 4.01 for Macintosh). Student's t-test was used for side comparisons within groups. p Values < 0.05 were considered statistically significant. For the behavioral tests the Kruskal-Wallis non parametric ANOVA with a multiple comparison post-hoc test at the 95% significance level was used.

Results

The infarct volume did not differ between the groups (20±3, 16±4, 17±6% of contralateral hemisphere in standard, early and delayed enriched, respectively). The experimental groups housed in the enriched environment performed significantly better than rats in standard environment on the rotating pole and the leg placement test (Fig. 2).

Rats with cortical lesions had a 5- to 6-fold ipsilateral increase in BrdU-labeled cells with no significant differences between the groups. The significance for difference to the sham group was p < 0.001 for the delayed enriched group and p < 0.01 for the early enriched and standard groups. On the side contralateral to the lesion the number of BrdU cells was increased 2- to 3-fold but the difference was not significant due to a large inter animal variation.

Triple labeling with BrdU, the glial marker GFAP and the two neuronal markers showed that most of the new cells were neurons. New astrocytes were mostly located in the subgranular zone and extended processes into the granular cell layer proper. Examples of BrdU/Calbindin and BrdU/GFAP co-labeling are shown in Fig. 3. About 80% of the cells co-labeled with NeuN and about 70% co-labeled with calbindin D28k (p < 0.001 for NeuN E7d; p < 0.01 for all the other groups) (Fig. 4). On the contralateral side the difference between sham and standard was significant for BrdU/NeuN co-labeled cells and between sham and early enriched

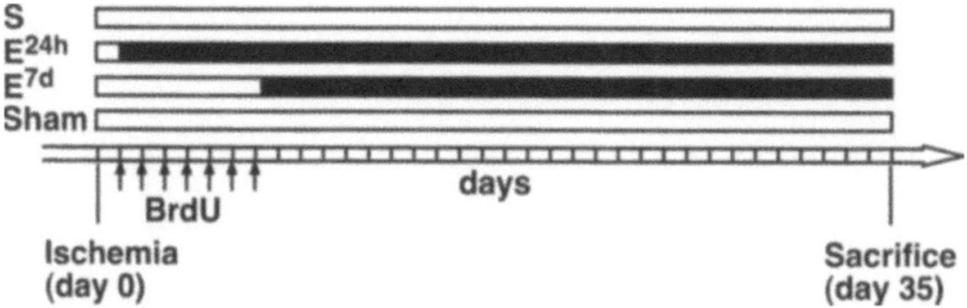

Fig. 1. The experimental design

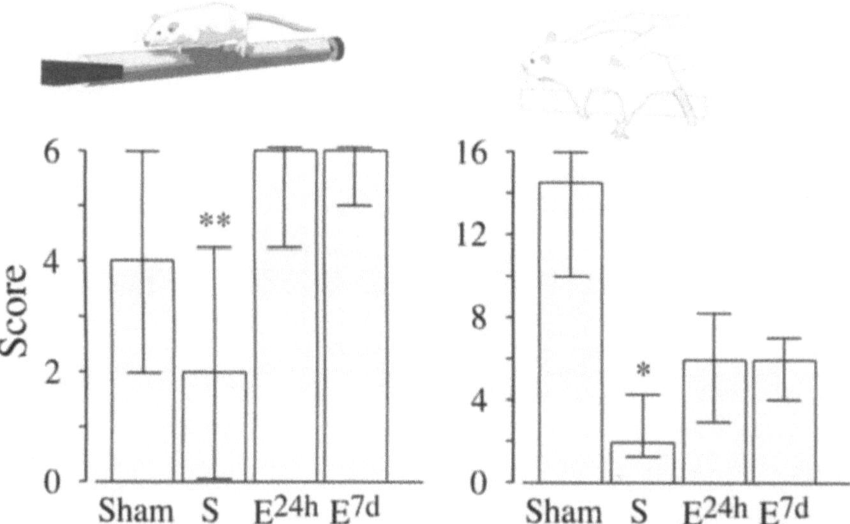

Fig. 2. Left side: Performance on a horizontal pole rotating 3 turns per minute in rats postoperatively housed in standard cages (S), or transferred to an enriched environment 24 h, E24h, or 7 days, E7d after a permanent focal cortical ischemia. The top score, 6, indicates that the rats manage the tests without any problems, ** p < 0.01 for difference between the standard and enriched groups. Right side: performance in a leg placement test in the same groups of rats. p < 0.05 for difference between rats housed in standard and enriched environment

for BrdU/Calbindin co-labeled cells. The standard and delayed enriched group had significantly more neurons on the ipsi-than on the contralateral side for both neuronal markers whereas the side difference in the early enriched group was not significant.

In contrast to the lack of environmental effect on neurogenesis, gliogenesis differed markedly with no increase in the standard but a 3- and 5-fold increase in newborn astrocytes in the early and delayed enriched groups, respectively (Fig. 5). The largest difference was seen between the standard and delayed enriched groups, with slightly less than 6% compared to 22% of the BrdU-labeled cells co-labeling for GFAP on the ipsilateral side. In addition, the ipsi- to contralateral difference

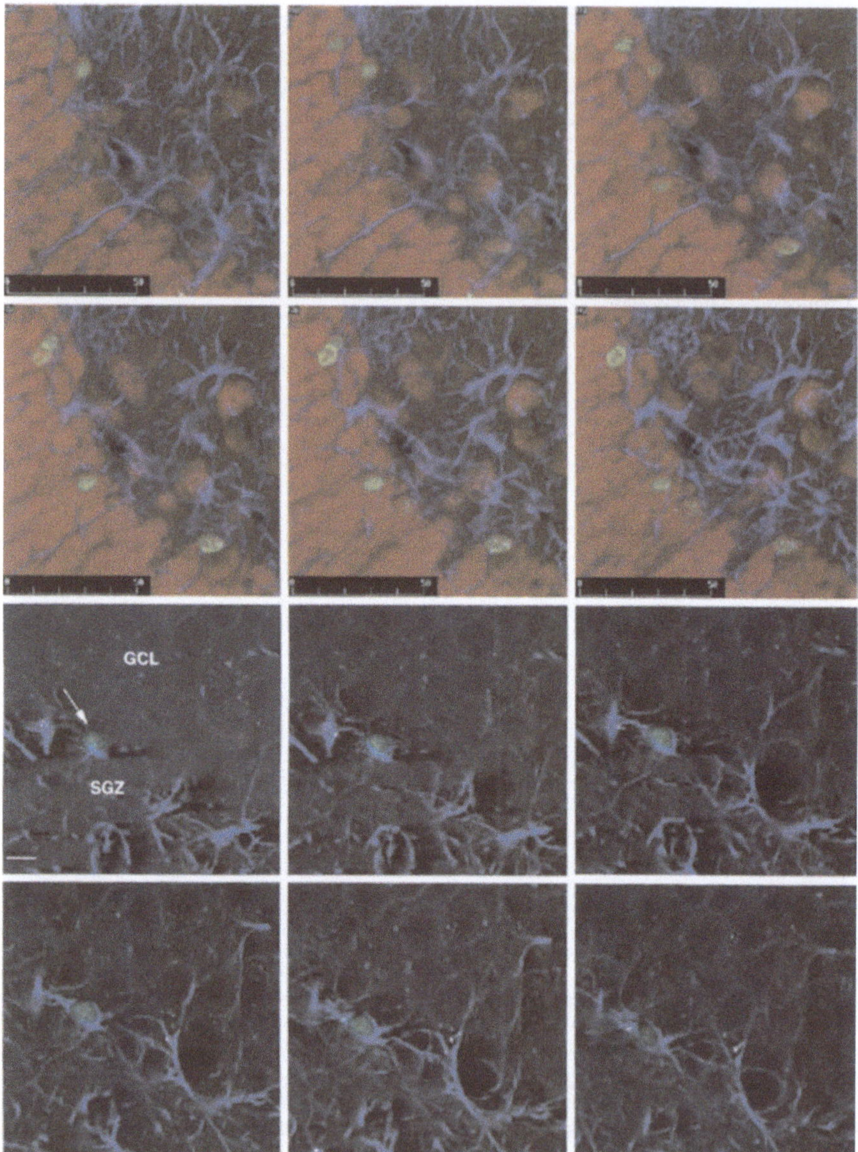

Fig. 3. Upper part: Triple immunofluorescence staining for BrdU (green), calbindin D28 and GFAP (blue) of the ipsilateral dentate gyrus with multiple cells co-labeled with BrdU and calbindin (green and red) identified with a Z-series confocal microscopy scan (step size 1.0 µm). In the lower part the arrow points to a BrdU-GFAP co-labeled new astrocyte in the SGZ identified with a Z-series confocal microscopy scan (step size 0.5 µm). The scale bar in the lower part corresponds to 10 µm. BrdU: bromodeoxyuridine; GFAP: glial fibrillary acidic protein; SGZ: subgranular zone; GCL: granular cell layer

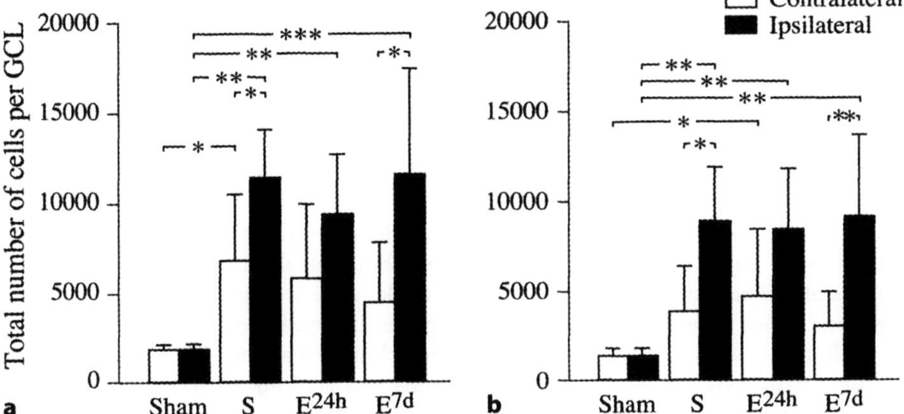

Fig. 4. The number of cells in the granular cell layer that co-labeled with BrdU and the neuronal markers NeuN (**a**) and calbindin D28k (**b**). S: Rats housed in standard housing after a distal ligation of the middle cerebral artery; E24h, E7d: rats transferred to an enriched environment 24 h, or 7 days after the ischemic insult. $^*p < 0.05$, $^{**}p < 0.01$, $^{***}p < 0.001$

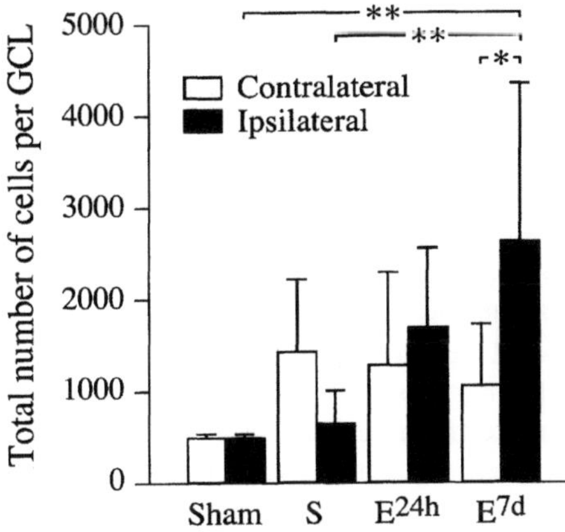

Fig. 5. The number of cells co-labeled with BrdU and the astrocyte marker GFAP. Note the striking difference on the ipsilateral side between S and E7d rats that were housed in the same standard environment during the BrdU administration. S: rats housed in standard housing after a distal ligation of the middle cerebral artery; E24h, E7d: rats transferred to an enriched environment 24 h and 7 days after the ischemic insult. $^*p < 0.05$; $^{**}p < 0.01$

was in the opposite direction. Whereas the standard group had 55% *less* astrocytes on the ipsi- than on the contralateral side (p = 0.054) the delayed enriched group has 253% *more* on the ipsilateral side (p < 0.05). The consequence of the low number of new astrocytes on the side ipsilateral to the infarct was a marked increase of the neuron to astrocyte ratio, 17.8, compared to 3.8 in the sham-operated rats. Early and delayed exposure to an enriched environment normalized the ratio to 5.5 and 4.3 respectively.

Compared to rats in standard housing, those introduced into enriched environmental enrichment 24 h after the arterial ligation had a significantly increased GCL volume on both sides whereas the volume in the delayed enriched rats was increased only on the contralateral side [35].

Discussion

Focal cortical ischemia induced a 5- to 6-fold increase in neurogenesis on the ipsilateral side and a less marked increase on the contralateral side as evaluated 5 weeks after the ischemic event and thus 4 weeks after 7 consecutive daily BrdU injections which labeled dividing cells during the first week post surgery.

Whereas postischemic environmental enrichment had no influence on hippocampal neurogenesis, the effect on gliogenesis was striking where a markedly perturbed neuron to glia ratio in rats housed in a standard environment was normalized by housing in enriched environments. The largest difference was seen between the standard and the delayed enriched groups that both had been housed in standard cages during the BrdU administration, thereby ruling out any possible influence on cell proliferation or early postischemic events. Our data suggest that enriched environment stimulates the differentiation of progenitor cells into astrocytes and/or selectively enhances the survival of progenitor cells committed to become astrocytes.

Considering that astrocytes have many functions essential for optimal neuronal function, the marked reduction of newly formed cells developing into astrocytes in the standard group is likely to provide a poor environment for newly formed neurons. In addition to being crucial for neuronal metabolism, intra- and extracellular homeostatic regulation, non-synaptic transmission, and being producers as well as targets for neurotrophins and other trophic factors, astrocytes take an active part in brain and synaptic plasticity [2, 5, 6, 17, 18, 24–26, 43, 50, 52, 53, 55, 61, 62]. During early brain development, neurons send out dendrites and axons to appropriate parts of the brain. However, most of their synapses are formed many days later and at about the same time that astrocyte matures [58]. TNF-a produced by astrocytes enhances synaptic efficacy by increasing surface expression of AMPA receptors, something that has been proposed to play a role in synaptic plasticity and modulation responses to neural injury [5]. Astrocytes increase their contact with synaptic elements in the visual cortex of rats reared in a complex environment [53]. Seasonal changes in astrocytes parallel neuronal plasticity in the song control area of the canary [26]. Spontaneous astrocytic calcium oscillations drive NMDA receptor-mediated neuronal excitation *in situ* [47]. It is thus reasonable to assume that the increase in gliogenesis in enriched compared to standard rats is beneficial.

Although no significant difference was observed between the early and delayed enriched group there was a consistent trend for less ipsi- to contralateral difference in the early enriched group. The fact that the volume of the granular cell layer was significantly increased compared to the standard group on both sides in the early enriched and only on the ipsilateral side in the delayed enriched group also suggests a difference between the groups related to the housing conditions during the first postoperative week. As no animals were killed directly after the end of BrdU administration we do not know if the initial progenitor cell proliferation differed between the groups. Ongoing studies are aimed at elucidating this point. As the early enriched rats clearly had a higher activity during the time of BrdU administration it seems logical that there could have been differences. Housing intact animals in an enriched environment has in most studies been shown to influence survival but not proliferation [32, 45, 60] although increased proliferation has been observed in one strain of mice [29].

Because physical exercise in the form of wheel running increases cell proliferation as well as net neuronal survival in mice [59], it has been proposed that physical exercise may be the most important component of housing in enriched environments [60]. However, whereas postoperative housing of animals in an enriched environment significantly improves functional outcome after focal brain ischemia [15, 20, 46], there is no evidence that postischemic running has a beneficial effect. In contrast, two studies on focal cerebral ischemia have indicated that rats housed in an enriched environment or social environment have a significant better outcome than rats housed in individual cages with free access to a running wheel [23, 51], which is in agreement with studies showing that motor learning but not repetitive physical exercise generates new synapses in cerebellar cortex [1]. It would clearly be of interest to look at postischemic cell proliferation in different types of stimulating environment.

Neurogenesis in normal and ischemic hippocampus can be prevented by injection of antagonists for the NMDA and AMPA glutamate receptors [4], and the NMDA receptor antagonist, MK801, has been reported to prevent neurogenesis after focal brain ischemia [3]. The periinfarct cortical area is hyperexcitable due to an imbalance between excitatory and inhibitory synaptic function [42, 44]. Whether that is beneficial or not in the early postischemic period stage is under discussion [64]. In studies on cell genesis in the perifocal region using a number of neuronal markers, BrdU was found to co-label with calbindin, calretinin, and parvalbumin and also with doublecortin but not with NeuN, MAP-2 and β-tubulin. Only calbindin was used for quantitative analysis since the other tracers co-localized with BrdU very rarely. Post-ischemic enrichment significantly decreased BrdU/calbindin co-labeling compared to standard housing [36].

The improvement in sensorimotor functional outcome after focal brain ischemia is most likely due to plastic changes in the remaining cortical and subcortical regions. The regions involved may depend on size and location of the lesion [21]. The behavioral tests used in this study evaluate sensorimotor functions, and we do not propose that the increased gliogenesis in hippocampus is directly related to outcome in our animals. The hippocampus is generally considered to be of main importance for spatial memory. However, in a recent review, it was suggested that the hippocampus may represent a bottleneck in processing new information and all learning is connected with memory [28]. We suggest that the current evidence for a major role of astrocytes in brain and synaptic plasticity justifies that more interest be directed towards gliogenesis in studies of post ischemic brain plasticity.

References

1. Anderson BJ, Li X, Alcantara AA, Isaacs KR, Black, Greenough WT (1994) Glial hypertrophy is associated with synaptogenesis following motor-skill learning, but not with angiogenesis following exercise. Glia 11:73–80
2. Araque A, Carmignoto G, Haydon PG (2001) Dynamic signaling between astrocytes and neurons. Annu Rev Physiol 63:795–813
3. Arvidsson A, Kokaia Z, Lindvall O (2001) N-methyl-D-aspartate receptor-mediated increase of neurogenesis in adult rat dentate gyrus following stroke. Eur J Neurosci 14:10–18
4. Bernabeu R, Sharp FR (2000) NMDA and AMPA/kainate glutamate receptors modulate dentate neurogenesis and CA3 Synapsin-I in normal and ischemic hippocampus. J Cereb Blood Flow Metab 20:1669–1680
5. Bezzi P, Domercq M, Vesce S, Volterra A (2001) Neuron-astrocyte cross-talk during synaptic transmission: physiological and neuropathological implications. Prog Brain Res 132:255–265
6. Chvatal A, Sykova E (2000) Glial influence on neuronal signaling. Prog Brain Res 125:199–216
7. Coggeshall RE, Lekan HA (1996) Methods for determining numbers of cells and synapses: a case for more uniform standards of review. J Comp Neurol 364:6–15
8. Dahlqvist P, Zhao L-R, Johansson I-M, Mattsson B, Johansson BB, Seckl JR, Olsson T (1999) Environmental enrichment alters NGFI-A mRNA expression after MCA-occlusion in rats. Neuroscience 93:527–535
9. Dash PK, Mach SA, Moore AN (2001) Enhanced neurogenesis in the rodent hippocampus following traumatic brain injury. J Neurosci Res 63:313–319
10. Dobrossy MD, Dunnett SB (2001) The influence of environment and experience on neural grafts. Nat Rev Neurosci 2:871–879
11. Eriksson PS, Perfilieva E, Björk-Eriksson T, Alborn A-M, Nordborg C, Peterson DA, Gage FH (1998) Neurogenesis in the adult human hippocampus. Nature Med 4:1313–1317
12. Goldman SA (1998) Adult neurogenesis: from canaries to the clinic. J Neurobiol 36:267–286
13. Gould E, Tanapat P (1997) Lesion-induced proliferation of neuronal progenitors in the dentate gyrus of the adult rat. Neuroscience 80:427–436
14. Gould E, Tanapat P, McEwen BS, Flugge G, Fuchs E (1998) Proliferation of granule cell precursors in the dentate gyrus of adult monkeys is diminished by stress. Proc Natl Acad Sci USA 95:3168–3171
15. Grabowski M, Sørensen J-C, Mattsson B, Zimmer J, Johansson BB (1995) Influence of an enriched environment and cortical grafting in functional outcome in brain infarcts of adult rats. Exp Neurol 133:1–7
16. Gundersen HJ, Bendtsen TF, Korbo L, Marcussen N, Moller, Nielsen K, Nyengaard JR, Pakkenberg B, Sorensen FB, Vesterby A, West MJ (1988) The new stereological tools: dissector, fractionator, nucleator and point sampled intercepts and their use in pathological research and diagnosis. Acta Pathol Microbiol Immunol Scand 96:857–881
17. Helmuth L (2001) Glia tell neurons to build synapses. Science 291:569–570
18. Hertz L, Hansson E, Rönnbäck L (2001) Signaling and gene expression in the neuron/glia unit during brain function and dysfunction: Holger Hydén in memoriam. Neurochem Int 39:227–252
19. Jin K, Minami M, Lan JQ, Mao XO, Batteur S, Simon RP, Greenberg DA (2001) Neurogenesis in dentate subgranular zone and rostral subventricular zone after focal cerebral ischemia in the rat. PNAS 98:4710–4715
20. Johansson BB (1996) Functional outcome in rats transferred to an enriched environment 15 days after focal brain ischemia. Stroke 27:324–326
21. Johansson BB (2000) Brain plasticity and stroke rehabilitation. The Willis Lecture. Stroke 31:223–230
22. Johansson BB, Belichenko PV (2002) Neuronal plasticity and dendritic spines: effect of environmental enrichment on intact and postischemic rat brain. J Cereb Blood Flow Metab 22:89–96
23. Johansson BB, Ohlsson A-L (1996) Environment, social interaction and physical activity as determinants of functional outcome after cerebral infarction in the rat. Exp Neurol 139:322–327
24. Jones TA, Greenough WT (1996) Ultrastructural evidence for increased contact between astrocytes and synapses in rats reared in a complex environment. Neurobiol Learn Mem 65:48–56
25. Jones TA, Hawrylak N, Greenough WT (1996) Rapid changes in GFAP immunoreactive astrocytes in rats reared in a complex environment. Brain Res 733:142–148
26. Kafitz KW, Guttinger HR, Muller CM (1999) Seasonal changes in astrocytes parallel neuronal plasticity in the song control area Hve of the canary. Glia 27:88–100
27. Kee NJ, Preston E, Wojtowicz JM (2001) Enhanced neurogenesis after transient global ischemia in the dentate gyrus of the rat. Exp Brain Res 136:313–320

28. Kempermann G (2002) Why new neurons? Possible function for adult hippocampal neurogenesis. J Neurosci 22:635–638
29. Kempermann G, Brandon EP, Gage FH (1998) Environmental stimulation of 129/SvJ mice causes increased cell proliferation and neurogenesis in the adult dentate gyrus. Curr Biol 8:939–942
30. Kempermann G, Gage FH (2002) Genetic influence on phenotypic differentiation in adult hippocampal neurogenesis. Brain Res Dev Brain Res 134:1–12
31. Kempermann G, Kuhn HG, Gage FH (1997) Genetic influence on neurogenesis in the dentate gyrus of adult mice. Proc Natl Acad Sci USA 94:10409–10414
32. Kempermann G, Kuhn HG, Gage FH (1997) More hippocampal neurons in adult mice living in an enriched environment. Nature 386:493–495
33. Kernie SG, Erwin TM, Parada LF (2001) Brain remodeling due to neuronal and astrocytic proliferation after controlled cortical injury in mice. Neurosci Res 66:317–326
34. Kolb B (1995) Brain plasticity and behavior. Hillside, NY: Lawrence Erlbaum
35. Komitova M, Perfilieva E, Mattsson B, Eriksson P, Johansson BB (2002) Effects of cortical ischemia and postischemic environmental enrichment on hippocampal cell genesis and differentiation in the adult rat. J Cereb Blood Flow Metab 22:852–860
36. Komitova M, Mattsson B, Eriksson PS, Johansson BB (2001) Post-ischemic housing influences cortical cell genesis. Abstract. Brain Research Interactive Symposium: Stem cells in the mammalian brain 132:P6–12
37. Lee J, Duan W, Long JM, Ingram DK, Mattson MP (2000) Dietary restriction increases the number of newly generated neural cells, and induces BDNF expression, in dentate gyrus of rats. J Mol Neurosci 15:99–108
38. Liu J, Bernabaeu R, Aigang L, Sharp FR (2000) Neurogenesis and gliogenesis in the postischemic brain. Neuroscientist 6:362–370
39. Liu J, Solway K, Messing RO, Sharp FR (1998) Increased neurogenesis in the dentate gyrus after transient global ischemia in gerbils. J Neurosci 18:7768–7778
40. Magavi SS, Leavitt BR, Macklis JD (2000) Induction of neurogenesis in the neocortex of adult mice. Nature 405:951–955
41. Mattsson B, Sørensen JC, Zimmer, Johansson BB (1997) Neural grafting to experimental neurocortical infarcts improves behavioral outcome and reduces thalamic atrophy in rats housed in an enriched but not in standard environment. Stroke 28:1225–1232
42. Mittmann T, Qu M, Zilles K, Luhmann HJ (1998) Long-term cellular dysfunction after focal cerebral ischemia; *in vitro* analysis. Neuroscience 85:15–27
43. Müller CM (1992) A role for glia in activity-dependent central nervous plasticity? Review and hypothesis. Internat Rev Neurobiol 34:215–218
44. Neumann-Haefelin T, Bosse F, Redecker C, Muller HW, Witte O (1999) Upregulation of GABA-receptor alpha1- and alpha2-subunit mRNA following ischemic cortical lesions in rats. Brain Res 816:234–237
45. Nilsson M, Perfilieva E, Johansson U, Orwar O, Eriksson PS (1999) Enriched environment increases neurogenesis in the adult rat dentate gyrus and improves spatial memory. J Neurobiol 15:569–578
46. Ohlsson A-L, Johansson BB (1995) The environment influences functional outcome of cerebral infarction in rats. Stroke 26:644–649
47. Parri PH, Gould TM, Crunelli V (2001) Spontaneous astrocytic Ca2+ oscillations *in situ* drive NMDA receptor-mediated neuronal excitation. Nature Neurosci 4:803–812
48. Perfilieva E, Risedal A, Nyberg J, Johansson BB, Eriksson PS (2001) Gender and strain influence on neurogenesis in dentate gyrus of young rats. J Cereb Blood Flow Metab 21:221–217
49. Peterson DA (2002) Stem cells in brain plasticity and repair. Curr Opin Pharmacol 2:34–42
50. Pfrieger FW, Barres BA (1996) New views on synapse-glia interactions. Curr Opin Neurobiol 6:615–621
51. Risedal A, Mattsson B, Dahlqvist P, Nordborg C, Olsson T, Johansson BB (2002) Environmental influences on functional outcome after a cortical infarct in the rat. Brain Res Bull, in press
52. Seil FJ (2001) Interactions between cerebellar Purkinje cells and their associated astrocytes. Histol Histopathol 16:955–968
53. Sirivaag AM, Greenough WT (1991) Plasticity of GFAP-immunoreactive astrocytes size and number in visual cortex of rats reared in complex environments. Brain Res 540:273–278
54. Sloviter RS (1989) Calcium-binding protein (calbindin-D28k) and parvalbumin immunocytochemistry: localization in the rat hippocampus with specific reference to the selective vulnerability of hippocampal neurons to seizure activity. J Comp Neurol 280:183–196
55. Smit AB, Syed NI, Schaap D, van Minnen J, Klumperman J, Kits KS, Lodder H, van der Schors RC, van Elk R, Sorgedrager B, Brejc K, Sixma TK, Geraerts WP (2001) A glia-derived acetylcholine-binding protein that modulates synaptic transmission. Nature 411:261–268

56. Snyder EY (1998) Neural stem-like cells: developmental lessons with therapeutic potential. The Neuroscientist 4:408–425
57. Takagi Y, Nozaki K, Takahashi J, Yodoi J, Ichikawa M, Hashimoto M (1999) Proliferation of neuronal precursor cells in the dentate gyrus is accelerated after transient forebrain ischemia in the mice. Brain Res 831:183–187
58. Ullian EM, Sapperstein SK, Christopherson KS, Barres BA (2001) Control of synapse number by glia. Science 291:657–661
59. Van Praag H, Kempermann G, Gage FH (1999) Running increases cell proliferation and neurogenesis in the adult mouse dentate gyrus. Nat Neurosci 2:266–270
60. Van Praag H, Kempermann G, Gage FH (2000) Neural consequences of environmental enrichment. Nature Rev Neurosci 1:191–198
61. Vernadakis A (1996) Glia-neuron intercommunications and synaptic plasticity. Prog Neurobiol 49:185–214
62. Wenzel J, Lammert G, Meyer U, Krug M (1991) The influence of long-term potentiation on the spatial relationship between astrocyte processes and potentiated synapses in the dentate gyrus neuropil of rat brain. Brain Res 560:122–131
63. Zeng J, Mattsson B, Schultz M, Johansson BB, Soerensen JC (2000) Expression of Zinc-positive cells and terminals in fetal neocortical homografts of adult rat depends on lesion type and rearing conditions. Exp Neurol 164:176–183
64. Zhao L-R, Mattsson B, Johansson BB (2000) Environmental influence on BDNF mRNA after focal brain ischemia in adult rats. Neurosci 97:177–184

VII
Abstract and Poster Presentations

Estrogen Protects Against Global Ischemia-Induced Neuronal Death and Prevents Activation of Apoptotic Signaling Cascades in the Hippocampal CA1

T. Jover, H. Tanaka, A. Calderone, K. Oguro, M. V. L. Bennett, A. M. Etgen, and R. S. Zukin

The importance of postmenopausal estrogen replacement therapy in affording protection against the selective and delayed neuronal death associated with cardiac arrest or cardiac surgery in women remains controversial. Here we report that exogenous estrogen at levels that are physiological in females affords protection against global ischemia-induced neuronal death and prevents activation of apoptotic signaling cascades in the hippocampal CA1 of male gerbils.

Global ischemia induced a marked increase in activated caspase-3 in CA1, evident at 6 h after ischemia. Global ischemia induced a marked upregulation of the proapoptotic neurotrophin receptor p75^{NTR} and a marked downregulation of the antiapoptotic trkA receptor in CA1, evident at 48 h. Global ischemia also induced a marked downregulation of mRNA encoding the AMPA receptor GluR2 subunit in CA1. Caspase-3, p75^{NTR}, trkA and GluR2 were not significantly changed in CA3 and dentate gyrus, indicating that the ischemia-induced changes in gene expression were region-specific. Exogenous estrogen attenuated the ischemia-induced increases in activated caspase-3 and blocked the increase in p75^{NTR} and the reduction in trkA in post-ischemic CA1 neurons, but did not prevent ischemia-induced downregulation of GluR2.

These findings demonstrate that long-term estrogen at physiological levels ameliorates ischemia-induced hippocampal injury and indicate that estrogen intervenes at the level of apoptotic signaling cascades to prevent onset of death in neurons otherwise "destined to die".

Dr. R. Suzanne Zukin, Albert Einstein College, Department of Neuroscience, 1300 Morris Park Avenue, Bronx, New York 10461, USA, Tel.: (718) 430-2160, Fax: (718) 430-8932, E-Mail: zukin@aecom.yu.edu

Maturation Phenomenon in Cerebral Ischemia V
A. M. Buchan et al. (Eds.)
© Springer-Verlag Berlin Heidelberg 2004

Ischemic Preconditioning Prevents Global Ischemia-Induced Downregulation of the AMPA Receptor GluR2 Subunit in the Hippocampal CA1 but does not Block Caspase-3 Activation nor Protect against GluR2 Antisense Knockdown

H. Tanaka, A. Calderone, T. Jover, S. Y. Grooms, R. S. Zukin, and M. V. L. Bennett

Animals subjected to sublethal transient global ischemia (preconditioning) exhibit neuroprotection against subsequent global ischemia-induced neuronal death in the hippocampal CA1 (ischemic tolerance). The molecular mechanisms underlying ischemic tolerance are, however, unclear.

Global ischemia modifies decreased GluR2 expression leading to an AMPAR subunit composition that promotes Ca^{2+}-permeability at CA1 synapses. Ischemic preconditioning induced a small, transient downregulation of GluR2 mRNA expression and prevented GluR2 mRNA and protein downregulation and neuronal death induced by a subsequent period of global ischemia lethal to CA1 neurons in naive animals. Sublethal ischemia and "sublethal" (and "lethal") GluR2 antisense, administered sequentially or together, acted synergistically to induce neuronal death.

This finding supports the concept that ischemic preconditioning acts at a step upstream and not downstream from suppression of GluR2 gene expression to afford neuroprotection and implicates transcriptional regulation of GluR2 expression in the adaptive mechanisms associated with ischemic tolerance. In addition, preconditioning prevented ischemia-induced upregulation of the pro-apoptotic neurotrophin receptor p75 and induction of terminal deoxynucleotidyl transferase-mediated UTP nick-end labeling (TUNEL) in CA1. In contrast, preconditioning did not block ischemia-induced activation of caspase-3 (or its activator, caspase-9), a protease implicated in the execution step of apoptosis.

This unexpected finding indicates that caspase-3 activation is not sufficient to induce neuronal death.

Dr. Michael Bennett, Albert Einstein College, Department of Neuroscience, 1300 Morris Park Avenue, Kennedy 720, Bronx, New York 10461, USA, Tel.: (718) 430-3679, Fax: (718) 430-8821, E-Mail: mbennett@aecom.yu.edu

Maturation Phenomenon in Cerebral Ischemia V
A. M. Buchan et al. (Eds.)
© Springer-Verlag Berlin Heidelberg 2004

Acute Recovery and Delayed Changes in Tissue Following Transient Cerebral Hypoxia-Ischemia: Impedance, Electron Microscopy and MR Imaging Studies in Rats

U. I. Tuor, M. Qiao, T. Foniok, B. Tomanek, and M. Del Bigio

A transient episode of cerebral hypoxia-ischemia will produce transient biochemical and cellular changes that may be followed by a delayed cell death. Magnetic resonance (MR) imaging techniques have demonstrated that despite a normalization of images post reperfusion, a secondary or delayed return of hyperintense areas in the T_2 and diffusion weighted (DW) images can appear several hours after the end of the hypoxic-ischemic insult [1, 2]. Reductions in extracellular space and cell swelling are known to occur during ischemia and are considered to underly some of these MR imaging changes. How well the extracellular space changes recover upon reperfusion is not known. Presently, we examine the pre, during and post hypoxic-ischemic changes in extracellular space in rats at two different ages because hypoxic-ischemic MR-imaging changes differ between them [2].

Changes in extracellular space were examined repeatedly prior to, during and after an episode of cerebral hypoxia-ischemia by measuring the impedance between electrodes within the ischemic hemisphere [3]. Impedance was measured in 1 week old (n = 5) and 4 week old (n = 5) rats and compared to respective control animals (n = 5/age group) [3]. Changes in extracellular space were also examined in 8 one week old rats using flash frozen tissue which was freeze substituted and examined with electron microscopic techniques [4]. Ultrastructural signs of cellular edema were investigated using electron microscopy in 26 additional animals perfusion fixed with glutaraldehyde. In 24 animals, multi-echo T_2 and DW MR images were also obtained prior to, during and one hour after hypoxia-ischemia using a 9.4 T system. Cerebral ischemia was produced by exposing sedated animals, which had undergone ligation of the right carotid artery while anesthetized, to 1.5–2 h (1 wk olds) or 30 min (4 wk olds) of hypoxia (8% O_2) and measurements were made prior to, during and 1 and/or 24 h after hypoxia.

The extracellular space, that was measured using impedance methods, decreased as early as 15 min after the start of hypoxia-ischemia in 4 week old brain whereas in 1 week old brain significant decreases did not occur until 30 min after the onset of hypoxia-ischemia. Irrespective of age, the extracellular space decreased during hypoxia-ischemia exhibiting a remarkable recovery to baseline levels within 1 h post hypoxia-ischemia. Similar patterns of change in the extracellular space were observed with the electron microscopy methods in one week old rats. The extracellular space decreased from $51 \pm 6\%$ in controls to $3 \pm 1\%$ during

Ursula Tuor, PhD, Senior Research Officer, NRC Research Professor (Adjunct), University of Calgary, Institute for Biodiagnostics (West), 3330 Hospital Drive NW, Calgary, Alberta, Canada, T2N 4N1
Tel.: (403) 221-3227, Fax: (403) 221-3230, E-Mail: ursula.tuor@nrc-cnrc.gc.ca

Maturation Phenomenon in Cerebral Ischemia V
A. M. Buchan et al. (Eds.)
© Springer-Verlag Berlin Heidelberg 2004

hypoxia-ischemia and enlarged to $30 \pm 21\%$ within 1 h post hypoxia-ischemia. Hyperintensities on DW images had a similar age-dependent difference in the onset and recovery of changes during and post-hypoxia. Ultrastructurally recovery was incomplete at 1 h with continued signs of swelling within cell constituents such as the endoplasmic reticulum and mitochondria and clumping of the nuclear chromatin in some cells. By 4–5 and 24 hours post-hypoxia there was marked subcellular changes including nuclear condensation and swelling of perivascular astrocytes and neurons.

The present results demonstrate that one of the early cellular alterations in cerebral tissue associated with a depletion of cerebral energy stores is a decrease in extracellular space. These decreases in extracellular space are well correlated to a decrease in the diffusibility of water as detected with diffusion-sensitive MR imaging and are consistent with the changes in extracellular space measured using freeze substitution and electron microscopic techniques. The perceived recovery of DW hyperintensities post-hypoxia-ischemia is associated with a reduction in the expanded extracellular space. However, ultrastructural abnormalities, some of which may reflect irreversible tissue damage are already visible within the first hour posthypoxia-ischemia. Thus the relatively gross measures of MR water diffusibility and tissue impedance techniques are detected as a recovery of the tissue from the ischemic insult whereas cellular abnormalities remain at the ultrastructural level. Determining the significance as to the reversibility of some of these ultrastructural changes requires further study.

Supported by the Canadian Institutes of Health Research.

References

1. Tuor UI, Kozlowski P, Del Bigio MR, Ramjiawan B, Su S, Malisza K, Saunders JK (1998) Exp Neurol 150:321–328
2. Ning G, Malisza KL, Del Bigio MR, Bascaramurty S, Kozlowski P, Tuor UI (1999) Pediatr Res 45:173–179
3. Goovaerts HG, Faes TJ, Valk-de-Roo GW et al. (1998) Physiol Meas 19:517–526
4. Palsgard E, Lindh U, Roomans GM (1994) Microsc Res Tech 28:254–258

A Selective Thrombin Inhibitor Prevents Thrombin-Induced Neuronal Cell Death and Mild Hypothermia Enhances its Neuroprotective Effects Following Transient Focal Ischemia in Rats

T. Kamiya, C. Ito, M. Ueda, K. Kato, S. Amemiya, T. Inaba, A. Terashi, and Y. Katayama

Background and Purpose

Thrombin is a multifunctional serine protease that has important roles in blood coagulation, the activation of platelets and the cleavage of fibrinogen to fibrin after ischemia. Recently, thrombin has been reported to induce apoptosis in neurons and astrocytes via activation of the thrombin receptor [1]. The aim of this study is, therefore, to determine whether a selective thrombin inhibitor, Argatroban, would prevent neuronal cell death, especially thrombin-induced apoptotic cell death, and whether extra-mild hypothermia (35 °C) would enhance the neuroprotective effect of a selective thrombin inhibitor following transient focal ischemia in rats.

Methods

Sprague-Dawley rats were subjected to MCAo using an intraluminal suture technique for 2 h [2]. The rats were reperfused for 24 h and decapitated for infarct and edema analysis [3]. Animals were randomly divided into the following four groups: (I) vehicle-treated, normothermic group: (II) vehicle-treated, mild hypothemic (35 °C) group: (III) Argatroban-treated, normothermic group: (IV) Argatroban-treated, mild hypothemic (35 °C) group. Argatroban-treated animals received a continuous injection of Argatroban (3.0 mg/kg) for 24 h after the onset of ischemia, while vehicle-treated groups received same dose of vehicle. During ischemia, temporal muscle and rectal temperatures were monitored and maintained at 37 °C in the normothermic animals and at 35 °C in the hypothermic animals. Neurological symptoms and survival rate were also tested just before decapitation.

Results

Infarct Volume: In group III, the cortical infarct volume (162 ± 28 mm^3) was significantly less than those in group I (205 ± 55 mm^3, $p < 0.05$). Moreover, Argatroban with hypothermia (group IV) decreased the cortical infarct volume (114 ± 27 mm^3) significantly compared with those of groups I and III ($p<0.05$), while there was no

Tatsushi Kamiya, Chikako Ito, Masayuki Ueda, Kengo Kato, Shimon Amemiya, Toshiki Inaba, Akiro Terashi and Yasuo Katayama, Second Department of Internal Medicine, Nippon Medical School, Tokyo, 113-8603, Japan.

Maturation Phenomenon in Cerebral Ischemia V
A.M. Buchan et al. (Eds.)
© Springer-Verlag Berlin Heidelberg 2004

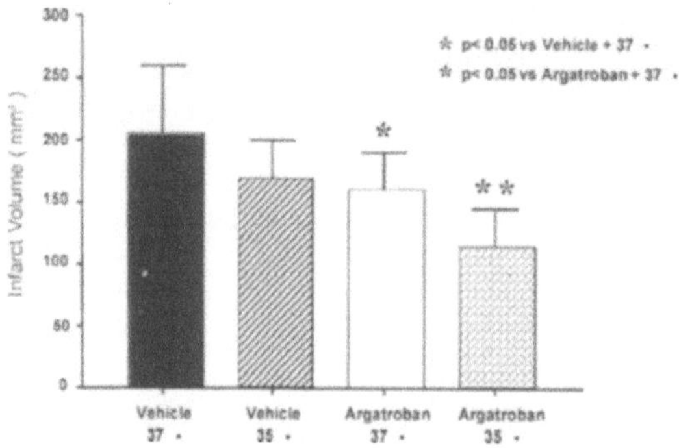

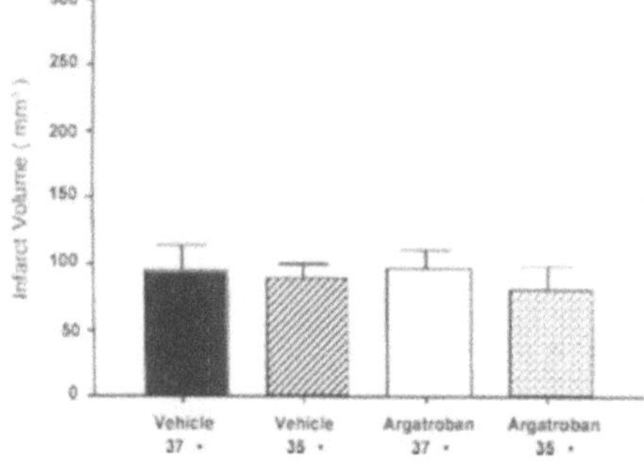

Fig. 1.

significant difference in the striatum among the groups. *Edema Volume*: The cortical edema volumes of group IV were significantly smaller than those in group I, II or III (p < 0.05). *Neurological Symptoms*: Argatroban improved neurological symptoms (posture and hemiplegia) significantly and also improved survival rate.

Conclusions

These results demonstrate that a selective thrombin inhibitor, Argatroban prevents neuronal cell death in the penumbra, involved throughout thrombin-induced apoptotic cell death. Furthermore, mild hypothermia enhances neuroprotective effects of Argatroban, suggesting that this combined therapy may be a new therapeutic strategy for the treatment of acute stroke.

References

1. Donovan FM et al (1997) J Neurosci 17:5316–5326
2. Katsumata T et al (2001) Brain Res 901:62–70
3. Jacewicz M et al (1992) J Cereb Blood Flow Metab 12:359–370

Zinc Dynamics in the Cerebral Cortex of Adult Mice Following Photothrombotic Stroke

S. Subramaniam and R. H. Dyck

Zinc is an important neuromodulatory transmitter that is co-released with gluta-mate at many central excitatory synapses. Growing evidence suggests that zinc is a key mediator and modulator of ischemic cell death in neurons. Vesicular release of zinc from presynaptic nerve terminals following ischemia is thought to be a key step in the excitotoxic injury of postsynaptic neurons. As a first step in determin-ing the role of zinc in post-ischemic neuronal injury we have assessed the distribu-tion of synaptic zinc in the motor cortex of adult mice at various intervals follow-ing induction of a photothrombotic stroke.

Methods

Photothrombotic stroke, which is based on photochemical damage to the vascular endothelium and subsequent platelet aggregation, is an effective means of inducing a stroke that is very similar to human thrombotic stroke. A photothrombotic stroke was induced by directing laser light (20 mW, 532 nm) for 5–20 min, through aper-tures of varying diameter (1 mm, 2 mm and 3 mm) onto the motor cortex of adult mice, immediately following iv administration of the photosensitive dye, Rose Ben-gal (20 mg/kg). The animals were allowed to survive up to 60 h then were adminis-tered sodium selenite (15 mg/kg, ip) 1 h prior to sacrifice in order to stain the pool of zinc contained in synaptic vesicles (Timm/Danscher method).

Results

The cortical infarct consisted of a cylindrical area, completely devoid of zinc stain-ing, that extended from the pial surface of the cortex through to the subcortical white matter. The infarcted area also included the corpus callosum and dorsal striatum when larger laser apertures were combined with longer laser exposure times (2 mm and 3 mm panels in figure below). At all time intervals assessed, the

Suresh Subramaniam and Richard H. Dyck, Dept of Neuroscience and Dept of Psychology, Univer-sity of Calgary, Calgary, AB

Suresh Subramaniam, MD, Rm 1162, 1403-29 Street NW, Calgary, Alberta, Canada, T2N 2T9, E-Mail: subramas@ucalgary.ca
Richard Dyck, PhD, Assistant Professor, University of Calgary, Psychology, 2500 University Drive, NW, Calgary, Alberta, Canada, T2N 1N4, Tel.: (403) 220-4206, Fax: (403) 282-8249, E-Mail: rdyck@-ucalgary.ca

Maturation Phenomenon in Cerebral Ischemia V
A. M. Buchan et al. (Eds.)
© Springer-Verlag Berlin Heidelberg 2004

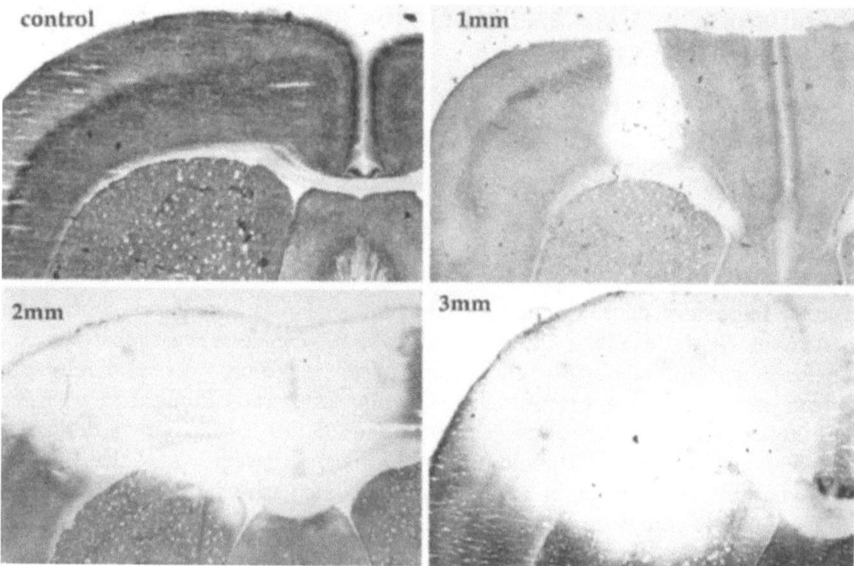

Fig. 1.

ischemic core was found to be essentially devoid of staining for synaptic zinc. However, we found that a narrow zone bordering the infarct exhibited significantly increased levels of staining for zinc, but only at survival periods less than 6 h. At longer survival times, zinc staining in the zone bordering the infarct appeared normal.

Conclusions

Although not yet fully understood, it is believed that decreased zinc staining in the ischemic zone reflects rapid release of zinc from synaptic vesicles in response to the ischemic insult. Furthermore, we believe that the increased zinc staining in the bordering zone at early survival times may represent the penumbral zone. As high extracellular levels of zinc are thought to be neurotoxic, therapeutic strategies aimed at manipulating the zinc levels in infarct and the salvageable penumbral zone could provide new targets for neuroprotective intervention.

Supported by the Canadian Stroke Network and a CIHR grant to RHD.

Persistent Neuroprotection against Focal Cerebral Ischemia Induced by Cortical Spreading Depression in Mice

H. Yanamoto, I. Nagata, J.-H. Xue, Z.-W. Zhang, K. Murao, K. Iihara, and H. Kikuchi

Introduction

A prolonged period of cortical spreading depression induces a potent resistance against temporary focal ischemia in rat brain [1]. The phenomenon of infarct tolerance was investigated in mice using a newly developed mouse neocortical infarction model.

Methods

In C57BL/6J mice, a prolonged period of CSD preconditioning was induced by intracerebral micro-infusion of 4 M KCl at a rate of 0.5 µl/h for 24 h via an osmotic mini-pump implanted under the skin of the back. After intervals of 3, 6, 9 or 12 days following the end of KCl-preconditioning, temporary neocortical focal ischemia was induced by permanent occlusion of the left middle cerebral artery, superimposed with 1 h-temporary occlusion of the bilateral common carotid arteries (2) (n = 7 each). After twenty-four hours, the brain was removed and cerebral infarcted volume was analyzed by TTC stain. Regional cerebral blood flow, before and during focal ischemia was monitored by a laser-Doppler flowmetry system in CSD- or vehicle (saline)-preconditioned groups. To examine the chronic effect of CSD-preconditioning against focal ischemia, the infarcted volumes were also analyzed 14 days after the induction of ischemia in groups with 6-day interval following CSD- or vehicle-preconditioning.

Results

In groups with 6- and 9-day interval following CSD, significantly smaller neocortical infarctions were demonstrated: 19 ± 6.5 mm^3 and 22 ± 6.3 mm^3, respectively, compared to the volume in the group without CSD (34 ± 10 mm^3). When the infarcted volume was analyzed 14 days after ischemia, the significant difference between groups with CSD- and vehicle-preconditioning was again demonstrated.

Hiroji Yanamoto [1,2], Izumi Nagata [2], Jing-Hui Xue [1], Zhi-wen Zhang [1], Kenichi Murao [2], Koji Iihara [2], Haruhiko Kikuchi [2]
[1] Laboratory for Cerebrovascular Disorders, [2] Department of Cerebrovascular Surgery, National Cardiovascular Center and Research Institute of NCVC, Suita, 565-8565 Japan

Maturation Phenomenon in Cerebral Ischemia V
A. M. Buchan et al. (Eds.)
© Springer-Verlag Berlin Heidelberg 2004

Conclusion

The phenomenon of infarct tolerance was demonstrated in mice. Neuroprotection by infarct tolerance in mouse neocortex persisted chronically.

References

1. Yanamoto H, Hashimoto N, Nagata I, Kikuchi H (1998) Infarct tolerance against temporary focal ischemia following spreading depression in rat brain. Brain Res 784:239–249
2. Yanamoto H, Nagata I, Niitsu Y, Sakai N, Zhang Z, Xue J-H, Kikuchi H (2000) A new mouse model of neocortical infarct caused by temporary focal ischemia. J Cereb Blood Flow Metab 21:516

Anatomical, Morphological and Behavioral Assessments of Four Different Models of Stroke

C. L. R. Gonzalez and Bryan Kolb

Introduction

Several models of focal cerebral injury have been developed to study the pathology and treatment of stroke. There is some controversy, however, over whether the behavioral and anatomical effects of the different lesion models are equivalent. In the present study we compared the anatomical and behavioral effects of three different models of permanent ischemia plus an additional group with ablations of the motor cortex by suction.

Methods

Thirty-five male Long-Evans rats were used in this study. Animals were randomly assigned to four different groups: motor cortex lesion by suction (n = 8), motor cortex lesion by devascularization (n = 8), medial cerebral artery occlusion (MCAO) only distal branches (MCAO-small, n = 10), and MCAO with both distal and proximal branches occluded (MCAO-large, n = 9). Two weeks before the surgeries and continuing for the duration of the behavioral testing, animals were food restricted to 18 g/day. Animals were trained pre-surgery to successfully open and consume 5 sunflower seeds and on a reaching task. At the end of the two weeks the total time to consume the 5 sunflower seeds and the number of shell pieces were recorded. For the tray reaching task the number of successful reaches (extension of the forepaw, grasp and consumption of the food) over the total (successful plus non-successful attempts) were recorded. In order to assess forepaw inhibition, animals were videotaped while swimming to a visible platform at the end of an aquarium two days before the surgeries. Tongue extension and forepaw use asymmetry were also assessed. All behavioral tasks used have been shown to be sensitive to motor deficits and in each case performance was videotaped. Behavior was assessed one week after the lesion and every other week for a total of nine weeks. At the end of the behavioral testing animals were sacrificed and the brains were processed for either Cresyl-Violet or Golgi-Cox.

Ms Claudia L. Gonzalez, Canadian Center for Behavioural Neuroscience, University of Lethbridge, Department of Psychology & Neuroscience, 4401 University Drive, Lethbridge, Alberta, Canada, T1K 3M4, Tel.: (403) 394-3993, Fax: (403) 329-2775, E-Mail: claudia.gonzalez@uleth.ca

Maturation Phenomenon in Cerebral Ischemia V
A. M. Buchan et al. (Eds.)
© Springer-Verlag Berlin Heidelberg 2004

Results

The lesion size was fairly consistent within and among the groups with suction, devascularization, and large MCAOs reducing brain weight about 17% in all groups. Animals with small MCAOs had smaller lesion size reducing brain weight up to 9% as compared to controls. One week after the surgeries all animals were severely impaired on all tasks and although they slowly improved over the nine weeks they never reached pre-operative baselines. Interestingly, no differences among groups were detected in any of the five tasks except in reaching where animals with small MCAOs tended to use both the impaired and non-impaired limbs, whereas animals with suction lesions, devascularization, or large MCAOs only used the non-impaired limb. Using the Golgi-Cox method, pyramidal cortical cells from an area adjacent to the lesion (Cg3, Layer III) were drawn in order to examine the level of subsequent neuronal plasticity following the different kinds of stroke. Sholl and branch order analyses were done for the apical and basilar trees of the cells contra- and ipsilateral to the lesion. Overall, there was no change in the apical tree of either side on the groups with MCAOs. There was, however, an increase on the apical tree in both contralateral and ipsilateral sides of the group with devascularization and a decrease on the ipsilateral side of the group with suction lesions. On the basilar tree contralateral to the lesion there was a slight decrease in the group with suction lesions but virtually no changes on the other three groups. Finally, there was a reduction on the basilar tree ipsilateral to the lesion in both groups with MCAOs and an even greater reduction on the group with suction lesions. The group with devascularization showed no changes from control and differed from the other three groups. A correlation between the morphological changes and the amount of cortical and striatal damage showed that the greater the cortical **or** striatal damage, the greater the atrophy on the basilar tree.

Conclusions

Two main conclusions could be drawn from these results: 1) The severity and duration of the behavioral effects of suction lesions, devascularization, and MCA occlusions is remarkably similar. 2) The method to induce the lesion produces striking differences in cortical plasticity only part of which can be explained by lesion extent. We are currently assessing whether or not the same pattern of morphological changes observed in the cortex are also present in the striatum.

Effects of Endurance Exercise on Recovery Following Focal Ischemic Infarct

N. Cooper, P. VandenBerg, S. Cooper, and J. Kleim

Introduction

Exercise has been shown to have several effects on the brain, including an increase in capillary density within the motor cortex. This adaptation is thought to occur in response to the increased metabolic demands associated with increased brain activity. Such changes may therefore provide a neural framework to support enhanced capacity for recovery from cortical ischemia. This study examined the effects of exercise on recovery in a rat model of focal cortical ischemia.

Methods

Adult male rats were allocated to either a Voluntary eXercise (VX) or Inactive Condition (IC). VX animals were housed for 30 days with unlimited access to running wheels while IC animals were housed in standard laboratory cages. VX animals exhibited a progressive increase in the distance traveled per day. Microelectrode stimulation was then used to derive high resolution maps of the forelimb representations within the motor cortex of animals from both conditions. In half of the animals, a focal ischemic infarct was then produced within 35% of the forelimb representation via electrocoagulation of the surface vasculature. All animals were allowed to recover for three weeks, during which a battery of behavioural tests were conducted to assess level of recovery. At the end of the recovery period, all animals were mapped again.

Results

Map and behavioural analyses are currently being completed.

Conclusions

Preliminary results suggest that running reduces damage from focal cortical ischemia.

Ms Natalie R. Cooper, University of Lethbridge, Department of Psychology & Neuroscience, 4401 University Drive West, Lethbridge, Alberta, Canada, T1K 3M4
Tel.: (403) 394-3995, Fax: (403) 329-2775, E-Mail: coopnr@uleth.ca
Jeffrey A. Kleim, PhD, AHFMR Medical Scholar, Canadian Center for Behavioural Neuroscience, Dept of Psychology & Neuroscience, University of Lethbridge, Lethbridge, Alberta, Canada, T1K 3M4
Tel.: (403) 329-2044, Fax: (403) 329-2775, http://www.psych.uleth.ca/jkleim/index.htm

Maturation Phenomenon in Cerebral Ischemia V
A. M. Buchan et al. (Eds.)
© Springer-Verlag Berlin Heidelberg 2004

Leukocyte Rolling and Adhesion in Pial Microvessels Following Cerebral Ischemia in the Mouse: Effects of Hypothermia

I. T. Sutcliffe, H. Smith, J. S. Hutchison, and D. B. Stanimirovic

Introduction

Neuroprotection by hypothermia has been demonstrated both clinically (1) and experimentally (2). Moderate hypothermia is most beneficial when applied immediately following the insult and for longer duration (2). Proposed mechanisms of action of hypothermia include lower glutamate concentrations, decreased BBB permeability, and reduction of inflammatory gene expression (3). We have recently shown that hypothermia inhibits rolling and adhesion of leukocytes in pial venules following systemic injection of IL-1β in C57Bl6 mice (4). In this study, we investigate the effects of hypothermia on leukocyte rolling and adhesion in pial microvessels induced by cerebral ischemia.

Methods

Forebrain ischemia was induced in C57/Bl6 mice by a 10 min bilateral common carotid artery (CCA) occlusion. Mice were anesthetized by i.p. injection of xylazine (10 mg/kg)/ketamine (150 mg/kg) mixture. Under a dissecting microscope, a thermister was placed intracranially in the left parietal bone for brain temperature monitoring. Both rectal and cerebral temperatures were recorded; rectal temperature was kept within 0.5 °C of the target temperature throughout the experimental period using a heated water mat and overhead heating lamp. Animals were divided into 3 experimental groups: (i) sham control, (ii) ischemia followed by a 4-h hypothermic (32 °C) reperfusion, and (iii) ischemia followed by normothermic (37 °C) reperfusion. Immediately before CCA occlusion, 1.5-h and 3.5-h after CCA occlusion, an open cranial window was made in the right parietal bone as described (4) and mice were placed under an intravital microscope (Olympus BHMJ

I. T. Sutcliffe [1,2], H. Smith [1,2], J. S. Hutchison [1], D. B. Stanimirovic [2]
[1] Division of Paediatric Intensive Care, [2] Institute for Biological Sciences, National Research Council, Ottawa, Ontario, Canada

Mr. Ian Sutcliffe, National Research Council of Canada, Institute for Biological Sciences, 1200 Montreal Road, Bldg M54, Ottawa, Ontario, Canada, K1A OR6, Tel.: (613) 993-3730, Fax: (613) 941-4475, E-Mail: ian.sutcliffe@nrc.ca
Dr. Danica Stanimirovic, Director, Neurobiology Program, Institute for Biological Sciences, National Research Council of Canada, 1200 Montreal Road,y Bldg M54, Ottawa, Ontario, Canada, K1A OR6, Tel.: (613) 993-3730, Fax: (613) 941-4475, E-Mail: danica.stanimirovic@nrc. ca

Maturation Phenomenon in Cerebral Ischemia V
A. M. Buchan et al. (Eds.)
© Springer-Verlag Berlin Heidelberg 2004

modular focusing mount and BH2-RFCA illuminator, Hitachi CCD video camera, Sony monitor and Panasonic WJ-810 time code generator). Regional CBF was monitored using laser-Doppler flowmetry. Acridine orange (17 µg/kg) was infused into the tail vein, and pial microcirculation was illuminated with a mercury vapor lamp through a 495 nm excitation filter for fluorescence or with a fibreoptic epi-illuminator for bright field recording. Videocassette recordings were made for 30 seconds each at 2–4 venules per animal. The number of leukocytes rolling and adhering to the walls of venules was determined during playback of the videotapes. The quantity of neutrophils entering the brain parenchyma at 24 h following cerebral ischemia was determined using myeloperoxidase assay and immunohistochemistry. Brain injury was assessed at 24 h and 72 h after ischemia by determining *in situ* end labeling (TUNEL), NeuN expression, immunohistochemical determination of GFAP expression, and blinded neuropathological scoring. CD18 expression and chemotaxis of freshly isolated human neutrophils, and ICAM-1 and IL-8 expression in human brain endothelial cells (HCEC) exposed to hypoxia, or hypoxia and hypothermia, were also determined.

Results

rCBF in pial microcirculation was reduced to <10% of basal at the end of 10 min ischemia, and recovered to 50% by 20 min of reperfusion. Four hours of postischemic hypothermia to 32 °C decreased neuronal loss (NeuN) and TUNEL labeling in hippocampal, striatal and cortical brain regions at 3 and 7 days following ischemia compared to mice maintained normothermic (37 °C). A marked increase in leukocyte rolling and adhesion in pial microcirculation was observed in ischemic mice following 0-h, 2-h and 4-h reperfusion at 37 °C compared to sham-operated animals. Hypothermia (32 °C) significantly decreased leukocyte rolling and adhesion at 2-h reperfusion. There was no significant infiltration of neutrophils into the brain parenchyma in this model. Hypothermia (32 °C) failed to affect either fMLP- or LPS-induced CD18 expression in neutrophils or neutrophil chemotaxis *in vitro* induced by hypoxia, but reduced the expression/release of IL-8 in hypoxic HCEC.

Conclusions

Postischemic hypothermia inhibits leukocyte rolling and adhesion and reduces neuronal damage in forebrain structures after global cerebral ischemia in C57/Bl6 mice. In-vitro studies suggest that attenuated rolling/adhesion is likely the consequence of reduced expression/production of chemokine(s) by cerebral endothelial cells.

References

1. Marion DW, Penrod LE, Kelsey SF et al (1997) N Engl J Med 336:540–546
2. Colbourne F, Corbett D, Zhao Z, Yang J, Buchan AM (2000) J Cereb Blood Flow Metab 20:1702–1708
3. Maron DW (2001) Curr Pharm Dev 7:1533–1536
4. Sutcliff IT, Smith HA, Stanimirovic D, Hutchison JS: J Cereb Blood Flow Metab 21:1310–1319

Early Hypoxic-Ischemic Changes in Blood Flow, T_1 Relaxation Time, Apparent Diffusion Coefficient and Water Content in Neonatal Rat Brain

M. Qiao, P. Latta, R. Buist, S. Bascaramurty, T. Foniok, E. McKenzie, B. Tomanek, and U. I. Tuor

Introduction

Technological developments in magnetic resonance imaging have provided the methods to non-invasively and repeatedly monitor cerebral blood flow (CBF) and water dynamics in human and experimental animals [1, 2]. Cerebral ischemia is associated with brain edema and tissue injury but the correlation between transient reductions in CBF and alterations in brain water are not well understood [3]. We determined how changes in brain water during and after transient hypoxia-ischemia (HI) are related to changes in CBF by measuring blood flow (artery spin-tagging imaging) and tissue water changes with magnetic resonance imaging techniques (T_1-weighted and diffusion-weighted imaging) in a neonatal model of cerebral hypoxia-ischemia (HI).

Methods

Cerebral HI was produced as described previously [4]. Briefly, the right carotid artery was isolated surgically (sham control) or occluded with subsequent exposure to 8% oxygen (HI) for 1.5 h in 23 one-week old Wistar rats. CBF, the apparent diffusion coefficient (ADC) and T_1 maps were acquired either in sham controls or in HI rats before, during and for 1 h after HI using a 9.4 T magnetic resonance system. Water content was assessed as wet/dry weights in subgroups of these animals. A paired t-test was used to compare differences in the mean values for CBF, ADC, T_1 and water content within the left and right hemispheres.

Results

In sham controls, there were no left-right differences in CBF, ADC, T_1 or water content. In HI rats, there was a modest reduction in CBF in the HI hemisphere ipsilateral to the carotid occlusion prior to the start of HI with no change in T_1 or ADC. CBF in the ipsilateral hemisphere declined further during HI and was partially restored at 1 h post HI. Marked increases in water content occurred in the ip-

Ms Min Qiao, Research Technical Officer, Institute of Biodiagnostics, National Research Council of Canada, 3330 Hospital Drive NW, Calgary, Alberta, Canada, T2N 4N1
Tel.: (403) 221-3225, Fax: (403) 221-3230, E-Mail: min.qiao@nrc-cnrc.gc.ca

Maturation Phenomenon in Cerebral Ischemia V
A. M. Buchan et al. (Eds.)
© Springer-Verlag Berlin Heidelberg 2004

silateral hemisphere during HI and remained elevated at 1 h post HI, corresponding well to the increases in T_1 and decreases in ADC observed at these same time points.

Conclusions

HI-induced cerebral edema in neonatal brain was consistently detected as an increase in T_1 and a reduction in the ADC and an increase in brain water indicative of an early occurrence of cellular edema [5]. Restoration of blood flow towards pre-hypoxia levels after HI did not reverse the brain edema in these animals (Funded by Canadian Institutes of Health Research).

References

1. Detre JA, Alsop DC (1999) Perfusion magnetic resonance imaging with continuous arterial spin labeling: methods and clinical applications in the central nervous system. Europ J Radiol 30:115–124
2. Lythgoe MF, Thomas DL, Calamante F, Pell GS, King MD, Busza AL, Sotak CH, Williams SR, Ordidge RJ, Gadian DG (2000) Acute changes in MRI diffusion, perfusion, T1, and T2 in a rat model of oligemia produced by partial occlusion of the middle cerebral artery. Mag Res Med 44:706–712
3. Qiao M, Malisza KL, Del Bigio MR, Tuor UI (2001) Correlation of Cerebral hypoxic-ischemic T2 changes with tissue alterations in water content and protein extravasation. Stroke 32:958–963
4. Tuor UI, Kozlowski P, Del Bigio MR, Ramjiawan B, Su S, Malisza K, Saunders JK (1998) Diffusion- and T2-weighted increases in magnetic resonance images of immature brain during hypoxia-ischemia: transient reversal posthypoxia. Exp Neurol 150:321–328
5. van Bruggen N, Roberts TPL, Cremer JE (1994) The application of magnetic resonance imaging to the study of experimental cerebral ischaemia. Cere Br Metabolism Review 6:180–210

Increased Expression of nNos Following Cortical Spreading Depression in Rat Brain

J.-H. Xue, H. Yanamoto, I. Nagata, Z.-W. Zhang, and H. Kikuchi

Introduction

In our previous study, a prolonged period (48 h) of cortical spreading depression (CSD) effectively induced resistance to focal ischemia in 12 days without affecting rCBF, and significantly reduced the volume of infarcted lesion in rats, a phenomenon called 'infarct tolerance' [1]. Recently, NOS has been implicated in the decrease of hydroxyl radical (OH) generation following oxidative stress, via NO production as an antioxidant defense mechanism [2]. To analyze the role of the antioxidant defense mechanism via NO, the expression of neuronal nitric oxide synthase (nNOS) was examined following a prolonged period of CSD in rat neocortex.

Methods

In 105 male Sprague-Dawley rats (8 weeks old), CSD was generated by continuous microinfusion of 4 M potassium chloride (KCl) to the primary somatosensory area in the left neocortex. The same amount of vehicle (saline) without CSD was used as control. At different intervals of 1, 3, 6, 9, 12, 18, or 24 days after the end of KCl or saline preconditioning, brains were removed under deep anesthesia, and Western blot and immunohistochemical analyses were performed using rabbit nNOS polyclonal antibody (KAP-NO003, Stressgen Biotech Co.) for the left neocortex.

Results

The levels of nNOS protein were significantly elevated in the neocortex at intervals of 1, 3, 6, 9 and 12 days following 48 h-CSD, compared to the same intervals in the controls (p < 0.01). The elevated nNOS protein level began to decrease at 18 days and returned to baseline level at 24 days. The immunoreactivity for nNOS gradually increased in a widely spread area in the ipsilateral neocortex for the initial 12 days, and remained elevated at 24 days following CSD.

Jing-Hui Xue[1], Hiroji Yanamoto[1,2], Izumi Nagata[2], Zhi-wen Zhang[1], Haruhiko Kikuchi[2]
[1] Laboratory for Cerebrovascular Disorders, [2] Department of Cerebrovascular Surgery, National Cardiovascular Center and Research Institute of NCVC, Suita, 565-8565 Japan

Maturation Phenomenon in Cerebral Ischemia V
A.M. Buchan et al. (Eds.)
© Springer-Verlag Berlin Heidelberg 2004

Conclusions

CSD significantly increased the levels of nNOS protein over 12 days in the neocortex, and the increase in nNOS-immunoreactivity proceeded in the period during the development of infarct tolerance, which indicated an upregulation of the defense mechanism via NO against focal ischemia in the brain with infarct tolerance.

References

1. Yanamoto H, Hashimoto N, Nagata I, Kikuchi H (1998) Infarct tolerance against temporary focal ischemia following spreading depression in rat brain. Brain Res 784:239–249
2. Andoh T, Lee SY, Chiueh CC (2000) Preconditioning regulation of bcl-2 and p66shc by human NOS1 enhances tolerance to oxidative stress. FASEB J 14:2144–2146

Regional Differences on Free Fatty Acid Accumulations between Upper and Lower Frontal Cortex in Rat Focal Ischemia

M. Kubota, M. Nakane, T. Nakagomi, A. Tamura, H. Hisaki, and N. Ueta

Purpose

We have already reported that the levels of FFAs in the global frontal cortex showed a biphasic increase at 1 and 24 h following rat permanent focal ischemia [1]. To evaluate the regional characteristic of frontal cortex (penumbra) on the alteration of membrane phospholipids after MCA occlusion, the rat frontal cortex was divided into upper and lower parts, and their free fatty acid compositions were examined.

Materials and Methods

The S-D rats were anesthetized with 2% halothane and the left MCA was permanently occluded at a proximal site using a micro bipolar coagulator according to the Tamura method. Each experimental group consisted of five rats and the durations of ischemia were 0, 0.5, 1, 3, 6 and 8 h, respectively. After treatment with microwaves, the left cerebral frontal cortex was divided into upper and lower parts. The former was rich in ischemic penumbra. The total lipids of both samples were extracted with 20 volumes of chloroform-methanol (2:1, by volume) in accordance with the Folch method. For the separation of FFA, the total lipid fraction was subjected to Bond Elut (NH_2) column, using 2% acetic acid in diethyl ether as the elution solvent. After evaporation of the elution solvent, FFA fraction was trimethylsilylated with TMS-PZ (trimethylsilylating reagent) (Tokyo Kasei, Japan) and subjected to GLC.

Results

In upper frontal cortex, FFA levels transiently increased at 0.5 and 1 h after MCA occlusion, and thereafter returned to pre-ischemic levels. The arachidonic acid, stearic acid and oleic acid were more prominent components in the transiently increased FFA (8-, 3- and 3-folds of pre-ischemic level, respectively). In lower frontal cortex, FFA levels progressively increased and reached a high value of 4138 µmol/

Masaru Kubota[1], Makoto Nakane[1], Tadayoshi Nakagomi[1], Akira Tamura[1], Harumi Hisaki[2], Nobuo Ueta[2]
Department of Neurosurgery[1] and Biochemistry[2], Teikyo University School of Medicine, Tokyo 173, Japan

Maturation Phenomenon in Cerebral Ischemia V
A. M. Buchan et al. (Eds.)
© Springer-Verlag Berlin Heidelberg 2004

kg wet weight (8-fold of pre-ischemic FFA levels) at 8 h, particularly in the level of PUFAs, with a high proportion of docosahexaenoic acid after 3 h of ischemia.

Couclusion

Until 1 h of ischemia, the transient accumulation of FFA in upper frontal cortex consisted mainly of stearic acid and arachidonic acid, which were thought to be derived from PIPs by the activation of PLC, and then returned to pre-ischemic levels. It indicates that sufficient levels of high-energy phosphate may be maintained to repair the degraded membrane phospholipids in this region within 8 h of permanent MCA occlusion. In contrast, in lower frontal cortex the time-dependent changes of docosahexaenoic acid, which might be thought to be the main acyl residues of PE, began to increase gradually after 3 h of ischemia. This finding supports that the degradation of membrane phospholipids may be associated with irreversible ischemic change in lower frontal cortex.

Reference

1. Narita K, Kubota M, Nakane M et al (2000) Therapeutic time window in the penumbra during permanent focal ischemia in rats. Neurol Res 22:393–400

Cerebral Ischemia and the Rat Motor Cortex:
Time Course of Dysfunction and the Effects
of Differential Rehabilitation Paradigms on Functional Recovery

P. M. VANDENBERG, C. D. GOERTZEN, N. R. COOPER, H. A. VAN DER LEE, and J. A. KLEIM

Introduction

Cortical ischemia has been shown to cause a loss of movement representations in the rat motor cortex that extends to regions beyond the infarct. It has been suggested that this dysfunction referred to as "diaschisis" contributes to the motor deficits observed following this loss of representation. We have shown that peri-infarct diaschisis and motor impairments can be overcome with motor rehabilitation. Further, both the physiological and behavioral recovery is dependent on skilled rehabilitation rather than simply increased use [1]. The present experiment investigated the time course of the representational dysfunction and whether the timing of skilled rehabilitative training following ischemia influenced motor representations in the rat.

Methods

In Part 1 of this experiment, standard intracortical microstimulation (ICMS) techniques were administered to male Long Evans rats to elicit forelimb movement representations within the caudal forelimb area (CFA) of the motor cortex ([2]. Half of the animals were then administered a focal ischemic infarct via bipolar coagulation of surface vasculature within approximately 30% of distal forelimb representation. The animals were then allowed to sit in for one hour and then the CFA remapped or they were allowed to recover for twenty-four hours before the CFA was remapped. The size of CFA movement representations were calculated for Map 1 and Map 2.

In Part 1, male Long Evans rats were trained on the skilled single pellet reach task and then ICMS was used to produce CFA motor maps. Half the animals were administered an ischemic insult. Then rats were allowed to sit in their home cages for two weeks or one month following ischemia before receiving ten days of rehabilitation in the form of skilled single pellet reach training. Following rehabilitation, CFA representations of the animals were remapped. The size of forelimb representation pre- and post-lesion was calculated. Also, motor skill represented by

Ms Penny M. VandenBerg, Canadian Center for Behavioral Neuroscience, Department of Psychology & Neuroscience, University of Lethbridge, 4401 University Drive West, Lethbridge, Alberta, Canada, T1K 3M4, Tel.: (403) 394-3995, Fax: (403) 329-2775, E-Mail: vandpm@uleth.ca

Maturation Phenomenon in Cerebral Ischemia V
A. M. Buchan et al. (Eds.)
© Springer-Verlag Berlin Heidelberg 2004

successful reach percent and sensory errors during rehabilitation were recorded for Day 1, 5, and 10.

Results

Part 1 results show at one hour, following a focal ischemic event, movement representations within the rat motor cortex directly damaged by ischemia is unresponsive to stimulation. Within the twenty-four hours following a focal ischemic event, areas outside the ischemic area also become unresponsive. Preliminary data from Part 2 seems to indicate that a two-week delay in rehabilitation does not decrease the size of movement representation retained after skilled rehabilitation. However, at one month intact areas within the motor cortex remain unresponsive. Reach percent post-ischemia seems to be lower after a one-month delay in rehabilitation.

Conclusions

Peri-infarct diaschitic dysfunction of movement representations within the rat motor cortex occurs within 24 h of ischemia. Preliminary data indicates that movement representations seem to be retained and behavioral recovery seems to be more complete if rehabilitation is administered before one month post-ischemia. It seems more effective to administer a rehabilitative training regimen immediately following ischemia rather than after a large delay.

References

1. Goertzen C, Yamagishi K, VandenBerg PM, Kleim JA (2001) Neural and behavioral compensation following ischemic infarct within motor cortex is dependent upon the nature of motor rehabilitation experience. Soc Neurosc Abs 27:761–810
2. Kleim JA, Barbay S, Cooper NR, Hogg TM, Reidel CN, Remple MS, Nudo RJ (2002) Motor learning-dependent synaptogenesis is localized to functionally reorganized motor cortex. Neurobiology of Learning and Memory 77:63–77

Rehabilitation-Induced Cortical Dysfunction Following Focal Ischemic Infarct

R. BRUNEAU, K. YAMAGISHI, and J. KLEIM

Introduction

We have previously shown that rehabilitative training following focal ischemia both improves motor ability and prevents loss of movement outside the infarct. However, we have also found that during the first week of rehabilitation all animals exhibit a behavioural relapse, for which the neural basis is unknown. The present study examines the functional integrity of the motor cortex and the distribution of GABA receptors at the time of relapse.

Methods

Rats were first trained on a skilled reaching task to establish a baseline measure of motor ability. Intracortical microstimulation was then used to produce "maps" of forelimb representations within the motor cortex. A focal ischemic infarct was then produced within 35% of the forelimb representation via electrocoagulation of the surface vasculature. Half of the animals then received rehabilitative training on the same skilled reaching task and reaching accuracy was monitored daily. When animals exhibited the relapse, they were immediately remapped. Non-rehab animals were pair matched with the stroke animals and were mapped at the same time.

Results

Results showed that in comparison with non-rehab controls, relapse animals exhibited a significant decrease in movement representations that extended well beyond the infarction. Analysis of GABA receptor density within the peri-infarct cortex is currently in progress.

Conclusions

Motor rehabilitation initiated early after stroke transiently exacerbates peri-infarct diaschisis and is associated with a relapse in motor performance.

Ms Rochelle Bruneau, Canadian Center for Behavioral Neuroscience, Department of Psychology and Neuroscience, University of Lethbridge, 4401 University Drive West, Lethbridge, Alberta, Canada, T1K 3M4, Tel.: (403) 329-2293, Fax: (403) 329-2775, E-Mail: rochelle.bruneau@uleth.ca

Maturation Phenomenon in Cerebral Ischemia V
A.M. Buchan et al. (Eds.)
© Springer-Verlag Berlin Heidelberg 2004

Behavioral Improvement by Activation of Endogenous Progenitors in the Rat Transient Forebrain Ischemia Model

H. Nakatomi, A. Tamura, N. Kawahara, T. Kirino, and M. Nakafuku

Introduction

The recent discovery of neuronal progenitors capable of producing new neurons in adult mammals raises the possibility of repairing damaged tissue by recruiting their latent regenerative potentials. Recently, we showed that activation of endogenous progenitors by intraventricular infusion of a cocktail of FGF-2 and EGF, 2–5 days after ischemia, led to massive regeneration of hippocampal neurons after transient global ischemia in the adult rat forebrain. In this model, we investigated whether regenerated neurons can contribute to the intervention of neurological deficits caused by ischemic insults.

Methods and Results

We used the Morris water-maze task to examine the impairment of hippocampus-involved learning and memory. When sham-operated control animals were trained twice a day for 5 consecutive days, they memorized the position of the hidden platform, thereby reaching it in a shorter escape latency during the training at 7 to 11 post-operative days. When untreated and growth factor-treated ischemic animals were subjected to the same task, both groups showed significantly longer escape latencies during the initial 3 days (6 sessions) than in the control group. These ischemic animals, however, were finally trained like the control animals during the last 2 days. To examine the function of regenerated neurons, separate subgroups of trained animals were subjected to the second block of training during 49 to 53 days after ischemia with two different task paradigms. In one paradigm, the acquisition of new memory was tested by placing the platform at a new position. The untreated ischemic animals again showed a deficiency in this task compared to the control. In contrast, the growth factor-infused animals could reach the new platform position faster than the untreated ischemic animals at all trial points.

Department of Neurosurgery[1], Teikyo University School of Medicine, Department of Neurobiology[2], and Neurosurgery[3], The University of Tokyo Graduate School of Medicine

Correspondence to Prof. Akira Tamura, Teikyo University School of Medicine, Department of Neurosurgery, 2-11-1 Kaga, Itabashi-ku, Tokyo, Japan, 173-8605
Tel.: 81-3-3964-2415, Fax: 81-3-5375-1716, E-Mail: nstamura@med.teikyo-u.ac.jp

Maturation Phenomenon in Cerebral Ischemia V
A.M. Buchan et al. (Eds.)
© Springer-Verlag Berlin Heidelberg 2004

The other set of animal subgroups was subjected to the task without changing the position of the platform. The control animals retained the memory of the original position, and thus could reach it as fast as they did at the end of the first block of training. Although the untreated ischemic animals exhibited severe impairment in this task, the growth factor-treatment remarkably improved this long-term memory-dependent cognitive performance. We observed no significant difference in either the swimming velocities or spontaneous locomotor activities among these animal groups.

Conclusions

Thus, we concluded that post-ischemic administration of growth factors significantly ameliorated the deficits in both the acquisition and retention of spatial memory.

VIII Round Table Discussion

5th International Workshop
Maturation Phenomenon in Cerebral Ischemia
Round Table Discussion

Alastair Buchan

This is the final session of the Fifth International Workshop Maturation Phenomenon in Cerebral Ischemia and we will have a Round Table. I've invited 3 of our prominent speakers and members to help chair the discussion and there are a number of topics. Perhaps we'll start with John Hallenbeck making some introductory remarks. We want to cover what we think are the important issues that have been presented in the various talks. We then want to open this up to some round table discussion to see what the problems are and what the promise might be.

John Hallenbeck

Well, I was just going to make a couple of points. I had kind of listed in general, the areas that we've been through in the course of the last 3 days and I could run through them but if you could run them through your mind I think you will realize that there are many different mechanisms, many different approaches and perspectives to the general problem of brain ischemia that have been discussed. Not only brain ischemia but one of the – I think – challenges in the field at this point is to try to integrate all these different concepts and mechanisms. Trying to even understand them all, represents a departure, I think, from the way things were in the 80's and early 90's when the models for progression of ischemic damage during the early hours of ischemia were far more monolithic. And there was greater density of investigators that were all focusing on similar mechanisms. This kind of challenge of integrating and understanding many different types of processes and mechanisms is, I think, only going to get worse and more challenging as this tidal wave comes in of genomics and proteomics with what has been termed the combinatorial explosion. The concepts we now have could get far more complex and intricate. So I think this is a central thing that we have to consider and address and it may be that in future meetings we will want to have people represented from other disciplines. That we will want to have people who are still in things like systems analysts, things that we don't really present at the present time so it gives us a discipline forecast.

Alastair Buchan

Just in listening to you John, I think of two big areas, and it's just like in clinical medicine, we worry about the diagnosis and we worry about the investigation. Here we've got models, and that is the diagnosis. There are a huge variety of mod-

Maturation Phenomenon in Cerebral Ischemia V
A. M. Buchan et al. (Eds.)
© Springer-Verlag Berlin Heidelberg 2004

els, which have different implications. I think we should try to get some discussion going on that. And then whereas we have picked out and historically we've looked at, I suppose, barbiturates and then calcium channels and then excitotoxicity and free radicals and then inflammation and then apoptosis, the reality is that these were things we could see, things that we could measure. A bit like measuring sugar in the urine in the 19th century. We've now got the chance to see pretty much everything that's going on but whether we can decipher it or not I don't know. We heard at Marburg – at the last meeting, a very elegant lecture on the way in which genomics and proteomics are going and the kinds of patterns that are emerging and maybe Ulrich, you could comment to some extent on how we're going to look and patterns, and whether different models might be a way of dissecting out the different patterns.

Ulrich Dirnagl

Well, I'm not quite sure whether I'm really the person to comment on that but we have ventured into transcript proteomics and we have learned a lot from that. And most of what I've learned was not so much about what ischemia... – it was much more about the limitations of these approaches. My personal belief is that we will see a lot coming out of the transcript proteomics of course. But at present the methodology we have available – or in part also the way we apply it is not really perfect. In our lab, and I think one of the problems lie, for example, if you want to expect patterns, which I think is something that the methods are very good in doing, we need extensive time kinetics and what we also need is we need tissues specifically. We need to laser dissect cells and we have to pool the data from the different models. I think that is extremely powerful but it only works if we can trust the data. I mean you can pool all kinds of transcript data from wherever but it's what you put in you will get out. But if there is junk at the *top* there is junk down the drain. We need very well defined models, we need to be sure that these transcriptors or proteoms that we put on the data base and so forth and that we exchange are extremely reliable and my feeling is that this is not the case at present. But if this is the case, I think it will be extremely powerful because then we can sort of merge a lot of, let's say protection models, where there is pure protection by preconditioning. Or we see manifest damage by *concarma* like in various ischemic models. And from this, at least what I think, we should be able to extract relevant signal transduction patterns. But as I said, I think we don't have the tools we need to do that. Also, if you think about biomathematic support, they are not really – they are emerging but they are not there. Lastly, I think all the work that has been done so far – and I think a lot has been presented here – is extremely relevant. Also then for the candidate approaches we have and the candidate data we already have, I think it is extremely important for the screening approaches because we can – it's kind of a calibration – and we can put that in like columns and put all the other things that come in by hypothesis not ignorance driven and so we have to bring that together but I think it might be another 3–5 years until we are there.

John MacManus

My two cents worth are to say that I do agree with this time frame. I think at the moment we are so mesmerized with the technology, we are taken with the hype of the technology. With the massive parallel transcriptomes analyses, and now the problems as I seen them are that; one, massive amounts of information are generated but in fact I don't think enough information is generated, chips that are available don't have the whole mouse genome, for example. There aren't very many large rat chips available yet. We are definitely going to need time to have these technologies in place. And another problem that I see is a particular situation that we are interested in of cerebral ischemia is one where I don't think looking at a particular conscriptome is in fact very useful because you have massive inhibition of protein synthesis. And, I think a lot of these messages that we see changed don't actually go anywhere. They don't produce a useful protein. We need to be very careful that the analysis we do make – that we follow them through. Do we get protein in the right place at the right time? Can we understand? Certainly the proteomics technologies are way behind the nucleic acid ones in terms of their abilities to do screening. Again, I think people are seduced by the idea of using mass spectrometry to analyze protein. But the problem, the bottleneck, is in the separation of protein by 2D gel electrophoresis methods that are available are pretty hopeless. Particularly to analyze the very small amount of proteins like some transcription factors are essentially the things that are controlling these methods. I absolutely agree with John's statement that systems analysts and systems scientists are going to have a huge impact on biology in the way that old fashioned bacteriologist had an effect on molecular biology. We are going to have to bring in people who essentially don't know any biology. I went to a talk at a meeting where Leroy Hood talked about his systems analysis approach, quoted papers in Science and where he looked at changes in genome yeast. Where you just take it from glucose into galactose. There is a huge change and shift in the metabolism of the yeast to do that apparently simple thing. And, the people on the papers were all engineers & physicists. Because I can teach anybody biology but it's very difficult to teach them the physics and the mathematics of systems analysis.

Alastair Buchan

Roger, any thoughts?

Roger Simon

Well someone, I won't say who, commented that as we move through these systems of biology it seems to turn into alphabet soup. And it's an alphabet that has way too many letters for many of us to think about. So our own approach was to get a hold of the weather systems computer people at one of the national laboratories. These guys, we told them that we have so many data pieces, maybe we have 30 000 data pieces in our microarray. That we have no idea how to deal with this and they said, well we can deal with that at lunch time because they can look at every pixel for global weather pattern simultaneously on the globe and watch it move over periods of time. So they took our data and we have a globe and now we have all of these patterns that sort through in regard to if gene A changes then what happens to gene B. But if gene A and B changes what happens to gene C. But if gene A and C change then what happens to gene E. This is I think, the kind of technology that we're going to need to step beyond at least what this mortal brain can deal with in regard to alphabet soup. My only fear is that we're going to pass this on to "Hal" the computer and it's going to be at that level. As a matter of fact, I made a mistake in hearing your Kings English presentation of your question a moment ago. I thought what you said was "what are we going to do with the patterns"? Which was what I was just speaking to. But what I thought you said is "what are we going to do with the patents"? That's another issue indeed. But I think this really completely open minded look to this informatics analysis is the link we absolutely have to make because alphabet soup is already beyond what my brain can handle.

Alastair Buchan

Thanks, Roger. Perhaps we'll hear from Mike Bennett next.

Michael Bennett

I would like to give as an example a really rather disturbing story from a simple invertebrate nervous system. Namely the stomatogastric ganglion of the crab. It has 7 neurons. And they've been working on this thing for 40 years and getting basically nowhere. And now, Eve Marter has done some simulations and found that if you want to get the output of these 7 neurons that gives a rhythmic beating output that causes the contractions of the gut. There are a number of different satisfactory solutions to that. If you consider cell A and B, say cell A can be exciting cell B and that will work for the pattern. Or alternatively cell B can excite cell A and give the same pattern. If you average them in together, it washes out. So the idea that single neurons aren't the same may in fact be true in terms of their gene expression. But they may still be giving the same result. That there are multiple solutions to give the same answer.

Alexander Baethmann

I would like to come back to the reason why we are here. The question I think is the general question is what is maturation phenomenon. What is the mechanism underlying the maturation phenomenon? I think the next pertinent problem then is what is the relationship with the clinical cause of an infarct or related issues? I think this has not been addressed very much so far. Maybe in the final minutes of this symposium we should ask the key players here, of everybody who likes to contribute, what his feeling or his current state of knowledge or the gain of this meeting, is that he has increased his understanding of the maturation phenomenon. And what is the contribution for progress on our clinical management and also our diagnosis is the first point. The second point is, yesterday, Peter Hossmann said there is actually no excellent stroke model, which has something to do with a clinical stroke. And I think this is really a challenge and I think there are so many here who work on MCA occlusion or all kinds of most in-vivo models, and they take this result without any objection? I think this should not be left not discussed otherwise all the animal experiments would be futile if Hossmann is right. I'm sure he meant it in a provocative way. Because also studying experiential stroke models or MCA occlusion models ... but we should discuss it. And really we should find here all the experts together what can we do to close the gap between the experimental models and the clinical situation?

Alastair Buchan

Thank you. John's suggestion as I understand it is to have two sections now. One would be to discuss the maturation phenomena and the other would be to spend the last 20 minutes discussing the models that we use to explore the chemistry, or the behavior, or whatever. When I introduced it, what I was trying to indicate was that over the 12–13 years that I've been doing experiments since I was at Cornell we've gone from one fashion to another. Very focused fashions. And there are fashions that were there before I was on the scene. The fashions come and go but then often link to single proteins such as the NMDA protein channel, the AMPA protein channel, and inflammatory cytokine or ICAM molecule. The reason we were having the discussion, I thought, was because I think all of us feel that there are so many things going on and perhaps people can correct me, but it's unlikely with respect that even something as telling us perhaps the Glur2 subunit, that there's going to be a single protein or a single change in the way in which the biology works that's going to address the cause and the treatment of the maturation of cell death. Perhaps Suzanne or Mike, do you think we'll come up with something as specific as a single genetic event, given what we've been discussing?

Suzanne Zukin

I think we already have a very clear message that there are multiple death cascades that are triggered following a global ischemic episode and I'm sure that's true for the other models as well. So I think we already have to look beyond that and ask when and where the different death cascades are triggered and what is the cross talk between them.

John Hallenbeck

As Suzanne said, there is a growing consensus that no single mechanism is sufficient to really control or determine progression of brain damage in ischemia. Whether in the early hours of focal ischemia or whether it's delayed by several days in the global ischemia maturation model. And maturation probably can be stretched to cover both of these processes. So the challenge then is that if you have a global factorial problem, is to try to transfer something that works. How do you deal with such a problem? I think the present state of the field is that those who are designing clinical trials are actively considering a combination. And those of us who are working at a basic level are looking for what might be called master switches that simultaneously counteract the deleterious mediators and mechanisms.

Alastair Buchan

So Pak, do you think there is a master switch?

Pak Chan

I hope so. And to me, I think the maturation of injury really are phenomena of reperfusion injury. But it's in a long delayed fashion. I think that maturation is a very good model for the degeneration and degenerative diseases in general. For those who work in Parkinson's disease and other disease, they don't have a good model. I think they should come to us and use this model as a delay – neurodegeneration model. I do agree with John's suggestion. There could be a master switch somewhere – somewhere in the process. To me I would suggest that the toxic state may be one of those master switch. The molecular switch – either the cell is going to die or the cell is going to survive. So, I've done so many studies involving with oxidative stress. So if you increase superoxide dismutase level, then push all the phenomena – caspase 3, cytochrome c, whatever you name it the mechanism, you push to the survival mode and if you have less SOD you increase the oxidative stress. You push the active side, you push the cells are going to die. I think that could be one of those molecules which could exist, I think it's really by our effort to find this kind of molecular switch. And that could be true. Maybe one single drug can affect many many pathways involving cell death. That drug could exist. I think nowadays we should not worry about the selectivity or specificity of a drug. If this drug can work on multiple pathways and either inhibit upstream or downstream I think that would be fine.

Alastair Buchan

So what I'm hearing is perhaps a drug might interfere and that would give you a clue. The other way of looking at this is what we were saying before. To see a pattern of genes or proteins that might point to something like a transcription factor such as NF-κB. So you've got a dichotomy there and my experience both in the lab and now with patients is that using drugs to explore this is very fraught.

Pak Chan

Yes, that's true, I agree with you. You still have to go through the basics. You still have to study the mechanisms first. So like oxidative stress – whether oxidative stress can play a role in the molecular master switch. And then develop a drug maybe SOD and ... or something to counteract this pattern of injury.

John MacManus

I do not believe in the master switch. I think that the decision whether to live or die is way too important for our biological systems to have evolved to have a black and white answer. As we go back to Roger's weather men there. There is a wonderful theory of chaos and butterflies fluttering and their wings have gone calm which will affect our weather here. Well this may affect our weather here but you can in no way predict it because it's a chaotic system. I believe in fact that biology is a chaotic system and so many networks and feedback systems been built into it that what we are seeing here is a result of suppressed cell death. That the death pathway is the default. We are quite lucky that there are many many many ways to have cell death.

Alastair Buchan

This is John's theory of snow in July. Which we have to tell you John, is something we do experience in Alberta!

John McManus

I just don't see that it's that simple that there is a master switch. I think we're dreaming in technicolor if we think it's that easy.

Fred Colbourne

I think we have the drug and it's called hypothermia. I think hypothermia also teaches us something about this issue of master switches. There are a number of mechanisms by which hypothermia reduces injury, but I cannot pinpoint one that is the reason why hypothermia is effective. I can pinpoint a number that no doubt contribute. Whether that's anti-inflammatory effects or whether that's $GluR_2$ mechanisms. There are lots. And I think ischemic injury in the brain whether it's global or focal or hemorrhagic is like a house burning down. But not one that's burning down because of a pot that's left on the stove that's started a fire in one location. It's a house burning down because somebody has set fires in many locations and ischemic injury is because of many master switches, many fires. We're going to have to take something that addresses many of those mechanisms. Right now the only thing I think we have like that is hypothermia. So I think if we can understand that a bit better then we can hone in on those 4 or 5 or a dozen master switches.

John Hallenbeck

To answer really quickly, maybe the term master switch, if it requires that there be a single switch, and I think if you used the English language you'd have to say a master switch is just one, might be the wrong concept. But beyond the view there could be targets in the cell where they could affect more than one process and this would be desirable and that was the concept.

Ulrich Dirnagl

From a totally different point of view concerning maturation and stroke. What I have found that has gone unnoticed somehow over the last one or two years is that from a committal standpoint the fact that there is maturation in stroke has now, at least with many clinicians, become accepted. This is the result of imaging studies because one of the arguments always were that stroke patients are getting better. I mean 95% of the stroke patients are getting better. And if you talk about maturation and the basic scientist they tell you our patients are not getting worse they are getting better. I think with perfusion imaging and some kind of a penumbra or whatever markers now, it is quite clear that there is a subset of patients where there is maturation. And, this maturation – and this also I think is good news I guess for the patient because this is a concept which we can salvage and its also good news for the basic scientist. Because, I think these clinical studies with the imaging method, they demonstrate quite the opposite from what we said before. That we don't have models that resemble anything that's happening clinically. In fact, there is a very nice registration of this subset of patients, obviously not the majority of patients, but a subset of patients who these resemble very much what we see in our models. And we have models with necroses right away and almost no maturation. I'm talking about focal ischemia. And we have models where there's a delay and there is maturation. So because with models then, I think it's very good that we have all these models because we just have to use them the right way and bring together the information we get from them. Because just as varietal and heterogenous the models are, the varietal and heterogenous the patients. So it's just different aspects of the model. So I'm not as pessimistic about all this stuff with modeling and also on the clinical side.

Guenter Mies

I think we're starting to think in molecular biology terms and I think that's a little bit far away because the animal models, in our hands at least, they all differ by basic physiological parameters. Namely perfusion, glucose metabolism, protein synthesis and for example if you do the thread model in the mouse and you keep the common carotid artery occluded then you reduce the baseline blood flow to 70% and that stays at this level. So when you put in your thread now you go down to 10% or 5%. When you reperfuse you only come back to the 70% because it prevented adequate reperfusion. This is what you have to know before you're able to interpret any other changes. You also have to know basic paramaters like inhibition of protein synthesis. That may not be the reason for the ischemic damage but it's a valuable marker of preceding ischemic injury. And I think we have to analyze the tissue, which we then take into further molecular biological analysis. We have to know more about this tissue and this was mentioned earlier here. It doesn't make sense if I make a low DNA chip from an area which is ATP depleted. Maybe there is interest to look into a necrotic tissue.

Alastair Buchan

What about models? On the one hand clinicians and people doing trials need data they can believe to go into trials. On the other, having accepted that consistency with an effect, the clinical situation is of course a complete lack of consistency and patients who have all manner of outcome. And you need huge sample sizes and thousands of patients. But the essential ingredient of using a model is surely to get as tight a consistency as you can either within the laboratory or hopefully across laboratories using the same technique. That's very difficult to do. But within one's own laboratory to get really really tight control of the physiological variables has always been something we were brought up with, but is rarely achieved. If we don't absolutely clamp to 0.1 of a degree centigrade over many many days, we see variability, which we accept, which we can counter if we deal with that. I worry about doing transcript screening experiments or gene chip experiments if people haven't got really really precise physiological control of what they are going to put in. I think that's going to create an enormous noise in the system that'll be hard to deal with.

Ulrich Dirnagl

I have just another comment. We're always talking about the brain. But for the outcome of the stroke patient the brain in fact may only be one variable. There are so many if we talk about variables that are measured in the clinical condition. If you think about pCO_2, pO_2, blood pressure, these are the obvious things, and glucose for example. But there are many other things. Like there is a very nice connection from the brain to the immune system. There are lots of changes after stroke in the immune system which in turn have effect on the brain again. So, in fact, it just came to my mind when you were talking about variability because that's just an additional variable that goes into that equation. And finally, in relating it back to the outcome of patients we have to not only think about the brain but also the person.

Maria Spatz

Well we are focused on the brain for a long time but I really think that we are ignoring the peripheral organs. Heart, lung and kidney, which do have an effect on the brain and the brain has an effect on those peripheral organs. So finally we have to think about the entire system. Not only the brain.

Alastair Buchan

Francesco, please.

Francesco Orzi

If we learn about the use of a number of events and clinical changes which occur in the penumbra are common to a number of neurogenerative conditions, particularly Parkinson's disease.

I wonder what the audience thinks about the use of toxins which interfere with the mitochondria in relation to this change particularly in Parkinson's disease.

Whether the use of these toxins could be somehow relevant or useful to our comprehension.

Alastair Buchan

Pak do you want to comment to that. So we're getting away from stroke and just setting up a model of cell death *in vivo*.

Pak Chan

We have used 3NP as a mitochondrial toxin. If you inject the 3NP, either IP or intracerebral injection into the brain and you can induce a delayed neuronal cell death, especially in the striatum area, this is very specific to the striatum. I think that's sort of trouble you know, to use this mitochondrial toxin to generate a model for neurodegeneration. It's similar to a stroke model. You injure the mitochondria so we use SOD2 knockout mechanism. You do a focal stroke or you do a global stroke. You see a similar event when you use a 3NP mitochondrial toxin. Also it causes a neurodegenerative mitochondrial injury, and a neuronal cell death. So it's a very similar, I think we can expand this model into a neurodegenerative disease model using a mitochondrial toxin.

John Hallenbeck

One other aspect of the model is that I was listening to John Harlen recently talk about his experiences. Harlen is from Seattle who for years and years has been a champion of the concept of reperfusion injury and many different issues. In many of the models he was able to show fairly robust protection by blocking adhesion molecules and this was carried into the clinic. Study after study failed in different tissues in the heart, in the brain, sometimes you could make the case that the agents that were used were toxic and so forth. But sometimes you couldn't make that case so he was forced to confront this discrepancy between the pre-clinical models which he'd done very carefully and the clinical failures. So what he did was to begin extending the ischemia times before reperfusion until he got to something that was more realistic in terms of endogenous thrombolysis, which doesn't occur in 2 hrs or 4 hrs or something like that. And his model started failing. So the problem I think for the investigator that has a family and kids and needs to feed them is do you continue having sensitive models that can see small effects, are designed to do this and publish and so forth and make an income or do you show what in preclinical models, the realistic thing, that no neuroprotective agent for 20 years has worked. It's kind of a problem but yet it's probably something that has to be considered.

Fred Colbourne

Personally, I don't think it's acceptable when one is interested in efficacy to test this in models of stroke that are 30 minutes for example in mouse. Because I don't think that would be something that would be urgent in treating the human population. I think that you would consider severe ischemia, more prolonged ischemia. Perhaps Dr. Hossmann is right in that we need to really do permanent ischemia models in addition to the temporary models. And maybe the clinicians can answer the question of whether what's necessary before you advance a therapy in a clinic.

Will you accept it in temporary models or do you require permanent ischemia as well.

Alastair Buchan

But you're never going to mimic a human stroke in a model and we're not trying to. And that's why I think the global model is very helpful, as I tried to illustrate on Monday to look at cell death. The focal models aren't there to replicate human stroke. They're there to show for a given ischemic insult, the intervention may ameliorate damage and the question then is can you translate it up through phase 1, 2 and 3. I think the clinicians make the mistake of thinking that the stroke in the rat is the same as a stroke in man and it's not. It's so much more complicated in man because they've got a lot of diseases and a lot of different organs. They've got a lot more things going on. There's a hemorrhagic component. It's just so completely different but the only clue you've got of efficacy is the rat data and you get evidence of safety from Phase 2. So when you actually press the button and do the Phase 3 study, that's when you've got rat efficacy and man's safety which is what Steven Warrach is trying to do. We are trying to stratify our patients and it becomes very important if we can produce some kind of biomarker and that's what we don't have. So those drug studies to come out of the animals and those surrogate markers or biomarkers that might come out of the animals, that's all you can get. You can't mimic the stroke of the human in a model, I don't think.

Fred Colbourne

I agree with that and I think that a focal model may not necessarily predict the results in humans any better than a global model. But, I think that it's important that we really consider the severity issue.

Alastair Buchan

But it's the best we've got. But what I really object to is when we go to the human trial and maybe Diane will speak, is I find it incredible that there are trials done when the animal data is negative. And then I don't feel we have an ethical right to put patients at risk. But companies are willing to say well the animal models are useless, therefore the negative animal data is a form of surrogate positivity.

Diane Stephenson

I agree. Companies are easily saying that animal models are not predictive. From my perspective, I'm thrilled to be here in this privileged group. But I have the challenge as a scientist in a company to try to persuade the pharmaceutical companies to say the games is not over – don't completely give up. One thing I've been very encouraged about in the last few years in stroke is the advances in imaging and the behavioral endpoints so that we don't just look at lesion size. But if we can monitor animals progressively over time in terms of infarct size and imaging and then in behavioral endpoints we at least try to match the animals, which is a little bit closer to what they do in humans. The question I have, is in at least two or three presentations at this meeting it's been clear that lesion size does not match what you see in behavior and that was from Claudia Gonzalez's talk and from Michael Chopp's talk. So I struggle with and I know you've been so helpful with so many discussions of what from a pharmaceutical perspective can we use as critical endpoints in our animal models because their resistance is growing more and more with all the failures that even the best pre-clinical package that you could ever have. You could have focal, you can have global data you can have post ischemic improvement both functionally, still isn't enough to convince higher management that you would go forward for a clinical trial.

Alastair Buchan

What John McManus would you say was a good measure at the cell level that would tell you that you'd make a difference? You spent a lot of time with the corollary. The reason why I ask you the question is that when I first met you, you were worrying about malignancy and you were working with the cancer biologists. Are there clues that we can get from the other side of the cell death spectrum? What would you look for?

John McManus

The other side? I've never gone over to the other side. I don't know. I hope you're not asking me to discuss necrosis and apoptosis. Cause I don't want to do that. I think the question from industry is one that is very important. I think it's not just important in industry, I think it's very important in research institutes. I think it's very important on grant selection committees. I think there's been a wave of testament and a wave of depression that's gone down through clinical trial failures to research managers asking why are you doing the research. In fact my managers have asked me that question. They have said exactly the same thing. You are doing measures of infarct, you're doing measures of cell death no matter how sophisticated they are by molecular biology. You are doing behavior and still nobody is going to be convinced. I have really no answer to you. There to me is no litmus paper that you could apply to cell death model in oncology or ischemia research.

Guenter Mies

Using rodents for focal ischemia we have shown the importance SF electrophysiology. In rodents Peri-infarct depolarizations increase the infarct.

Fred Colbourne

Perhaps it's my age, I'm not so pessimistic that our models cannot predict the clinical situation. I think that a reason for a lot of the failures – all of the failures I think is because a lot of them – our work hasn't done a few key things, which are, assess the major physiological variables, to assess that particular drug in multiple models over extended survival times and look at multiple behavioral measures in addition to cell death. I don't think you can rely on either one and I think if you ask yourselves the question, which of those therapies were properly assessed I don't think any of them have.

Alastair Buchan

I will follow that up by saying I think that with the cardiac arrest papers coming out, although Dr. Spatz could say well in fact it's because hypothermia and cardiac arrest protected the heart and the kidneys which made them better but there is no histological measure in those patients. But more or less your experiment was translated to man and we've demonstrated that we can predict from, in that case rat 4-VO to global ischemia. And Steven Warrach has shown that some of the data from the citocholine experiments has been translated to some of the patients using imaging as a surrogate and shown a reduction in the amount of focal injury with an intervention in a clinical situation. So I think there is beginning to be the evidence of translation but I don't think we have any sense of what the mechanism of that intervention did at a fundamental level. So I think we're beginning to see translation of outcome but I worry it's not taking the mechanism with it.

Roger Simon

Before we just kind of bash everything everyone's been doing over the last 20 years I suggest that maybe there's another way to look at this. I don't think there are any negative trials because the trials have not been set up to recapitulate the animal experiments. And this brings me back to a conversation that was had at one of the stroke meetings a very long time ago by an argument between Myron Ginsberg and, I think Hossmann, no it was Dieter Heiss arguing "is the window in a rodent similar to or a microcosm of the therapeutic window in a human?" I may have the argument wrong but I believe Ginsberg said that if you have a 30 minute window in a rodent, that's the window and I think Heiss said that what he thought from his pet scan experiments it was much broader in the human. I think that the Ginsberg view is correct. So we have all of these studies of neuroprotectants that have been done over the last twenty years. None of them show a time window of efficacy of more than 90 minutes or so. And so that experiment has never been done so I don't think it's true that we've done all these pre-clinical experiments and all of them have failed. As a matter of fact, clinical trial of the pre-clinical experiments has never been done. And then one other comment about the models. I think that all of these models are relevant and I think the answer is there and I think that we know the answer is there because it's so pleuri-potential. I think that if what one's looking for is the protection phenomenon, the protection phenomenon can be seen in focal ischemia, it can be seen in global ischemia. It can be seen in endotoxemia, it can be seen in epilepsy. The tolerence in epilepsy tolerizes against ischemia, the tolerance in ischemia tolerizes against epilepsy. The same thing is true in endotoxin. The same thing is true in ischemia tolerizing against traumatic brain injury. So the challenge then is to pick out this piece of the biology that's the protective bit and not be side tracked by hypoxia of blood flow, spreading depression, or whatever the elements are. I think all of these models are relevant, I think the answer is there. And the answer is not just true for brain. It's true for kidney, it's true for heart, it's true for intestine, and there is even a phenomenon in the middle ear called toughening in which a brief loud sound is given and protects against cochlear injury from an otherwise intolerable loud sound and BCL2 up-regulation is seen in the tolerance.

Alastair Buchan

Thank-you, we've only got a few minutes left, is there anybody who wants to bring up a new topic?

Minoru Tanaka

I think one of the reasons for discrepancy between animals, stroke data and clinical trials is the significant and difference between rodent brain and human brain. In the human brain white matter is proportionally more important. On the other hand, in the rodent brain there is very little white matter. So there is a difference in the mechanism of damage between grey matter and white matter. We have to focus much more on white matter damage.

Alastair Buchan

That's very important. A number of the companies have tried to separate out lacunar or white matter stroke from cortical stroke and the imaging studies are beginning to give us a sense of which drugs might be used for what kind of injury. So I think that we are sort of getting there but it begs the question with the rats being predictive of humans. But the rats could at least predict that you're going to get protection of cortex. One of the things that I'm always disturbed by is not the gray/white differentiation, it's the cortical/striatal differences and the posterior circulation/anterior circulation. There are differences clinically which we're not trying to replicate in the models but it's not just the gray/white ganglionic vs. cortex differences, it's also the different compartments. It's very clear that in the rat it's unusual to get a big swollen hemisphere with hemorrhagic conversion and herniation. But in humans we face that all the time and I think compartmentalization is something that's not easily replicated by animal models. I think imaging has got to be able to tell us which compartment, the edema, the inflammation, the injury is in and until we work at different compartmental levels we could get into a lot of difficulties.

So we are looking for any other final remarks. Fred........

Fred Colbourne

I'd like to say thanks to you and your staff particularly Amley Wilson for organizing this conference. I think you've done a wonderful job. To you and to Ito again I say thanks for everything.

Umeo Ito

Before I formally declare the Fifth International Workshop of the Maturation Phenomenon in Cerebral Ischemia closed, I would like to express my sincere gratitude to all of you who contributed a great deal to the creativity of the workshop and I do hope that our partnerships will continue forever. We will try to publish your very important contributions with a summary of the very active Round Table discussions as a proceedings for this workshop as early as possible. The Advisory Board members decided the next Sixth International Workshop will be held in May, 2005 in Mainz Germany by Professor Kempski's group. I would like to introduce him to speak about it now.

Oliver Kempski

Thank you for the surprise. First of all, I would also like to thank all of the organizers here in Banff. You, Professor Ito, for this wonderful meeting. The environment alone is worth coming here but the meeting was even better and so when I invite you to come to Mainz I hope we can match that. I'm sure we cannot match the environment but as I've said recently we have some wine there and some castles and so I think at least a tour will be good. And so, I hope you will all come so at least the meeting may also become at least a little bit as good as this one was. If you all come it may even get better. Thank you.

Umeo Ito

Finally, I would like to extend a heartily appreciation to Professor Buchan and his group to bring about such a great success of this meeting.

So now, I formally declare the Fifth International Workshop on Maturation Phenomenon in Cerebral Ischemia formally closed. Thank you very much.

Subject Index

MIX
Papier aus verantwortungsvollen Quellen
Paper from responsible sources
FSC® C105338

If you have any concerns about our products,
you can contact us on
ProductSafety@springernature.com

In case Publisher is established outside the EU,
the EU authorized representative is:
Springer Nature Customer Service Center GmbH
Europaplatz 3, 69115 Heidelberg, Germany

Printed by Libri Plureos GmbH
in Hamburg, Germany